Mastering MATLAB

THE MATLAB CURRICULUM SERIES

To facilitate the use of MATLAB throughout the curriculum, Prentice-Hall has developed the following series of MATLAB-related titles:

Marcus, *Matrices and MATLAB: A Tutorial*
0–13–562901–2 © 1993

Etter, *Engineering Problem Solving with MATLAB*
0–13–293069–2 © 1993

Ogata, *Solving Control Engineering Problems with MATLAB*
0–13–182213–6 © 1994

Hanselman/Kuo, *MATLAB Tools for Control System Analysis and Design, Second Edition*
0–13–202293–1 © 1995

Garcia, *Numerical Methods for Physics*
0–13–151986–7 © 1994

Burrus et al, *Computer-Based Exercises for Signal Processing*
0–13–219825–8 © 1994

Polking, *MATLAB Manual for Ordinary Differential Equations*
0–13–133944–3 © 1995

Hanselman/Littlefield, *Mastering MATLAB: A Comprehensive Tutorial and Reference*

Mastering MATLAB

A Comprehensive Tutorial and Reference

Duane Hanselman
Bruce Littlefield

The MATLAB® Curriculum Series

Prentice-Hall
Upper Saddle River, New Jersey 07458

Library of Congress Cataloging-in-Publication Data

Hanselman, Duane C.
    Mastering MATLAB : a comprehensive tutorial and reference / Duane
Hanselman, Bruce Littlefield.
       p.   cm.
    Includes bibliographical references and index.
    ISBN 0–13–191594–0
    1. MATLAB.   2. Numerical analysis—Data processing.
I. Littlefield, Bruce. II. Title.
QA297.H298   1996
519.4'0285'5369—dc20   95–39898
                                               CIP

Acquisitions editor: Tom Robbins
Production editor: Rose Kernan
Copy editor: BookMasters, Inc.
Cover designer: Joe Sengotta
Buyer: Donna Sullivan
Editorial assistant: Phyllis Morgan

© 1996 by Prentice-Hall, Inc.
Simon & Schuster/A Viacom Company
Upper Saddle River, NJ 07458

The Author and publisher of this book have used their best efforts in preparing this book. These
efforts include the development, research, and testing of the theories and programs to determine
their effectiveness. The author and publisher make no warranty of any kind, expressed or implied,
with regard to these programs or the documentation contained in this book. The author and publisher shall not be liable in any event for incidental or consequential damages in connection with,
or arising out of, the furnishing, performance, or use of these programs.

Printed in the United States of America

10 9 8 7 6 5 4 3 2 1

ISBN 0-13-191594-0

Prentice-Hall International (UK) Limited, London
Prentice-Hall of Australia Pty. Limited, Sydney
Prentice-Hall Canada, Inc., Toronto
Prentice-Hall Hispanoamericana, S.A., Mexico
Prentice-Hall of India Private Limited, New Delhi
Prentice-Hall of Japan, Inc., Tokyo
Simon & Schuster Asia Pte. Ltd., Singapore
Editora Prentice-Hall do Brasil, Ltda., Rio de Janeiro

Contents

v

Preface

This text is about MATLAB. If you use MATLAB or are considering using MATLAB, this text is for you. This text represents an alternative to learning MATLAB on your own with or without the help of the documentation that comes with MATLAB. The informal style of this text makes it easy to read. As the title suggests, this text provides the tools you need to master MATLAB. As a programming language and data visualization tool, MATLAB offers a rich set of capabilities to solve problems in engineering, scientific, computing, and mathematical disciplines. The fundamental goal of this text is to help you increase your productivity by showing you how to use these capabilities efficiently. Because of the interactive nature of MATLAB, the material is generally presented in the form of examples that you can duplicate by running MATLAB as you read this text.

This text covers only those topics that are of use to a general audience. The material presented applies equally to all computer platforms including Unix workstations, Macintoshes, and PCs. Beyond the functions that are part of standard MATLAB itself, only the Symbolic Toolbox is discussed. None of the other, more-specialized toolboxes are discussed. Moreover, aspects of MATLAB that are machine-dependent are not discussed, such as the writing of MEX-files.

This text develops a number of function M-files that extend the capabilities of MATLAB. In the text, they demonstrate various MATLAB capabilities and programming techniques. In the text they are collectively called the *Mastering MATLAB Toolbox*. These M-files are available on floppy disk free of charge from The MathWorks, Inc. by sending in the postcard inside this text. As an alternative, they are available on The MathWorks FTP site. For information regarding this approach see Chapter 23. You can contact the MathWorks directly by writing to: The MathWorks, Inc., 24 Prime Parkway, Natick, MA 01760; phone: (508) 653-1415; fax: (508) 653-2997; e-mail: info@mathworks.com; WWW: http://www.mathworks.com

Since MATLAB continues to evolve as a software tool, this text focuses on MATLAB version 4.2c. For the most part, the material applies equally to all 4.X versions of

MATLAB. When appropriate, distinctions between versions are made. Moreover, likely changes to be seen in MATLAB version 5 are noted.

We, the authors, encourage you to give us feedback on this text. What are the best features of the text? Which areas need more work? Which topics should be left out? What topics should be added? We can be reached at the e-mail address: `mm@eece.maine.edu.`

ACKNOWLEDGMENTS

We would like to thank Don Fowley for recognizing our abilities and making this project possible. In addition, we would like to thank the folks at The MathWorks, Inc. for their co-operation and assistance in this project as well as with the tutorial sections of *The Student Edition of MATLAB.*

I, Bruce Littlefield, would like to thank my wife Hazel for her enormous patience and understanding during the development of this book. She has offered unflagging support and encouragement despite losing her husband for far too many nights and weekends.

I, Duane Hanselman, would like to acknowledge the love and dedication of my wife Pamela and that of our children Ruth, Sarah, and Kevin.

TRADEMARK INFORMATION

	MATLAB M-files **Referenced for** **The HELP Text**
Internet World Wide Web Telnet	

Products/
Companies

Macintosh
Unix
CompuServe
Delphi
America Online
NetScape
Mosaic
Microsoft Windows
X Window System
VMS
Handle Graphics
Tektronix
The MathWorks
MATLAB

MATLAB M-files
Used in
Examples

```
fliplr
linspace
toc
tic
nargchk
humps
gcf
gca
gco
```

```
dialog
questdlg
helpdlg
warndlg
errordlg
uigetfile
uiputfile
uisetcolor
uisetfont
sqrt
conj
size
zoom
ginput
polar
bar
stairs
hist
rose
stem
errorbar
compass
feather
del2
findobj
orient
waitforbuttonpress
```

Mastering MATLAB

Getting Started

1.1 INTRODUCTION

This text assumes that you have some familiarity with matrices and computer programming. Matrices are at the heart of MATLAB since all data in MATLAB are stored as matrices. Besides common matrix algebra operations, MATLAB offers array operations that allow one to manipulate sets of data in a wide variety of ways quickly. In addition to its matrix orientation, MATLAB offers programming features similar to those of other computer programming languages. And finally, MATLAB offers graphical user interface (GUI) tools that allow one to use MATLAB as an application development tool. This combination of matrix data structures, programming features, and GUI tools makes MATLAB an extremely powerful tool for solving problems in many fields. In this text each of these aspects of MATLAB will be discussed in detail. To facilitate learning, detailed examples are presented. Many of these examples illustrate construction of M-file functions in the *Mastering MATLAB Toolbox* that accompanies this text.

1.2 TYPOGRAPHICAL CONVENTIONS

The following conventions are used throughout this text:

Bold	Important terms and facts
Bold Italics	New terms
Bold Initial Caps	Keyboard key names, menu names, and menu items
`Constant Width`	User input, function and file names, commands, and screen displays
Italics	Window names, book titles, toolbox names, company names, example text, and mathematical notation

Basic MATLAB Features

Running MATLAB creates one or more windows on your computer monitor. Of these, the **Command** window is the primary place where you interact with MATLAB. The prompt » is displayed in the *Command* window, and when the *Command* window is active, a cursor (most likely blinking) should appear to the right of the prompt. This cursor and the MATLAB prompt signify that MATLAB is waiting to perform a mathematical operation.

2.1 SIMPLE MATH

Just like a calculator, MATLAB can do simple math. Consider the following simple example: Mark goes to the office supply store and buys 4 erasers at 25 cents each, 6 memo pads at 22 cents each, and 2 rolls of tape at 99 cents each. How many items did Mark buy, and how much did they cost?

To solve this using your calculator, you enter

$$4 + 6 + 2 = 12 \text{ items} \qquad 4 \cdot 25 + 6 \cdot 22 + 2 \cdot 99 = 430 \text{ cents}$$

In MATLAB this can be solved in a number of different ways. First, the above calculator approach can be taken.

```
» 4+6+2
ans =
     12

» 4*25 + 6*22 + 2*99
ans =
     430
```

Note that MATLAB doesn't care about spaces for the most part and that multiplication takes precedence over addition. Note also that MATLAB calls the result **ans** (short for *answer*) for both computations.

As an alternative, the above can be solved by storing information in ***MATLAB variables.***

```
» erasers=4
erasers =
     4

» pads=6
pads =
     6

» tape=2;

» items=erasers+pads+tape
items =
     12

» cost=erasers*25+pads*22+tape*99
cost =
     430
```

Here we created three MATLAB variables **erasers**, **pads**, and **tape** to store the number of each item. After entering each statement, MATLAB displayed the results except in the case of **tape**. The semicolon at the end of the line » **tape=2;** tells MATLAB to evaluate the line but not tell us the answer. Finally, rather than calling the results **ans**, we told MATLAB to call the number of items purchased **items** and the total price paid **cost**. At each step MATLAB remembered past information. Because MATLAB remembers things, let's ask what the average cost per item was.

```
» average_cost=cost/items
average_cost =
     35.8333
```

Because *average cost* is two words and MATLAB variable names must be one word, the underscore was used to create the single MATLAB variable `average_cost.`

In addition to addition and multiplication, MATLAB offers the following basic arithmetic operations.

Operation	Symbol	Example
addition, a + b	+	`5+3`
subtraction, a − b	−	`23-12`
multiplication, a · b	*	`3.14*0.85`
division, a ÷ b	/ or \	`56/8=8\56`
power, a^b	^	`5^2`

The order in which these operations are evaluated in a given expression is given by the usual rules of precedence that can be summarized as follows: **expressions are evaluated from left to right with the power operation having the highest order of precedence, followed by multiplication and division having equal precedence, followed by addition and subtraction having equal precedence.** Parentheses can be used to alter this usual ordering, in which case evaluation initiates within the innermost parentheses and proceeds outward.

2.2 THE MATLAB WORKSPACE

As you work in the *Command* window, MATLAB remembers the commands you enter as well as the values of any variables you create. These commands and variables are said to reside in the ***MATLAB Workspace*** and can be recalled whenever you wish. For example, to check the value of `tape` all you have to do is ask MATLAB for it by entering its name at the prompt.

```
» tape
tape =
     2
```

If you can't remember the name of a variable, you can ask MATLAB for a list of the variables it knows by using the MATLAB command `who`.

```
» who

Your variables are:

ans               average_cost      tape          items
erasers           pads              cost

leaving 3241192 bytes of memory free.
```

Note that MATLAB doesn't tell you the value of all the variables, it just gives you their names. To find their values you must enter their names at the MATLAB prompt. Just like a calculator, there's only so much room to store variables. The last line of the **who** command response says that there is lots of room left! Each piece of data takes 8 bytes, so there is room to create 405,142 more pieces of data. How much workspace room you have depends on the computer you are using.

To recall previous commands, MATLAB uses the **Cursor** keys, ←→↑↓, on your keyboard. For example, pressing the ↑ key once recalls the most recent command to the MATLAB prompt. Repeated pressing scrolls back through prior commands one at a time. In a similar manner pressing the ↓ key scrolls forward through commands. Moreover, entering the first few characters of a known previous command at the prompt, and then pressing the ↑ key, immediately recalls the most recent command having those initial characters. At any time the ← and → keys can be used to move the cursor within the command at the MATLAB prompt. In this manner the command can be edited. Depending on the computer you are using, there are other command line editing features. Please see your *MATLAB User's Guide* for further information.

2.3 SAVING AND RETRIEVING DATA

In addition to remembering variables, MATLAB can save and load data from files on your computer. On some platforms, the **Save Workspace As . . .** menu item in the **File** menu opens a standard file dialog box for saving all current variables. Similarly, the **Load Workspace . . .** menu item in the **File** menu opens a dialog box for loading variables from a previously saved workspace. Saving variables does not delete them from the MATLAB workspace. Loading variables of the same name as those found in the MATLAB workspace changes the variable values to those loaded from the file.

If the **File** menu approach is not available or does not meet your needs, MATLAB provides two commands, **save** and **load**, which offer more flexibility. In particular, the **save** command allows you to save one or more variables in the file format of your choice. For example,

```
» save
```

stores all variables in binary format in the file **matlab.mat**.

```
» save data
```

saves all variables in binary format in the file `data.mat`.

> » `save data erasers pads tape`

saves the variables `erasers`, `pads`, and `tape` in binary format in the file `data.mat`.

> » `save data erasers pads tape -ascii`

saves the variables `erasers`, `pads`, and `tape` in 8 digit ASCII format in the file `data`. ASCII formated files are editable using any common text editor. Note that ASCII files do not get the extension `.mat`.

> » `save data erasers pads tape -ascii -double`

saves the variables `erasers`, `pads`, and `tape` in 16 digit ASCII format in the file `data`.

For further information on these commands, see the *MATLAB Reference Guide*. For even greater flexibility, MATLAB provides numerous low level file I/O functions that mimic those found in the C programming language. These functions give one the ability to save and load files in any format and are briefly described in Chapter 15.

2.4 NUMBER DISPLAY FORMATS

When MATLAB displays numerical results, it follows several rules. By default, if a result is an integer, MATLAB displays it as an integer. Likewise, when a result is a real number, MATLAB displays it with approximately four digits to the right of the decimal point. If the significant digits in the result are outside of this range, MATLAB displays the result in scientific notation similar to scientific calculators. You can override this default behavior by specifying a different numerical format using the **Numerical Format** menu item in the **Options** menu if available or by typing the appropriate MATLAB command at the prompt. Using the variable `average_cost` from the above example, these numerical formats are

MATLAB Command	average_cost	Comments
`format long`	`35.83333333333334`	16 digits
`format short e`	`35.833e+01`	5 digits plus exponent
`format long e`	`35.83333333333334e+01`	16 digits plus exponent
`format hex`	`4041eaaaaaaaaaab`	hexadecimal
`format bank`	`35.83`	2 decimal digits

MATLAB Command	average_cost	Comments
`format +`	`+`	positive, negative, or zero
`format rat`	`215/6`	rational approximation
`format short`	`35.8333`	default display

It is important to note that MATLAB does not change the internal representation of a number when different display formats are chosen; only the display changes.

2.5 ABOUT VARIABLES

Like any other computer language, MATLAB has rules about variable names. Earlier it was noted that variable names must be a single word containing no spaces. More specifically, MATLAB variable naming rules are

Variable Naming Rules	Comments/Examples
Variable names are case sensitive.	`Items, items, itEms,` and **`ITEMS`** are all different MATLAB variables.
Variable names can contain up to 19 characters; characters beyond the 19th are ignored.	`howaboutthisvariablename`
Variable names must start with a letter, followed by any number of letters, digits, or underscores. Punctuation characters are not allowed since many have special meaning to MATLAB.	`how_about_this_variable_name` `X51483` `a_b_c_d_e`

In addition to these naming rules, MATLAB has several special variables. They are

Special Variables	Value
ans	The default variable name used for results
pi	The ratio of the circumference of a circle to its diameter
eps	The smallest number such that when added to one creates a number greater than one on the computer
flops	Count of floating point operations
inf	Which stands for infinity, e.g., 1/0
NaN	Which stands for Not-a-Number, e.g., 0/0
i (and) j	$i = j = \sqrt{-1}$
nargin	Number of function input arguments used
nargout	Number of function output arguments used
realmin	The smallest usable positive real number
realmax	The largest usable positive real number

As you create variables in MATLAB, there may be instances where you wish to redefine one or more variables. For example,

```
» erasers=4;

» pads=6;

» tape=2;

» items = erasers+pads+tape
items =
      12

» erasers = 6
erasers =
      6

» items
items =
      12
```

Here, using the first example again, we found the number of items Mark purchased. Afterward we changed the number of erasers to 6, overwriting its prior value of 4. In doing so, the value of **items** has not changed. Unlike a spreadsheet, MATLAB does *NOT* recalculate the number of items based on the new value of **erasers**. When MATLAB performs a calculation, it does so using the values it knows at the time the requested command is evaluated. In the above example, if you wish to recalculate the number of items, the total cost, and the average cost, it is necessary to recall the appropriate MATLAB commands and ask MATLAB to evaluate them again.

The special variables given on page 9 follow this guideline also. When you start MATLAB, they have the values given above. If you change their values, the original special values are lost until you clear the variables or you restart MATLAB. With this in mind, avoid redefining special variables, unless absolutely necessary.

Variables in the MATLAB workspace can be unconditionally deleted by using the command **clear**. For example,

```
» clear erasers
```

deletes just the variable **erasers**.

```
» clear cost items
```

deletes both **cost** and **items**.

```
» clear
```

deletes all variables in the workspace! You are not asked to confirm this command. All variables are cleared and cannot be retrieved!

Needless to say the **clear** command is dangerous and should be used with caution. Thankfully, there is seldom a need to clear all variables from the workspace.

2.6 COMMENTS AND PUNCTUATION

All text after a percent sign (%) is taken as a comment statement.

```
» erasers=4 % Number of erasers.
erasers =
     4
```

The variable **erasers** is given the value 4 and MATLAB simply ignores the percent sign and all text following it. This feature makes it easy to document what you are doing.

Multiple commands can be placed on one line if they are separated by commas or semicolons.

```
» erasers=4, pads=6; tape=2
erasers =
     4
```

```
                    tape =
                       2
```

Commas tell MATLAB to display results; semicolons suppress printing.

```
         » average_cost=cost/...
            items
         average_cost =
              35.8333
```

As shown above, a succession of three periods tells MATLAB that the rest of a statement appears on the next line. Statement continuation as shown above works if the succession of three periods occurs between variable names or operators. That is, a variable name cannot be split between two lines.

```
         » average_cost=cost/it...
      ems
      ??? age_cost=cost/items
                            |
      Missing operator, comma, or semi-colon.
```

Likewise, comment statements cannot be continued.

```
         » % Comments cannot be continued...
         » either
      ??? Undefined function or variable either.
```

You can interrupt MATLAB at any time by pressing **Ctrl-C** (pressing the **Ctrl** and **C** keys simultaneously) on a PC. Pressing ⌘. (pressing the ⌘ and . keys simultaneously) on a Macintosh does the same thing. See your *MATLAB User's Guide* for information on interrupting MATLAB on other computer platforms.

2.7 COMPLEX NUMBERS

One of the most powerful features of MATLAB is that it does not require any special handling for complex numbers. Complex numbers are formed in MATLAB in several ways. Examples of complex numbers include:

```
         » c1=1-2i % the appended i signifies the imaginary part
         c1 =
            1.0000 - 2.0000i

         » c1=1-2j % j also works
         c1 =
            1.0000 - 2.0000i
```

```
» c2=3*(2-sqrt(-1)*3)
c2 =
   6.0000 - 9.0000i

» c3=sqrt(-2)
c3 =
        0 + 1.4142i

» c4=6+sin(.5)*i
c4 =
   6.0000 + 0.4794i

» c5=6+sin(.5)*j
c5 =
   6.0000 + 0.4794i
```

In the last two examples, the MATLAB default values of $i=j=\sqrt{-1}$ are used to form the imaginary part. Multiplication by i or j is required in these cases since $sin(.5)i$ and $sin(.5)j$ have no meaning to MATLAB. Termination with the characters i and j, as shown in the first two examples above, only works with simple numbers, not expressions.

Some programming languages require special handling for complex numbers wherever they appear. In MATLAB no special handling is required. Mathematical operations for complex numbers are written the same as those for real numbers.

```
» c6=(c1+c2)/c3 % from the above data
c6 =
  -7.7782 - 4.9497i

» check_it_out=i^2 % sqrt(-1) squared must be -1!
check_it_out =
  -1.0000 + 0.0000i

» check_it_out=real(check_it_out) % strip off imaginary part
check_it_out =
     -1
```

In general, operations on complex numbers lead to complex numbers. Thus, even though $i^2 = -1$ is strictly real, MATLAB keeps the zero imaginary part. As shown, the MATLAB function **real** extracts the real part of a complex number.

As a final example of complex arithmetic, consider the Euler (sounds like *Oiler*) identity which relates the *polar* form of a complex number to its *rectangular* form.

$$M\angle\theta \equiv M \cdot e^{j\theta} = a + bi$$

where the polar form is given by a *magnitude* M and an *angle* θ, and the rectangular form is given by a + bi. The relationships among these forms are

$$M = \sqrt{a^2 + b^2}$$

$$\theta = \tan^{-1}(b/a)$$

$$a = M \cos \theta$$

$$b = M \sin \theta$$

In MATLAB the conversion between polar and rectangular forms makes use of the functions `real`, `imag`, `abs`, and `angle`.

```
» c1
c1 =
    1.0000 - 2.0000i

» mag_c1=abs(c1)
mag_c1 =
    2.2361

» angle_c1=angle(c1)
angle_c1 =
   -1.1071

» deg_c1=angle_c1*180/pi
deg_c1 =
  -63.4349

» real_c1=real(c1)
real_c1 =
    1

» imag_c1=imag(c1)
imag_c1 =
   -2
```

The MATLAB function **abs** computes the magnitude of complex numbers or the absolute value of real numbers, depending upon which one you give it. Likewise the MATLAB function **angle** computes the angle of a complex number in radians.

2.8 MATHEMATICAL FUNCTIONS

A partial list of the common functions that MATLAB supports is shown in the table on pages 14–15. Most of these functions are used the same way you would write them mathematically.

```
» x=sqrt(2)/2
x =
    0.7071
```

```
» y=asin(x)
y =
      0.7854

» y_deg=y*180/pi
y_deg =
      45.0000
```

These commands find the angle where the sine function has a value of $\sqrt{2}/2$. **Note that MATLAB only works in radians.** Other examples include:

```
» y=sqrt(3^2 + 4^2) % show 3-4-5 right triangle relationship
y =
     5

» y=rem(23,4) % remainder function, 23/4 has a remainder of 3
y =
     3

» x=2.6, y1=fix(x), y2=floor(x), y3=ceil(x), y4=round(x)
x =
     2.6000
y1 =
     2
y2 =
     2
y3 =
     3
y4 =
     3
```

See the *MATLAB Reference Guide* for more specific information on these functions. Further illustration of these functions will appear throughout this text.

Common Functions

abs(x)	Absolute value or magnitude of complex number
acos(x)	Inverse cosine
acosh(x)	Inverse hyperbolic cosine
angle(x)	Four quadrant angle of complex
asin(x)	Inverse sine
asinh(x)	Inverse hyperbolic sine
atan(x)	Inverse tangent

`atan2(x,y)`	Four quadrant inverse tangent
`atanh(x)`	Inverse hyperbolic tangent
`ceil(x)`	Round toward plus infinity
`conj(x)`	Complex conjugate
`cos(x)`	Cosine
`cosh(x)`	Hyperbolic cosine
`exp(x)`	Exponential: e^x
`fix(x)`	Round toward zero
`floor(x)`	Round toward minus infinity
`gcd(x,y)`	Greatest common divisor of integers **x** and **y**
`imag(x)`	Complex imaginary part
`lcm(x,y)`	Least common multiple of integers **x** and **y**
`log(x)`	Natural logarithm
`log10(x)`	Common logarithm
`real(x)`	Complex real part
`rem(x,y)`	Remainder after division: `rem(x,y)` gives the remainder of **x/y**
`round(x)`	Round toward nearest integer
`sign(x)`	Signum function: return sign of argument, e.g., `sign(1.2)=1, sign(-23.4)=-1, sign(0)=0`
`sin(x)`	Sine
`sinh(x)`	Hyperbolic sine
`sqrt(x)`	Square root
`tan(x)`	Tangent
`tanh(x)`	Hyperbolic tangent

2.9 SCRIPT FILES

For simple problems, entering your requests at the MATLAB prompt is fast and efficient. However, as the number of commands increases or in the case when you wish to change the value of one or more variables and reevaluate a number of commands, typing at the MATLAB prompt quickly becomes tedious. MATLAB provides a logical solution to this problem. It allows you to place MATLAB commands in a simple text file and then tell MATLAB to open the file and evaluate commands exactly as it would if you had typed them at the MATLAB prompt. These files are called *script* files or simply *M-files.* The term *script* symbolizes the fact that MATLAB simply reads from the script found in the file. The term *M-file* recognizes the fact that script file names must end with the extension `.m`, e.g., `example1.m`.

To create a script M-file on a Macintosh or PC, choose **New** from the File menu and select **M-file.** This procedure brings up a text editor window where you enter MATLAB commands. On other platforms, it is convenient to open a separate terminal window and use your favorite text editor in that window to generate script M-files. The figure below shows the commands from the office supply store example considered at the beginning of this chapter.

```
% example1 script file
erasers=4;
pads=6;
tape=2;
items=erasers+pads+tape % number of items purchased
cost=erasers*25+pads*22+tape*99 % total cost
average_cost=cost/items
```

After this file is saved as the M-file **example1.m** on your disk, MATLAB will execute the commands in **example1.m** by simply typing **example1** at the MATLAB prompt.

```
» example1
items =
      12
cost =
     430
average_cost =
     35.8333
```

When MATLAB interprets the **example1** statement above, it follows the hierarchy described in the next section of this chapter. In brief, MATLAB prioritizes current MATLAB variables and built-in MATLAB commands ahead of M-file names. Thus, if **example1** is not a current MATLAB variable or a built-in MATLAB command (it isn't), MATLAB opens the file **example1.m** (if it can find it) and evaluates the commands found there just as if they were entered directly at the *Command* window prompt. As a result, commands within the M-file have access to all variables in the MATLAB workspace and all variables created in the M-file become part of the workspace. Normally, the commands read in from the M-file are not displayed as they are evaluated. The **echo on** command tells MATLAB to display or echo commands to the *Command* window as they are read and evaluated. You can probably guess what the **echo off** command does. Similarly, the command **echo** by itself toggles the echo state.

With this M-file feature it is simple to answer *what if?* questions. For example, one can repeatedly open the **example1.m** M-file, change the number of **erasers**, **pads**, or **tape**, save the file, and then ask MATLAB to reevaluate the commands in the file. The power of this capability cannot be overstated. Moreover, by creating M-files your commands are saved on disk for future MATLAB sessions.

Script files are also convenient for entering large arrays that may, for example, come from laboratory measurements. By using a text editor to enter one or more arrays, the editing capabilities of the editor make it is easy to correct mistakes without having to type the whole array in again. As above, this approach also saves the data on disk for future use.

The utility of MATLAB comments is readily apparent when using script files as shown in **example1.m**. Comments allow you to document the commands found in a script file so that they are not forgotten when viewed in the future. In addition, the use of semicolons at the end of lines to suppress the display of results allows you to control script file output so that only important results are shown.

Because of the utility of script files, MATLAB provides several functions that are particularly useful when used in M-files. They are

---

M-File Functions

disp(ans)	Display results without identifying variable names
echo	Control the *Command* window echoing of script file commands
input	Prompt user for input
keyboard	Give control to keyboard temporarily (Type **return** to quit)
pause	Pause until user presses any keyboard key
pause(n)	Pause for **n** seconds
waitforbuttonpress	Pause until user presses mouse key, or keyboard key

---

When a MATLAB command is not terminated in a semicolon, the results of the command are displayed in the *Command* window with the variable name identified. For a prettier display it is sometimes convenient to suppress the variable name. In MATLAB this is accomplished with the command **disp**.

```
» erasers=4 % traditional way to display result
erasers =
     4
» erasers=6;

» disp(erasers) % display result without variable name
     6
```

Rather than repeatedly edit a script file when computations for a variety of cases are desired, the `input` command allows one to prompt for input as a script file is executed. For example, reconsider the `example1.m` script file with modifications.

```
% example script file, modified to include 'input' function.

erasers=4;
pads=6;
tape=input('Enter number of rolls of tape > '); % prompt quote
items=erasers+pads+tape
cost=erasers*25+pads*22+tape*99
average_cost=cost/items
```

Running this script file produces

```
» example1

Enter number of rolls of tape > 5
items =
     15
cost =
    727
average_cost =
    48.4667
```

In response to the prompt, the number 5 was entered and the **Return** or **Enter** key was pressed. The remaining commands were evaluated with `tape=5`. The function input accepts any valid MATLAB expression for input. For example, running the script file again and providing different input gives

```
» example1

Enter number of roles of tape > round(sqrt(13))+3
items =
     17
cost =
    925
average_cost =
    54.4118
```

In this case, the number of roles of tape was set equal to the result of evaluating the expression `round(sqrt(13))+3`.

The `keyboard` command gives control to the keyboard and lets the user enter commands as desired. When the command `return` is entered, evaluation of commands from the script file resumes. Again consider the above script file, modified to include the `keyboard command`.

```
% example script file, modified to include 'keyboard' command.

erasers=4;
pads=6;
tape=2;
keyboard
items=erasers+pads+tape
cost=erasers*25+pads*22+tape*99
average_cost=cost/items
```

Running this script file produces

```
» example1

K» tape=7
tape =
        7
K» who


Your variables are:

erasers     pads        tape

leaving 455808 bytes of memory free.

K» return
items =
      17
cost =
    925
average_cost =
    54.4118
```

When the command **keyboard** is exe~uted, the prompt **K»** appears saying that control has been passed to the keyboard temporaril, In this example, the number of rolls of tape was changed to 7, the **who** command was issued showing the variables in existence, and the **return** command was entered passing control back to the script file.

2.10 FILE MANAGEMENT

MATLAB provides several file management commands that allow you to list file names, view and delete M-files, and show and change the current directory or folder. In addition, you can view and modify MATLAB's search path. A summary of these commands is given in the following table.

File Management Functions

cd	Show present working directory or folder
p=cd	Return present working directory in **p**
cd path	Change to directory or folder given by **path**
chdir	Same as **cd**
chdir path	Same as **cd path**
delete test	Delete the M-file **test.m**
dir	List all files in the current directory or folder
ls	Same as **dir**
matlabroot	Return directory path to MATLAB executable program
path	Display or modify MATLAB's search path
pwd	Same as **cd**
type test	Display the M-file **test.m** in the *Command window*
what	Return a listing of all M-files and MAT-files in the current directory or folder
which test	Display the directory path to **test.m**

Of the commands in the above table, the **path** command controls MATLAB's Search Path. This path is a list of directories where MATLAB files are located. Use of MATLAB's search path is described as follows.

MATLAB Search Path

In general, when you enter » `cow,` MATLAB does the following:

(1) it checks to see if `cow` is a **variable** in the MATLAB workspace; if not,

(2) it checks to see if `cow` is a **built-in function;** if not,

(3) it checks to see if a MEX-file `cow.mex` exists in the **current directory;** if not,

(4) it checks to see if an M-file named `cow.m` exists in the **current directory;** if not,

(5) it checks to see if `cow.mex` or `cow.m` exists anywhere on the **MATLAB Search Path,** by searching the path in the order in which it is specified.

Whenever a match is found, MATLAB accepts it and acts accordingly. The above path searching procedure is also followed when the `load` command is issued. First MATLAB looks in the current directory, and then it begins searching the MATLAB Search Path for the desired data file.

If you have M-files, MAT-files, or MEX-files stored in a directory not on the MATLAB Search Path and not in the current directory, MATLAB cannot find them. There are two solutions to this problem: (1) make the desired directory the current directory using the `cd` or `chdir` commands from the above table, or (2) add the desired directory to the MATLAB Search Path.

This last approach is easily accomplished using the path command.

```
» path
      MATLABPATH

    hd:Applications:MATLAB:Toolbox:matlab:general
    hd:Applications:MATLAB:Toolbox:matlab:ops
    hd:Applications:MATLAB:Toolbox:matlab:lang
    hd:Applications:MATLAB:Toolbox:matlab:elmat
    hd:Applications:MATLAB:Toolbox:matlab:specmat
    hd:Applications:MATLAB:Toolbox:matlab:elfun
    hd:Applications:MATLAB:Toolbox:matlab:specfun
    hd:Applications:MATLAB:Toolbox:matlab:matfun
    hd:Applications:MATLAB:Toolbox:matlab:datafun
    hd:Applications:MATLAB:Toolbox:matlab:polyfun
    hd:Applications:MATLAB:Toolbox:matlab:funfun
    hd:Applications:MATLAB:Toolbox:matlab:sparfun
```

```
hd:Applications:MATLAB:Toolbox:matlab:plotxy
hd:Applications:MATLAB:Toolbox:matlab:plotxyz
hd:Applications:MATLAB:Toolbox:matlab:graphics
hd:Applications:MATLAB:Toolbox:matlab:color
hd:Applications:MATLAB:Toolbox:matlab:sounds
hd:Applications:MATLAB:Toolbox:matlab:strfun
hd:Applications:MATLAB:Toolbox:matlab:iofun
hd:Applications:MATLAB:Toolbox:matlab:demos
hd:Applications:MATLAB:Toolbox:local
hd:Applications:MATLAB:extern
hd:Applications:MATLAB
```

Issuing path with no arguments shows the current MATLAB Search Path. Note that the above display changes depending on your computer platform and MATLAB configuration.

```
» path(path,'hd:Applications:MATLAB:Toolbox:MM')
```

appends **hd:Applications:MATLAB:Toolbox:MM** to the current path above.

If you have one or more directories that you wish to permanently search, it is convenient to place **path** commands that append the desired directories in the script file **startup.m** as described in section 2.12.

2.11 COMMAND WINDOW CONTROL

Depending on your hardware setup, you may not have a menu bar or scroll bars on the *Command* window. For this reason, MATLAB provides several commands for managing the *Command* window. These commands include:

Command Window Control Commands

clc	Clear the *Command* window
diary	Save *Command* window text to a file
home	Move cursor to upper left corner
more	Page the *Command* window

The command **more** keeps the *Command* window from scrolling text beyond view. The command **diary** creates a file containing *Command* window text as it is generated, but does not save *Figure* windows.

2.12 MATLAB AT STARTUP

When MATLAB starts up, it executes two script M-files, `matlabrc.m` and `startup.m`. Of these, `matlabrc.m` comes with MATLAB and generally should not be modified. The commands in this M-file set the default *Figure* window size and placement as well as a number of other default features. On some platforms, the default MATLAB Search Path is set in `matlabrc.m`. On all platforms, commands in `matlabrc.m` check for the existence of the script M-file `startup.m` on the MATLAB Search Path. If it exists, the commands in it are executed.

The optional M-file `startup.m` typically contains commands that add personal default features to MATLAB. For example, it is very common to put one or more `path` commands in `startup.m` to append additional directories to the MATLAB Search Path. Similarly, the default number display format can be changed, e.g., `format compact`. If you have a grayscale monitor, the command `graymon` is useful for setting default grayscale graphics features. Further, if you want plots to have a white background rather than the default black, the command `whitebg` can be placed in `startup.m`. Since `startup.m` is a standard script M-file, there are no restrictions as to what commands can be placed in it. However, it's probably not wise to place the `quit` command in `startup.m`!

2.13 ON-LINE HELP

You probably have the sense that MATLAB has many more commands than you can possibly remember. To help you find commands, MATLAB provides assistance through its extensive **on-line help** capabilities. These capabilities are available in three forms: the MATLAB command `help`, the MATLAB command `lookfor`, and interactively using help from the menu bar.

The help Command

The MATLAB `help` command is the simplest way to get help if you know the topic you want help on. Typing `help topic` displays help about that topic if it exists.

```
» help sqrt

SQRT    Square root.
    SQRT(X) is the square root of the elements of X. Complex
    results are produced if X is not positive.

    See also SQRTM.
```

Here we received help on MATLAB's square root function. On the other hand,

```
» help cows

cows not found.
```

simply says that MATLAB knows nothing about **cows**.

Note in the above **sqrt** example that **SQRT** is capitalized in the help text. However, when used **sqrt** is never capitalized. In fact, because MATLAB is case sensitive, **SQRT** is unknown and produces an error.

```
» SQRT(2)
??? SQRT(
        |
Missing operator, comma, or semi-colon.
```

To summarize, function names are capitalized in the help text solely to aid readability; but in use, functions are called using lower case characters.

The **help** command works well if you know the exact topic you want help on. Since many times this isn't true, **help** provides guidance to direct you to the exact topic you want by simply typing **help** with no **topic**.

```
» help

HELP topics:

MATLAB :general     - General purpose commands.
MATLAB :ops         - Operators and special characters.
MATLAB :lang        - Language constructs and debugging.
MATLAB :elmat       - Elementary matrices and matrix manipulation.
MATLAB :specmat     - Specialized matrices.
MATLAB :elfun       - Elementary math functions.
MATLAB :specfun     - Specialized math functions.
MATLAB :matfun      - Matrix functions - numerical linear algebra.
MATLAB :datafun     - Data analysis and Fourier transform functions.
MATLAB :polyfun     - Polynomial and interpolation functions.
MATLAB :funfun      - Function functions - nonlinear numerical methods.
MATLAB :sparfun     - Sparse matrix functions.
MATLAB :plotxy      - Two-dimensional graphics.
MATLAB :plotxyz     - Three dimensional graphics.
MATLAB :graphics    - General purpose graphics functions.
MATLAB :color       - Color control and lighting model functions.
MATLAB :sounds      - Sound processing functions.
MATLAB :strfun      - Character string functions.
MATLAB :iofun       - Low-level file I/O functions.
```

```
MATLAB :demos          - The MATLAB Expo and other demonstrations.
Toolbox:local          - Local function library.


For more help on directory/topic, type "help topic".
```

Your display may differ slightly from the above. In any case, this display describes categories from which you can ask for help. For example, » **help general** returns a list (too long to show here) of general MATLAB topics that you can use the » **help topic** command to get help on.

While the **help** command allows you to access help, it is not the most convenient way to do so unless you know the exact topic you are seeking help on. When you are uncertain about the spelling or existence of a topic, the other two approaches to obtaining help are often more productive.

The lookfor Command

The **lookfor** command provides help by searching through all the first lines of MATLAB help topics and M-files on the MATLAB Search Path and returning those that contain a key word you specify. **Most important is that the key word need not be a MATLAB command.** For example,

```
» lookfor complex

CONJ       Complex conjugate.
IMAG       Complex imaginary part.
REAL       Complex real part.
CDF2RDF    Complex diagonal form to real block diagonal form.
RSF2CSF    Real block diagonal form to complex diagonal form.
CPLXPAIR   Sort numbers into complex conjugate pairs.
```

The key word **complex** is not a MATLAB command, but was found in the help descriptions of six MATLAB commands. You may find more or less. Given this information, the **help** command can be used to display help about a specific command.

```
» help conj
CONJ Complex conjugate.
    CONJ(X) is the complex conjugate of X.
```

In summary, the **lookfor** command provides a way to find MATLAB commands and help topics given a general key word.

Menu-driven Help

As an alternative to getting help in the *Command* window, menu-driven help may be available from the **Menu Bar** depending on the computer you are using. On a Macintosh, menu-driven help is available by selecting **About MATLAB . . .** from the **Apple** menu or by selecting **MATLAB Help** from the **Balloon Help** menu. On a PC, menu-driven help is available by selecting **Table of Contents . . .** or **Index . . .** from the **Help** menu. Doing so creates a *MATLAB Help* window that lets you double-click to select any topic or function from the displayed list. The *MATLAB Help* window uses the standard *MS Windows* help format that allows you to search for topics, set bookmarks, annotate topics, and print help screens.

For more assistance on menu-driven help, see the *MATLAB User's Guide*.

3

Arrays

All of the computations considered to this point have involved single numbers called *scalars.* Operations involving scalars are the basis of mathematics. At the same time, when one wishes to perform the same operation on more than one number at a time, repeated scalar operations are time-consuming and cumbersome. To solve this problem, MATLAB defines operations on data arrays.

3.1 SIMPLE ARRAYS

Consider the problem of computing values of the sine function over one half of its period, namely: y = sin(x) over $0 \le x \le \pi$. Since it is impossible to compute sin(x) at all points over this range (there are an infinite number of them!), we must choose a finite number of points. In doing so we are *sampling* the function. To pick some number, let's evaluate sin(x) every 0.1π in this range, i.e., let x = 0, 0.1π, 0.2π, ... , 1.0π. If you were using a scientific calculator to compute these values, you would start by making a list or *array* of the values of x. Then you would enter each value of x into your calculator, find its sine, and write down the result as the second *array* y. Perhaps you would write them in an organized fashion as follows:

x	0	0.1π	0.2π	0.3π	0.4π	0.5π	0.6π	0.7π	0.8π	0.9π	π
y	0	0.31	0.59	0.81	0.95	1.0	0.95	0.81	0.59	0.31	0

As shown, x and y are ordered lists of numbers, i.e., the first value or element in y is associated with the first value or element in x, the second element in y is associated with the second element in x, and so on. Because of this ordering, it is common to refer to individual values or elements in x and y with subscripts, e.g., x_1 is the first element in x, y_5 is the fifth element in y, x_n is the nth element in x.

MATLAB handles arrays in a straightforward and intuitive way. Creating arrays is easy—just follow the visual organization given above.

```
» x=[0 .1*pi .2*pi .3*pi .4*pi .5*pi .6*pi .7*pi .8*pi .9*pi pi]
x =
  Columns 1 through 7
         0    0.3142    0.6283    0.9425    1.2566    1.5708    1.8850
  Columns 8 through 11
    2.1991    2.5133    2.8274    3.1416

» y=sin(x)
y =
  Columns 1 through 7
         0    0.3090    0.5878    0.8090    0.9511    1.0000    0.9511
  Columns 8 through 11
    0.8090    0.5878    0.3090    0.0000
```

To create an array in MATLAB, all you have to do is start with a left bracket, enter the desired values separated by spaces (or commas), and then close the array with a right bracket. Notice that finding the sine of the values in **x** follows naturally. MATLAB understands that you want to find the sine of each element in **x** and place the results in an associated array called **y**. This fundamental capability makes MATLAB different from other computer languages.

Since spaces separate array values, complex numbers entered as array values cannot have embedded spaces unless expressions are enclosed in parentheses. For example, [1 -2i 3 4 5+6i] contains 5 elements whereas the identical arrays [(1 -2i) 3 4 5+6i] and [1-2i 3 4 5+6i] contain four.

3.2 ARRAY ADDRESSING

Now since **x** in the above example has more than one element, namely it has 11 values separated into columns, MATLAB gives you the result back with the columns identified. As shown above, **x** is an array having one row and eleven columns, or in mathematical jargon it is a ***row vector,*** a ***one-by-eleven array,*** or simply an *array* of length 11.

In MATLAB, individual array elements are accessed using ***subscripts,*** e.g., **x(1)** is the first element in **x**, **x(2)** is the second element in **x**, and so on. For example,

```
» x(3) % The third element of x
ans =
    0.6283

» y(5) % The fifth element of y
ans =
    0.9511
```

To access a block of elements at one time, MATLAB provides ***colon notation.***

```
» x(1:5)
ans =
    0      0.3142      0.6283      0.9425      1.2566
```

This is the first through fifth elements in **x**. **1:5** says start with 1 and count up to 5.

```
» y(3:-1:1)
ans =
    0.5878      0.3090          0
```

This is the third, second, and first elements in reverse order. **3:-1:1** says start with 3, count down by 1, and stop at 1.

```
» x(2:2:7)
ans =
    0.3142      0.9425      1.5708
```

This is the second, fourth, and sixth elements in **x**. **2:2:7** says start with 2, count up by 2, and stop when you get to 7. In this case adding 2 to 6 gives 8, which is greater than 7 so the eighth element is not included.

```
» y([8 2 9 1])
ans =
    0.8090      0.3090      0.5878      0
```

Here we used another array **[8 2 9 1]** to extract the elements of the array **y** in the order we wanted them! The first element taken is the eighth, the second is the second, the third is the ninth, and the fourth is the first. In reality **[8 2 9 1]** itself is an array that addresses the desired elements of **y**.

3.3 ARRAY CONSTRUCTION

Earlier we entered the values of **x** by typing each individual element in **x**. While this is fine when there are only 11 values in **x**, what if there were 111 values? Using the colon notation, two other ways of entering **x** are

```
» x=(0:0.1:1)*pi
x =
  Columns 1 through 7
        0    0.3142    0.6283    0.9425    1.2566    1.5708    1.8850
  Columns 8 through 11
    2.1991    2.5133    2.8274    3.1416

» x=linspace(0,pi,11)
x =
  Columns 1 through 7
        0    0.3142    0.6283    0.9425    1.2566    1.5708    1.8850
  Columns 8 through 11
    2.1991    2.5133    2.8274    3.1416
```

In the first case above, the colon notation (`0:0.1:1`) creates an array that starts at 0, increments or counts by 0.1, and ends at 1. Each element in this array is then multiplied by π to create the desired values in **x**. In the second case, the MATLAB function **linspace** is used to create **x**. This function's arguments are described by

$$linspace(first\_value,last\_value,number\_of\_values)$$

Both of these array creation forms are common in MATLAB. The colon notation form allows you to directly specify the increment between data points, but not the number of data points. **linspace** on the other hand allows you to directly specify the number of data points, but not the increment between the data points.

Both of the above array creation forms create arrays where the individual elements are linearly spaced with respect to each other. For the special case where a logarithmically spaced array is desired, MATLAB provides the **logspace** function

```
» logspace(0,2,11)
ans =
  Columns 1 through 7
    1.0000    1.5849    2.5119    3.9811    6.3096   10.0000   15.8489
  Columns 8 through 11
   25.1189   39.8107   63.0957  100.0000
```

Here we created an array starting at 10^0, ending at 10^2, and containing 11 values. The function arguments are described by

$$logspace(first\_exponent,last\_exponent,number\_of\_values)$$

Though it is common to begin and end at integer powers of ten, **logspace** works equally well with nonintegers.

Sometimes an array is required that is not conveniently described by a linearly or logrithmically spaced element relationship. There is no uniform way to create these arrays. However, array addressing and the ability to combine expressions can help eliminate the need to enter individual elements one at a time.

```
» a=1:5,b=1:2:9
a =
      1      2      3      4      5
b =
      1      3      5      7      9
```

creates two arrays. Remember that multiple statements can appear on a single line if they are separated by commas or semicolons.

```
» c=[b a]
c =
      1    3    5    7    9    1    2    3    4    5
```

creates an array **c** composed of the elements of **b** followed by those of **a**.

```
» d=[a(1:2:5) 1 0 1]
d =
      1      3      5      1      0      1
```

creates an array **d** composed of the first, third, and fifth elements of **a** followed by three additional elements.

The simple array construction features of MATLAB are summarized in the table below.

Simple Array Construction

`x=[2 2*pi sqrt(2) 2-3j]`	Create row vector **x** containing elements specified
`x=first:last`	Create row vector **x** starting with **first**, counting by one, ending at or before **last**
`x=first:increment:last`	Create row vector **x** starting with **first**, counting by **increment**, ending at or before **last**

Simple Array Construction

`x=linspace(first,last,n)`	Create row vector **x** starting with `first`, ending at `last`, having **n** elements
`x=logspace(first,last,n)`	Create logarithmically-spaced row vector **x** starting with 10^{first}, ending at 10^{last}, having **n** elements

3.4 ARRAY ORIENTATION

In the above examples, arrays contained one row and multiple columns. As a result of this row orientation, they are commonly called *row vectors.* It is also possible for an array to be a *column vector,* having one column and multiple rows. In this case, all of the above array manipulation and mathematics apply without change. The only difference is that results are displayed as columns rather than rows.

Since the array creation functions illustrated above all create row vectors, there must be some way to create column vectors. The most straightforward way to create a column vector is to specify it element by element, separating values with semicolons.

```
» c=[1;2;3;4;5]
c =
         1
         2
         3
         4
         5
```

Based on this example, **separating elements by spaces or commas specifies elements in different columns, whereas separating elements by semicolons specifies elements in different rows.** The semicolon serves two purposes in MATLAB. At the end of a statement or between statements on a single line, it suppresses printing. When it appears *within* an array construction statement, it means *start a new row.*

To create a column vector using the colon notation `start:increment:end`, or the functions `linspace` and `logspace`, one must *transpose* the resulting row into a column using the MATLAB transpose operator (`'`).

```
» a=1:5
a =
        1     2     3     4     5
```

creates a row vector using the colon notation format.

```
» b=a'
b =
     1
     2
     3
     4
     5
```

uses the transpose operator to change the row vector **a** into the column vector **b**.

```
» c=b'
c =
    1    2    3    4    5
```

applies the transpose again and changes the column back into a row.

In addition to the simple transpose above, MATLAB also offers a transpose operator with a preceding dot. In this case the ***dot-transpose operator*** is interpreted as the non-complex conjugate transpose. When an array is complex, the transpose (') gives the complex conjugate transpose, i.e., the sign on the imaginary part is changed as part of the transpose operation. On the other hand, the dot-transpose (.') transposes the array but does not conjugate it.

```
» c=a.'
c =
     1
     2
     3
     4
     5
```

shows that .' and ' are identical for real data.

```
» d=a+i*a
d =
  Columns 1 through 4
   1.0000 + 1.0000i   2.0000 + 2.0000i   3.0000 + 3.0000i   4.0000 + 4.0000i
  Column 5
   5.0000 + 5.0000i
```

creates a simple complex row vector from the array **a** using the default value `i=sqrt(-1)`.

```
» e=d'
e =
   1.0000 - 1.0000i
   2.0000 - 2.0000i
```

```
              3.0000 - 3.0000i
              4.0000 - 4.0000i
              5.0000 - 5.0000i
```

creates a column vector **e** that is the *complex conjugate* transpose of **d**.

```
        » f=d.'
        f =
              1.0000 + 1.0000i
              2.0000 + 2.0000i
              3.0000 + 3.0000i
              4.0000 + 4.0000i
              5.0000 + 5.0000i
```

creates a column vector **f** that is the *noncomplex conjugate* transpose of **d**.

If an array can be a row vector or a column vector, it makes intuitive sense that arrays could just as well have both multiple rows and multiple columns. **That is, arrays can also be in the form of** *matrices.* Creation of matrices follows that of row and column vectors. Commas or spaces are used to separate elements in a specific row, and semicolons are used to separate individual rows.

```
        » g=[1 2 3 4;5 6 7 8]
        g =
              1      2      3      4
              5      6      7      8
```

Here **g** is an array or matrix having 2 rows and 4 columns, i.e., it is a *2 by 4 matrix* or it is a *matrix of dimension 2 by* **4**. The semicolon tells MATLAB to start a new row between the 4 and 5.

```
        » g=[1 2 3 4
        5 6 7 8
        9 10 11 12]
        g =
              1      2      3      4
              5      6      7      8
              9     10     11     12
```

In addition to semicolons, pressing the **Return** or **Enter** key while entering a matrix also tells MATLAB to start a new row.

```
        » h=[1 2 3;4 5 6 7]
        ??? All rows in the bracketed expression must have the same
        number of columns.
```

MATLAB strictly enforces the fact that all rows must contain the same number of columns.

3.5 SCALAR-ARRAY MATHEMATICS

In the first array example discussed previously, the array **x** is multiplied by the scalar π. Other simple mathematical operations between scalars and arrays follow the same natural interpretation. Addition, subtraction, multiplication, and division by a scalar simply apply the operation to all elements of the array.

```
» g-2
ans =
       -1     0     1     2
        3     4     5     6
        7     8     9    10
```

subtracts 2 from each element in **g**.

```
» 2*g-1
ans =
        1     3     5     7
        9    11    13    15
       17    19    21    23
```

multiplies each element in **g** by two and subtracts one from each element of the result. Note that scalar-array mathematics uses the same order of precedence used in scalar expressions to determine the order of evaluation.

3.6 ARRAY-ARRAY MATHEMATICS

Mathematical operations between arrays are not quite as simple as those between scalars and arrays. Clearly, array operations between arrays of different sizes or dimensions are difficult to define and of even more dubious value. **However, when two arrays have the same dimensions, addition, subtraction, multiplication, and division apply on an element-by-element basis in MATLAB.** For example,

```
» g % recall previous array
g =
        1     2     3     4
        5     6     7     8
        9    10    11    12

» h=[1 1 1 1;2 2 2 2;3 3 3 3] % create new array
h =
        1     1     1     1
        2     2     2     2
        3     3     3     3
```

```
» g+h % add h to g on an element-by-element basis
ans =
     2     3     4     5
     7     8     9    10
    12    13    14    15

» ans-h % subtract h from the previous answer to get g back
ans =
     1     2     3     4
     5     6     7     8
     9    10    11    12

» 2*g-h % multiplies g by 2 and subtracts h from the result
ans =
     1     3     5     7
     8    10    12    14
    15    17    19    21
```

Note that array-array mathematics also uses the same order of precedence used in scalar expressions to determine the order of evaluation.

Element-by-element multiplication and division work similarly but use slightly unconventional notation:

```
» g.*h
ans =
     1     2     3     4
    10    12    14    16
    27    30    33    36
```

Here we multiplied the arrays **g** and **h** element by element using the ***dot multiplication*** symbol **.*** . The dot preceding the standard asterisk multiplication symbol tells MATLAB to perform element-by-element array multiplication. Multiplication without the dot signifies ***matrix multiplication,*** which will be discussed later. For this particular example, matrix multiplication is not defined.

```
» g*h
??? Error using ==> *
Inner matrix dimensions must agree.
```

Array division, or dot division, requires use of the dot symbol also.

```
» g./h
ans =
    1.0000    2.0000    3.0000    4.0000
    2.5000    3.0000    3.5000    4.0000
    3.0000    3.3333    3.6667    4.0000
```

```
» h.\g
ans =
        1.0000      2.0000      3.0000      4.0000
        2.5000      3.0000      3.5000      4.0000
        3.0000      3.3333      3.6667      4.0000
```

As with scalars, division is defined using both the forward and backward slashes. In both cases, the array *below* the slash is divided into the array *above* the slash.

Division without the dot is the ***matrix division*** operation, which is an entirely different operation.

```
» g/h

Warning: Rank deficient, rank = 1 tol =    5.3291e-15
ans =
        0           0      0.8333
        0           0      2.1667
        0           0      3.5000
» h\g

Warning: Rank deficient, rank = 1 tol =    3.3233e-15
ans =
      2.7143    3.1429      3.5714    4.0000
        0           0           0         0
        0           0           0         0
        0           0           0         0
```

Matrix division gives results that are not necessarily the same size as **g** and **h**. Matrix operations are discussed later.

Array powers are defined in several ways. As with multiplication and division, ^ is reserved for matrix powers and .^ is used to denote element-by-element powers.

```
» g,h
g =
        1       2       3       4
        5       6       7       8
        9      10      11      12
h =
        1       1       1       1
        2       2       2       2
        3       3       3       3
```

recalls the arrays used earlier.

```
» g.^2
ans =
        1        4        9       16
       25       36       49       64
       81      100      121      144
```

squares the individual elements of **g**.

```
» g.^-1 % MATLAB doesn't like this syntax
??? g.^-
        |
Missing variable or function.

» g.^(-1)   % However, with parentheses it works!
ans =
    1.0000    0.5000    0.3333    0.2500
    0.2000    0.1667    0.1429    0.1250
    0.1111    0.1000    0.0909    0.0833
```

finds the reciprocal of each element in **g**.

```
» 2.^g
ans =
        2        4        8       16
       32       64      128      256
      512     1024     2048     4096
```

raises 2 to the power of each element in the array **g**.

```
» g.^h
ans =
        1        2        3        4
       25       36       49       64
      729     1000     1331     1728
```

raises the elements of **g** to the corresponding elements in **h**. In this case the first row is unchanged; the second row is squared, and the third row is cubed.

```
» g.^(h-1)
ans =
        1        1        1        1
        5        6        7        8
       81      100      121      144
```

shows that scalar and array operations can be combined.

The following table summarizes basic array operations.

Element-by-Element Array Mathematics

Illustrative data:	$a = [a_1\ a_2\ \ldots\ a_n]$, $b = [b_1\ b_2\ \ldots\ b_n]$, $c = $ <a scalar>
Scalar addition	$a+c = [a_1+c\ a_2+c\ \ldots\ a_n+c]$
Scalar multiplication	$a*c = [a_1*c\ a_2*c\ \ldots\ a_n*c]$
Array addition	$a+b = [a_1+b_1\ a_2+b_2\ \ldots\ a_n+b_n]$
Array multiplication	$a.*b = [a_1*b_1\ a_2*b_2\ \ldots\ a_n*b_n]$
Array right division	$a./b = [a_1/b_1\ a_2/b_2\ \ldots\ a_n/b_n]$
Array left division	$a.\backslash b = [a_1\backslash b_1\ a_2\backslash b_2\ \ldots\ a_n\backslash b_n]$
Array powers	$a.\wedge c = [a_1\wedge c\ a_2\wedge c\ \ldots\ a_n\wedge c]$
	$c.\wedge a = [c\wedge a_1\ c\wedge a_2\ \ldots\ c\wedge a_n]$
	$a.\wedge b = [a_1\wedge b_1\ a_2\wedge b_2\ \ldots\ a_n\wedge b_n]$

3.7 ARRAY MANIPULATION

Since arrays and matrices are fundamental to MATLAB, there are many ways to manipulate them in MATLAB. Once matrices are formed, MATLAB provides powerful ways to insert, extract, and rearrange subsets of them by identifying subscripts of interest. Knowledge of these features is a key to using MATLAB efficiently. To illustrate the matrix and array manipulation features of MATLAB, consider the following examples.

```
» A=[1 2 3;4 5 6;7 8 9]
A =
        1       2       3
        4       5       6
        7       8       9

» A(3,3)=0 % set element in 3rd row, 3rd column to zero
A =
        1       2       3
        4       5       6
        7       8       0
```

changes the element in the third row and third column to zero.

```
» A(2,6)=1 % set element in 2nd row, 6th column to one
A =
     1    2    3    0    0    0
     4    5    6    0    0    1
     7    8    0    0    0    0
```

places 1 in the second row, sixth column. Since **A** does not have six columns, the size of **A** is increased as necessary and filled with zeros so that the matrix remains rectangular.

```
» A=[1 2 3;4 5 6;7 8 9]; % restore original data

» B=A(3:-1:1,1:3)
B =
     7    8    9
     4    5    6
     1    2    3
```

creates a matrix **B** by taking the rows of **A** in reverse order.

```
» B=A(3:-1:1,:)
B =
     7    8    9
     4    5    6
     1    2    3
```

does the same as the above example. Here the **final single colon means take all columns.** That is, **:** is short for **1:3** in this example because **A** has three columns.

```
» C=[A B(:,[1 3])]
C =
     1    2    3    7    9
     4    5    6    4    6
     7    8    9    1    3
```

creates **C** by appending all rows in the first and third columns of **B** to the right of **A**.

```
» B=A(1:2,2:3)
B =
     2    3
     5    6
```

creates **B** by extracting the first two rows and last two columns of **A**.

```
» C=[1 3]
C =
     1    3
```

```
» B=A(C,C)
B =
        1       3
        7       9
```

uses the array **C** to index the matrix **A** rather than specifying them directly using the colon notation `start:increment:end` or `start:end`. In this example, **B** is formed from the first and third rows and first and third columns of **A**.

```
» B=A(:)
B =
            1
            4
            7
            2
            5
            8
            3
            6
            9
```

builds **B** by stretching **A** into a column vector, taking its columns one at a time in order.

```
» B=B.'
B =
        1     4     7     2     5     8     3     6     9
```

illustrates the dot-transpose operation introduced earlier.

```
» B=A
B =
        1       2       3
        4       5       6
        7       8       9

» B(:,2)=[]
B =
        1       3
        4       6
        7       9
```

redefines **B** by throwing away all rows in the second column of the original **B**. **When you set something equal to the empty matrix [], it gets deleted, causing the matrix to collapse to what remains.** Note that you must delete whole rows or columns so that the result remains rectangular.

```
» B=B.'
B =
            1       4       7
            3       6       9
```

illustrates the transpose of a matrix. In general the ith row becomes the ith column of the result, so the original 3-by-2 matrix becomes a 2-by-3 matrix.

```
» B(2,:)=[]
B =
            1       4       7
```

throws out the second row of **B**.

```
» A(2,:)=B
A =
            1       2       3
            1       4       7
            7       8       9
```

replaces the second row of **A** with **B**.

```
» B=A(:,[2 2 2 2]) % this manipulation gets used OFTEN!
B =
      2       2       2       2
      4       4       4       4
      8       8       8       8
```

creates **B** by duplicating all rows in the second column of **A** four times. At *The Mathworks Inc.*, this is commonly called *Tony's Trick*. In its most common form, a vector is duplicated to form a matrix.

```
» B=[1 4 7] % pick a new B to work with
B =
            1       4       7

» C=B([1 1 1 1],:)
C =
            1       4       7
            1       4       7
            1       4       7
            1       4       7
```

Here **C** is formed by duplicating each row of **B** four times.

```
» A % show A again
A =
     1      2      3
     1      4      7
     7      8      9
```

```
» A(2,2)=[]
???  In an assignment A(matrix,matrix) = B, the number of rows in B
and the number of elements in the A row index matrix must be the same.
```

shows that you can only throw out entire rows or columns. MATLAB does not know how to collapse a matrix when partial rows or columns are thrown out.

```
» B=A(4,:)
???  Index exceeds matrix dimensions.
```

Since **A** does not have a fourth row, MATLAB doesn't know what to do and says so.

```
» B(1:2,:)=A
???  In an assignment A(matrix,:) = B, the number of columns in A and
B must be the same.
```

shows that you can't squeeze one matrix into another of a different size.

```
» B(3:4,:)=A(2:3,:)
B =
     1      4      7
     0      0      0
     1      4      7
     7      8      9
```

But you can place the second and third columns of **A** into the same size area of **B**. Since the second through fourth rows of **B** did not exist, they are created as necessary. Moreover, the second row of **B** is unspecified, so it is filled with zeros.

```
» G(1:6)=A(:,2:3)
G =
     2      4      8      3      7      9
```

creates a row vector **G** by extracting all rows in the second and third columns of **A**. Note that the shapes of the matrices are different on opposite sides of the equal sign.

Sometimes it is more convenient to address matrix elements with a *single index*. When a single index is used in MATLAB, the index counts elements down the columns starting with the first. For example,

```
» D=[1 2 3 4;5 6 7 8; 9 10 11 12] % new data
D =
     1     2     3     4
     5     6     7     8
     9    10    11    12

» D(2) % second element
ans =
     5

» D(5) % fifth element (3 in first column plus 2 in second column)
ans =
     6

» D(12) % last element in matrix
ans =
    12

» D(4:7) % fourth through seventh elements
ans =
     2     6    10     3
```

There is a common vector-matrix manipulation that is easily written using the above array manipulation properties. Consider the following data:

```
» D=[1 2 3 4;5 6 7 8; 9 10 11 12] % same data as above
D =
     1     2     3     4
     5     6     7     8
     9    10    11    12

» v=[2; 4; 8] % a column vector
v =
     2
     4
     8
```

In many situations it is desirable to perform a mathematical operation between **v** and each column of **D**. (Or, if **v** were a row vector of length 4, we could perform an operation between **v** and each row of **D**.) For example, if we wanted to subtract **v** from each column of **D**, we could write

```
» E=[D(:,1)-v D(:,2)-v D(:,3)-v D(:,4)-v]
E =
    -1     0     1     2
     1     2     3     4
     1     2     3     4
```

Here we took each column of **D** separately and subtracted **v** from it. While this works, it is cumbersome to write. A faster approach is to duplicate **v** in order to create a matrix the size of **D**, then perform the mathematical operation. For example,

```
» E=D-[v v v v]
E =
       -1        0        1        2
        1        2        3        4
        1        2        3        4
```

This is faster than the previous approach, but still not necessarily the best approach, especially if one does not know how many columns **D** has. The best approach uses *Tony's Trick*.

```
» E=D-v(:,[1 1 1 1])
E =
       -1        0        1        2
        1        2        3        4
        1        2        3        4
```

Here **v(:,[1 1 1 1])** takes all rows of **v** in the first column and duplicates them to form four columns. When the number of columns in **D** is unknown, the functions **size** and **ones**, which will be described later in this chapter, can be used.

```
» c=size(D,2)
c =
    4

» E=D-v(:,ones(1,c))
E =
       -1        0        1        2
        1        2        3        4
        1        2        3        4
```

Here **size** returned the number of columns in **D**, and **ones(1,c)** created **[1 1 1 1]** as in the previous example.

In addition to addressing matrices based on their subscripts, arrays containing zeros and ones, called *logical arrays,* can also be used **if the size of the array is equal to that of the array it is addressing.** In this case, True (1) elements are retained and False (0) elements are discarded. Logical arrays are usually created through the use of a relational operator. A more thorough discussion of relational and logical operators can be found in Chapter 5 of this text.

```
» x=-3:3 % Create data
x =
     -3 -2 -1 0 1 2 3
```

```
» abs(x)>1
ans =
       1    1    0    0    0    1    1
```

returns ones where the absolute value of **x** is greater than 1.

```
» y=x(abs(x)>1)
y =
       -3   -2    2    3
```

creates **y** by taking those values of **x** where **x's** absolute value is greater than 1.

```
» y=x([1 1 1 1 0 0 0])
y =
       -3 -2 -1 0
```

creates **y** by selecting only the first four values, discarding others. Compare this result with

```
» y=x([1 1 1 1])
y =
       -3    -3    -3    -3
```

Here **y** is created by taking the first element of **x** four times. In the former example, the array [1 1 1 1 0 0 0] is the same size as **x**, thus **y** is created using the 0-1 logical array approach described here. In the latter example, the array [1   1   1   1] is smaller than **x**, so **y** is created using [1   1   1   1] as a numerical index as described earlier in this section. If this difference in usage is too subtle and confusing, try several examples on your own.

```
» y=x([1 0 1 0])
??? Index into matrix is negative or zero.
```

produces an error because [1   0   1   0] is not the same size as **x**, and **0** is not a valid array subscript.

```
» x(abs(x)>1)=[]
x =
       -1    0    1
```

throws out values of **x** where **abs(x)>1**. This example illustrates that relational operations can appear on the left-hand side of an assignment statement also.

Relational operations work on matrices as well as vectors.

```
» B=[5 -3;2 -4]
B =
       5      -3
       2      -4
```

```
» x=abs(B)>2
x =
        1      1
        0      1
```

Likewise, 0-1 logical array extraction works for matrices as well.

```
» y=b(abs(B)>2)
y =
        5
       -3
        4
```

However, the results are converted into a column vector, since there is no way to define a matrix having only three elements.

The above array-addressing techniques are summarized in the following table.

Array Addressing

`A(r,c)`	Addresses a subarray within **A** defined by the index vector of desired rows in **r** and index vector of desired columns in **c**
`A(r,:)`	Addresses a subarrray within **A** defined by the index vector of desired rows in **r** and all columns
`A(:,r)`	Addresses a subarray within **A** defined by all rows and the index vector of desired columns in **c**
`A(:)`	Addresses all elements of **A** as a column vector taken column by column
`A(i)`	Addresses a subarray within **A** defined by the single index vector of desired elements in **i**, as if **A** was the column vector, `A(:)`
`A(x)`	Addresses a subarray within **A** defined by the logical array **x**. **x** must contain only the values **0** and **1**, and must be the same size as **A**

3.8 SUBARRAY SEARCHING

Many times it is desirable to know the indices or subscripts of those elements of an array that satisfy some relational expression. In MATLAB this task is performed by the function `find`, which returns the subscripts where a relational expression is True.

```
» x=-3:3
x =
     -3    -2    -1     0     1     2     3

» k=find(abs(x)>1)
k =
      1     2     6     7
```

finds those indices where **abs(x)>1**.

```
» y=x(k)
y =
     -3    -2     2     3
```

creates *y* using the indices in **k**.

The *find* function also works for matrices.

```
» A=[1 2 3;4 5 6;7 8 9]
A =
      1     2     3
      4     5     6
      7     8     9

» [i,j]=find(A>5)
i =
      3
      3
      2
      3
j =
      1
      2
      3
      3
```

Here the indices stored in **i** and **j** are the associated row and column indices, respectively, where the relational expression is True. That is, **A(i(1),j(1))** is the first element of **A** where **A>5**, and so on.

Note that when a MATLAB function returns two or more variables, they are enclosed by square brackets on the left-hand side of the equal sign. This syntax is different from the array-manipulation syntax discussed above where **[i,j]** on the right-hand side of the equal sign builds a new array with **j** appended to the right of **i**.

The above concepts are summarized in the following table.

Array Searching

`i=find(x)`	Return indices of the array **x** where its elements are nonzero
`[r,c]=find(X)`	Return row and column indices of the array **X** where its elements are nonzero

---

3.9 ARRAY SIZE

Earlier in the discussion of scalars, the **who** command was illustrated as a command that displays the names of all user-created variables. In the case of arrays, it is also important to know their sizes. In MATLAB the command **whos** provides this additional information.

```
» whos
         Name       Size      Elements      Bytes      Density      Complex

           a      1 by 5             5         40         Full          No
           b      5 by 1             5         40         Full          No
           c      1 by 5             5         40         Full          No
           d      1 by 5             5         80         Full          Yes
           e      5 by 1             5         80         Full          Yes
           f      5 by 1             5         80         Full          Yes
           g      3 by 4            12         96         Full          No

Grand total is 42 elements using 456 bytes
leaving 2316084 bytes of memory free.
```

In addition to the names and sizes of the variables, **whos** identifies the total number of elements in each variable, the total number of bytes occupied, whether the variable is full or sparse, and whether the data contain complex data. By default, matrices are full. Sparse matrices are discussed in Chapter 4.

In those cases where the size of a matrix or vector is unknown, but is needed for some manipulation, MATLAB provides two utility functions **size** and **length**.

```
» A=[1 2 3 4;5 6 7 8]
A =
       1       2       3       4
       5       6       7       8
```

```
» s=size(A)
s =
      2     4
```

With one output argument, the **size** function returns a row vector whose first element is the number of rows and whose second element is the number of columns.

```
» [r,c]=size(A)
r =
      2
c =
      4
```

With two output arguments, **size** returns the number of rows in the first variable and the number of columns in the second variable.

```
» r=size(A,1) % number of rows
r =
      2

» c=size(A,2) % number of columns
c =
      4
```

Called with two arguments, **size** returns either the number of rows or columns.

```
» length(A)
ans =
      4
```

returns the number of rows or the number of columns, whichever is larger.

```
» B=pi:0.01:2*pi;

» size(B)
ans =
      1    315
```

shows that **B** is a row vector, and

```
» length(B)
ans =
    315
```

returns the length of the vector.

```
» size([])
ans =
       0     0
```

shows that the empty matrix does indeed have zero size.

The above concepts are summarized in the table below.

Array Size

`whos`	Display variables that exist in the workspace and their sizes
`s=size(A)`	Return a two-element vector **s**, whose first element is the number of rows in **A** and whose second element is the number of columns in **A**
`[r,c]=size(A)`	Return two scalars **r** and **c** containing the number of rows and columns in **A**, respectively
`r=size(A,1)`	Return the number of rows in **A** in the variable **r**
`c=size(A,2)`	Return the number of columns in **A** in the variable **c**
`n=length(A)`	Return `max(size(A))` in the variable **n**

3.10 ARRAY MANIPULATION FUNCTIONS

Because of their general utility, MATLAB offers several functions that automate common array manipulation functions. They include

---

Array Manipulation Functions

`flipud(A)`	Flip a matrix upside down
`fliplr(A)`	Flip a matrix left to right
`rot90(A)`	Rotate a matrix counterclockwise 90 degrees
`reshape(A,m,n)`	Return an **m** by **n** matrix whose elements are taken columnwise from **A**. **A** must contain **m*n** elements

Array Manipulation Functions

`diag(A)`	Extract the diagonal of the matrix **A** as a column vector
`diag(v)`	Create a diagonal matrix with the vector **v** on its diagonal
`tril(A)`	Extract lower triangular part of the matrix **A**
`triu(A)`	Extract upper triangular part of the matrix **A**

Further information regarding these functions can be found in the *MATLAB Reference Guide* and by using on-line help.

3.11 M-FILE EXAMPLES

In this section two functions in the *Mastering MATLAB Toolbox* are illustrated. These functions illustrate the array manipulation concepts demonstrated in this chapter and illustrate how M-file functions are written. For further information regarding M-file functions, see Chapter 8.

Before discussing the internal structure of the M-file functions, consider what they do.

```
» A=[1 2 3 4;2 4 6 8;3 6 9 12] % new data
A =
     1     2     3     4
     2     4     6     8
     3     6     9    12

» shiftud(A,1) % shift A DOWN 1 row, fill blank row with zeros
ans =
     0     0     0     0
     1     2     3     4
     2     4     6     8

» shiftud(A,-1) % shift A UP 1 row, fill blank row with zeros
ans =
     2     4     6     8
     3     6     9    12
     0     0     0     0

» shiftud(A,-1,1) % circularly shift A UP 1 row
ans =
     2     4     6     8
     3     6     9    12
     1     2     3     4
```

```
» shiftud(A,2,1) % circularly shift A DOWN 2 rows
ans =
     2     4     6     8
     3     6     9    12
     1     2     3     4


» shiftud(A,2,0) % shift A DOWN 2 rows, fill blank rows with zeros
ans =
     0     0     0     0
     0     0     0     0
     1     2     3     4


» shiftlr(A,1) % shift A RIGHT 1 row, fill blank row with zeros
ans =
     0     1     2     3
     0     2     4     6
     0     3     6     9


» shiftlr(A,-2) % shift A LEFT 2 rows, fill blank rows with zeros
ans =
     3     4     0     0
     6     8     0     0
     9    12     0     0


» shiftlr(A,-2,0) % same as above
ans =
     3     4     0     0
     6     8     0     0
     9    12     0     0


» shiftlr(A,-2,1) % circularly shift A LEFT 2 rows
ans =
     3     4     1     2
     6     8     2     4
     9    12     3     6
```

The functions **shiftud** and **shiftlr** shift and scroll matrix rows and columns, respectively. The first input argument is the matrix to be manipulated. The second is the shift amount, with positive integers being down or right and negative integers being up or left. If a third input argument is present and logical True, the shift is performed in a circular sense. The body of the M-file function **shiftud.m** is given below.

```
function y=shiftud(a,n,cs)
%SHIFTUD Shift or Circularly Shift Matrix Rows.
% SHIFTUD(A,N) with N>0 shifts the rows of A DOWN N rows.
% The first N rows are replaced by zeros and the last N
% rows of A are deleted.
%
% SHIFTUD(A,N) with N>0 shifts the rows of A UP N rows.
% The last N rows are replaced by zeros and the first N
% rows of A are deleted.
%
% SHIFTUD(A,N,C) where C is nonzero performs a circular
% shift of N rows, where rows circle back to the other
% side of the matrix. No rows are replaced by zeros.

% Copyright (c) 1996 by Prentice-Hall, Inc.

if nargin<3, cs=0; end % if no third argument, default is False
cs=cs(1);                    % make sure third argument is a scalar
[r,c]=size(a);               % get dimensions of input
dn=(n>=0);                   % dn is True if shift is down
n=min(abs(n),r);             % limit shift to less than rows

if n==0|(cs&n==r)            % simple no shift case
y=a;
elseif ~cs&dn                % no circular and down
y=[zeros(n,c); a(1:r-n,:)];
elseif ~cs&~dn               % no circular and up
y=[a(n+1:r,:); zeros(n,c)];
elseif cs&dn        % circular and down
y=[a(r-n+1:r,:); a(1:r-n,:)];
elseif cs&~dn                % circular and up
y=[a(n+1:r,:); a(1:n,:)];
end
```

The first line in the file identifies the function to MATLAB and specifies the input and output arguments. The next contiguous set of comment lines is the text that is displayed when » **help shiftud** is typed. The first help line, called the H1 line, is the line searched by the **lookfor** command. The first command in the function checks for the existence of the third input argument. If it does not exist, a default value of 0 or False is used. Next, in case the user passed a matrix in **cs**, only the first value is used. The rest of the function is a straightforward application of control flow and array manipulation.

The function **shiftlr** is even simpler than **shiftud**.

```
function y=shiftlr(a,n,cs)
% SHIFTLR Shift or Circularly Shift Matrix Columns.
% SHIFTLR(A,N) with N>0 shifts the columns of A to the RIGHT N columns.
% The first N columns are replaced by zeros and the last N
% columns of A are deleted.
%
% SHIFTLR(A,N) with N>0 shifts the columns of A to the LEFT N
% columns.
% The last N columns are replaced by zeros and the first N
% columns of A are deleted.
%
% SHIFTLR(A,N,C) where C is nonzero performs a circular
% shift of N columns, where columns circle back to the
% other side of the matrix. No columns are replaced by zeros.
%

% Copyright (c) 1996 by Prentice-Hall, Inc.

if nargin<3,cs=0;end
y=(shiftud(a.',n,cs)).';
```

Here **shiftlr** simply transposes the input, calls **shiftud**, and then transposes the result.

4

Matrix Operations and Functions

4.1 SETS OF LINEAR EQUATIONS

Originally MATLAB was written to simplify the matrix and linear algebra computations that appear in many applications. One of the most common linear algebra problems is the solution of a linear set of equations. For example, consider the set of equations

$$\begin{bmatrix} 1\ 2\ 3 \\ 4\ 5\ 6 \\ 7\ 8\ 0 \end{bmatrix} \cdot \begin{bmatrix} x_1 \\ x_2 \\ x_3 \end{bmatrix} = \begin{bmatrix} 366 \\ 804 \\ 351 \end{bmatrix}$$

$$A \cdot x = b$$

where the mathematical multiplication symbol ($\cdot$) is now defined in the matrix sense as opposed to the array sense discussed earlier. In MATLAB, this matrix multiplication is denoted with the asterisk notation *****. The above equations define the matrix product of the matrix A and the vector x as being equal to the vector b. The existence of solutions to the above equation is a fundamental problem in linear algebra. Moreover, when a solution does exist, there are numerous approaches to finding the solution, such as Gaussian elimination,

LU factorization, or direct use of A^{-1}. Analytically, the solution is written as $x = A^{-1} \cdot b$. It is beyond the scope of this text to discuss the many analytical and numerical issues of linear or matrix algebra. We only wish to demonstrate how MATLAB can be used to solve problems like the one above.

To solve the above problem it is necessary to enter A and b:

```
» A=[1 2 3;4 5 6
7 8 0]
A =
1       2       3
4       5       6
7       8       0

» b=[366;804;351]
b =
    366
    804
    351
```

As discussed earlier, the entry of the matrix **A** shows the two ways that MATLAB distinguishes between rows. The semicolon between the 3 and the 4 signifies the start of a new row as does the new line between the 6 and the 7. The vector **b** is a column because each semicolon signifies the start of a new row.

With a background in linear algebra, it is easy to show that this problem has a unique answer whenever the ***determinant*** of the matrix **A** is nonzero:

```
» det(A)
ans =
    27
```

Since this is true, MATLAB can find the solution of $A \cdot x = b$ in two ways, one of which is preferred. The less favorable but more straightforward method is to take $x = A^{-1} \cdot b$ literally.

```
» x=inv(A)*b
x =
    25.0000
    22.0000
    99.0000
```

Here **inv(A)** is a MATLAB function that computes A^{-1}, and the matrix multiplication operator *****, without a preceding dot, is matrix multiplication. The preferable solution is found using the matrix left division operator.

```
» x=A\b
x =
```

```
25.0000
22.0000
99.0000
```

This equation utilizes an LU factorization approach and expresses the answer as the left division of A into b. The left division operator has no preceding dot as this is a matrix operation, not an element-by-element array operation. There are many reasons this second solution is preferable. Of these, the simplest is that the latter method requires fewer multiplications and divisions and as a result is faster. In addition, this solution is generally more accurate for larger problems. In either case, if MATLAB cannot find a solution or cannot find it accurately, it displays an error message.

　　If you've studied linear algebra rigorously, you know that when the number of equations and number of unknowns differ, a single unique solution usually does not exist. However, with further constraints a practical solution usually can be found. In MATLAB, when all redundant equations have been removed and there are more equations than unknowns, i.e., the overdetermined case, **use of a division operator / or \ automatically finds the solution that minimizes the squared error in A · x − b.** This solution is of great practical value and is called the *least squares solution.* Consider the example,

```
» A=[1 2 3;4 5 6;7 8 0;2 5 8] % 4 equations in 3 unknowns
A =
       1       2       3
       4       5       6
       7       8       0
       2       5       8

» b=[366 804 351 514]' % a new r.h.s. vector
b =
     366
     804
     351
     514

» x=A\b % compute the least squares solution
x =
   247.9818
  -173.1091
   114.9273

» res=A*x-b % this residual has the smallest norm.
res =
  -119.4545
    11.9455
     0.0000
    35.8364
```

On the other hand, when there are fewer equations than unknowns, i.e., the underdetermined case, an infinite number of solutions exists. Of these solutions, MATLAB computes two in a straightforward way. Use of the division operator gives a solution that has a maximum number of zeros in the elements of x. Alternatively, **computing x=pinv(A)*b gives a solution where the length or norm of x is smaller than all other possible solutions.** This solution, based on the *pseudoinverse,* also has great practical value and is called the **minimum norm solution.** Consider the example,

```
» A=A' % create 3 equations in 4 unknowns
A =
       1       4       7       2
       2       5       8       5
       3       6       0       8

» b=b(1:3) % new r.h.s. vector
b =
   366
   804
   351

» x=A\b % solution with maximum zero elements
x =
          0
   -165.9000
     99.0000
    168.3000

» xn=pinv(A)*b % find minimum norm solution
xn =
     30.8182
   -168.9818
     99.0000
    159.0545

» norm(x) % Euclidean norm with zero elements
ans =
   256.2200

» norm(xn) % minimum norm solution has smaller norm!
ans =
   254.1731
```

4.2 MATRIX FUNCTIONS

In addition to the solution of linear sets of equations, MATLAB offers numerous matrix functions that are useful for solving numerical linear algebra problems. Thorough discussion of these functions is beyond the scope of this text. A brief description of many of the matrix functions is given below. For further information, use on-line help or consult the *MATLAB Reference Guide.*

Matrix Functions

`balance(A)`	Scale to improve eigenvalue accuracy
`cdf2rdf(A)`	Complex diagonal form to real block diagonal form
`chol(A)`	Cholesky factorization
`cond(A)`	Matrix condition number
`condest(A)`	1-norm matrix condition number estimate
`d=eig(A),` `[V,D]=eig(A)`	Eigenvalues and eigenvectors
`det(A)`	Determinant
`expm(A)`	Matrix exponential
`expm1(A)`	M-file implementation of **expm**
`expm2(A)`	Matrix exponential using Taylor series
`expm3(A)`	Matrix exponential using eigenvalues and eigenvectors
`funm(A,'fun')`	Compute general matrix function
`hess(A)`	Hessenberg form
`inv(A)`	Matrix inverse
`logm(A)`	Matrix logarithm
`lscov(A,b,V)`	Least squares with known covariance
`lu(A)`	Factors from Gaussian elimination
`nnls(A,b)`	Nonnegative least squares
`norm(A)`	Matrix and vector norms:
`norm(A,1)`	1-norm,
`norm(A,2)`	2-norm (Euclidean),
`norm(A,inf)`	Infinity,
`norm(A,p)`	P-norm (vectors only),
`norm(A,'fro' )`	F-norm
`null(A)`	Null space
`orth(A)`	Orthogonalization
`pinv(A)`	Pseudoinverse
`poly(A)`	Characteristic polynomial
`polyvalm(A)`	Evaluate matrix polynomial
`qr(A)`	Orthogonal-triangular decomposition
`qrdelete(Q,R,j)`	Delete column from qr factorization
`qrinsert(Q,R,j,x)`	Insert column in qr factorization

`qz(A,B)`	Generalized eigenvalues
`rank(A)`	Number of linearly independent rows or columns
`rcond(A)`	Reciprocal condition estimator
`rref(A)`	Reduced row echelon form
`rsf2csf(U,T)`	Real schur form to complex schur form
`schur(A)`	Schur decomposition
`sqrtm(A)`	Matrix square root
`svd(A)`	Singular value decomposition
`trace(A)`	Sum of diagonal elements

4.3 SPECIAL MATRICES

MATLAB offers a number of special matrices; some of them are general utilities, while others are matrices of interest to specialized disciplines. The general utility matrices include

```
» a=[1 2 3;4 5 6];
» b=find(a>10)
b =
        []
```

Here **b** is the empty matrix. MATLAB returns the empty matrix when an operation leads to no result. In the above example, there are no indices of **a** that are greater than 10. Empty matrices have zero size, but their variable names do exist in the workspace.

```
» zeros(3) % a 3-by-3 matrix of zeros
ans =
        0        0        0
        0        0        0
        0        0        0

» ones(2,4) % a 2-by-4 matrix of ones
ans =
        1        1        1        1
        1        1        1        1

» ones(3)*pi
ans =
        3.1416      3.1416      3.1416
        3.1416      3.1416      3.1416
        3.1416      3.1416      3.1416
```

an example of creating a 3–by–3 matrix with all elements equal to π.

```
» rand(3,1)
ans =
    0.2190
    0.0470
    0.6789
```

a 3–by–1 matrix of uniformly distributed random numbers between zero and one.

```
» randn(2)
ans =
    1.1650    0.0751
    0.6268    0.3516
```

a 2–by–2 matrix of normally distributed random numbers with zero mean and unit variance. The algorithms for **rand** and **randn** can be found in S. K. Park and K. W. Miller, *Random Number Generators: Good ones are hard to find,* Comm. ACM, vol. 32, no. 10, Oct. 1988, pp. 1192–1201.

```
» eye(3)
ans =
    1    0    0
    0    1    0
    0    0    1
```

a 3–by–3 identity matrix. (Or maybe it should be spelled *eye*dentity!)

```
» eye(3,2)
ans =
    1    0
    0    1
    0    0
```

a 3–by–2 identity matrix.

In addition to explicitly specifying the size of matrix, you can also use the function **size** to create a matrix the same size as another.

```
» A=[1 2 3;4 5 6];
» ones(size(A))
ans =
    1    1    1
    1    1    1
```

a matrix of ones the same size as **A**.

These and other special matrices include those given in the table below.

Special Matrices

[]	The empty matrix
compan	Companion matrix
eye	Identity matrix
gallery	Several small test matrices
hadamard	Hadamard matrix
hankel	Hankel matrix
hilb	Hilbert matrix
invhilb	Inverse Hilbert matrix
magic	Magic square
ones	Matrix containing all ones
pascal	Pascal triangle matrix
rand	Uniformly distributed random matrix with elements between 0 and 1
randn	Normally distributed random matrix with elements having zero mean and unit variance
rosser	Symmetric eigenvalue test matrix
toeplitz	Toeplitz matrix
vander	Vandermonde matrix
wilkinson	Wilkinson eignenvalue test matrix
zeros	Matrix containing all zero elements

4.4 SPARSE MATRICES

In many practical applications matrices are generated that contain only a few nonzero elements. As a result, these matrices are said to be sparse. For example, circuit simulation and finite element analysis programs routinely deal with matrices containing fewer than 1% nonzero elements. If a matrix is large, e.g., `max(size(A)) > 100`, and has a high percentage of zero elements, it is both wasteful of computer storage space to store the zero elements and wasteful of computational power to perform arithmetic operations using the zero elements. To eliminate the storage of zero elements, it is common to store only the nonzero elements of a matrix along with two sets of indices that identify the row and column positions of those elements. Similarly, to eliminate arithmetic operations on

the zero elements, special algorithms have been developed to solve typical matrix problems, such as solving a set of linear equations, wherein operations involving zeros are
minimized.

The techniques used to optimize sparse matrix computations are complex in implementation as well as in theory. Fortunately, MATLAB hides this complexity. In MATLAB
sparse matrices are stored in variables just as regular full matrices are. Moreover, most computations with sparse matrices use the same syntax as that used for full matrices. In this text,
only the creation of sparse matrices and the conversion to and from sparse matrices will be
illustrated. For more in-depth information consult the *MATLAB User's Guide*. In general,
operations on full matrices produce full matrices and operations on sparse matrices produce
sparse matrices. In addition, operations on a mixture of full and sparse matrices generally
produces sparse matrices, unless the operation makes the result too densely populated with
nonzeros to make sparse storage efficient.

Sparse matrices are created using the MATLAB function **sparse**. For example,

```
» As=sparse(1:10,1:10,ones(1,10))
As =
    (1,1)        1
    (2,2)        1
    (3,3)        1
    (4,4)        1
    (5,5)        1
    (6,6)        1
    (7,7)        1
    (8,8)        1
    (9,9)        1
    (10,10)      1
```

creates a 10–by–10 identity matrix. In this usage **sparse(i,j,s)** creates a sparse matrix whose k^{th} nonzero element is **s(k)**, and **s(k)** appears in the row **i(k)** and column
j(k). Note the difference in how sparse matrices are displayed.

```
» As=sparse(eye(10))
As =
    (1,1)        1
    (2,2)        1
    (3,3)        1
    (4,4)        1
    (5,5)        1
    (6,6)        1
    (7,7)        1
    (8,8)        1
    (9,9)        1
    (10,10)      1
```

creates the 10–by–10 identity matrix again, this time by converting the full matrix
eye(10) to sparse format.

```
» A=full(As)
A =
     1    0    0    0    0    0    0    0    0    0
     0    1    0    0    0    0    0    0    0    0
     0    0    1    0    0    0    0    0    0    0
     0    0    0    1    0    0    0    0    0    0
     0    0    0    0    1    0    0    0    0    0
     0    0    0    0    0    1    0    0    0    0
     0    0    0    0    0    0    1    0    0    0
     0    0    0    0    0    0    0    1    0    0
     0    0    0    0    0    0    0    0    1    0
     0    0    0    0    0    0    0    0    0    1
```

converts the sparse matrix back to its full form.

To compare sparse matrix storage to full matrix storage, consider the following example.

```
» B=eye(200); % FULL 200-by-200 identity matrix

» Bs=sparse(B); % Sparse 200-by-200 identity matrix

» whos
```

Name	Size	Elements	Bytes	Density	Complex
B	200 by 200	40000	320000	Full	No
Bs	200 by 200	200	3200	0.0050	No

Here the sparse matrix **Bs** contains only 0.5% nonzero elements and requires 3200 bytes of storage. On the other hand, **B**, the same matrix in full matrix form, requires two orders of magnitude more bytes of storage!

Other sparse matrix functions include the following:

Sparse Matrix Functions

`colmmd`	Reorder by column minimum degree
`colperm`	Reorder by ordering columns based on nonzero count
`condest`	Estimate 1-norm matrix condition
`dmperm`	Reorder by Dulmage-Mendelsohn decomposition
`find`	Find indices of nonzero entries
`gplot`	Graph theory plot of sparse matrix
`nnz`	Number of nonzero entries
`nonzeros`	Nonzero entries
`normest`	Estimate 2-norm
`nzmax`	Storage allocated for nonzeros
`randperm`	Random permutation
`spalloc`	Allocate memory for nonzeros
`spaugment`	Form least squares augmented system
`spconvert`	Convert sparse to external format
`spdiags`	Sparse matrix formed from diagonals
`speye`	Sparse identity matrix
`spones`	Replace nonzeros with ones
`spparms`	Set sparse matrix routine parameters
`sprandn`	Sparse random matrix
`sprandsym`	Sparse symmetric random matrix
`sprank`	Structural rank
`spy`	Visualize sparse structure
`symbfact`	Symbolic factorization analysis
`symmmd`	Reorder by symmetric minimum degree
`symrcm`	Reorder by reverse Cuthill-McKee algorithm

Relational and Logical Operations

In addition to traditional mathematical operations, MATLAB supports relational and logical operations. You may be familiar with these if you've had some experience with other programming languages. The purpose of these operators and functions is to provide answers to True/False questions. One important use of this capability is controlling the flow, or order of execution, of a series of MATLAB commands (usually in an M-file) based on the results of True/False questions.

As inputs to all relational and logical expressions, MATLAB considers *any nonzero number to be True and zero to be False*. The output of all relational and logical expressions produces ***One for True*** and ***Zero for False***.

5.1 RELATIONAL OPERATORS

MATLAB relational operators include all common comparisons.

Relational Operator	Description
<	less than
<=	less than or equal to
>	greater than
>=	greater than or equal to
==	equal to
~=	not equal to

MATLAB relational operators can be used to compare two arrays of the same size, or to compare an array to a scalar. In the latter case, the scalar is compared with all elements of the array, and the result has the same size as the array. Some examples include

```
» A=1:9,B=9-A
A =
     1     2     3     4     5     6     7     8     9
B =
     8     7     6     5     4     3     2     1     0

» tf=A>4
tf =
     0     0     0     0     1     1     1     1     1
```

finds elements of **A** that are greater than 4. Zeros appear in the result where $A \leq 4$, and ones appear where $A > 4$.

```
» tf=(A==B)
tf =
     0     0     0     0     0     0     0     0     0
```

finds elements of **A** that are equal to those in **B**. **Note that = and == mean two different things: == compares two variables and returns ones where they are equal and zeros where they are not; = on the other hand is used to assign the output of an operation to a variable.**

```
» tf=B-(A>2)
tf =
     8     7     5     4     3     2     1     0    -1
```

finds where **A>2** and subtracts the resulting vector from **B**. This example shows that since the outputs of logical operations are numerical arrays of ones and zeros, they can be used in mathematical operations, too.

```
» B=B+(B==0)*eps
B =
  Columns 1 through 7
     8.0000      7.0000      6.0000      5.0000      4.0000      3.0000      2.0000
  Columns 8 through 9
     1.0000      0.0000
```

is a demonstration of how to replace zero elements in an array with the special MATLAB number **eps**, which is approximately 2.2e-16. This particular expression is sometimes useful to avoid dividing by zero as in

```
» x=(-3:3)/3
x =
    -1.0000     -0.6667     -0.3333          0     0.3333     0.6667     1.0000

» sin(x)./x

Warning: Divide by zero
ans =
     0.8415      0.9276      0.9816        NaN     0.9816     0.9276     0.8415
```

computing the function sin(x)/x gives a warning because the fifth data point is zero. Since sin(0)/0 is undefined, MATLAB returns **NaN** (meaning not-a-number) at that location in the result. Try again, after replacing the zero with **eps**:

```
» x=x+(x==0)*eps;

» sin(x)./x
ans =
     0.8415      0.9276      0.9816     1.0000     0.9816     0.9276
     0.8415
```

Now sin(x)/x for x=0 gives the correct limiting answer.

5.2 LOGICAL OPERATORS

Logical operators provide a way to combine or negate relational expressions. MATLAB logical operators include:

	Logical Operator	Description
	&	AND
	|	OR
	~	NOT

Some examples of the use of logical operators are

```
» A=1:9;B=9-A;

» tf=A>4
tf =
      0    0    0    0    1    1    1    1    1
```

finds where **A** is greater than 4.

```
» tf=~(A>4)
tf =
      1    1    1    1    0    0    0    0    0
```

negates the above result, i.e., swaps where the ones and zeros appear.

```
» tf=(A>2)&(A<6)
tf =
      0    0    1    1    1    0    0    0    0
```

returns ones where **A** is greater than 2 AND less than 6.

Finally, the above capabilities make it easy to generate arrays representing signals with discontinuities or signals that are composed of segments of other signals. The basic idea is to multiply with ones those values that you wish to keep in an array, and to multiply all other values with zeros. For example,

```
» x=linspace(0,10,100);   % create data

» y=sin(x);               % compute sine

» z=(y>=0).*y;            % set negative values of sin(x) to zero

» z=z + 0.5*(y<0);        % where sin(x) is negative add 1/2
```

```
» z=(x<=8).*z;              % set values past x=8 to zero

» plot(x,z)

» xlabel('x'),ylabel('z=f(x)'),title('A Discontinuous Signal')
```

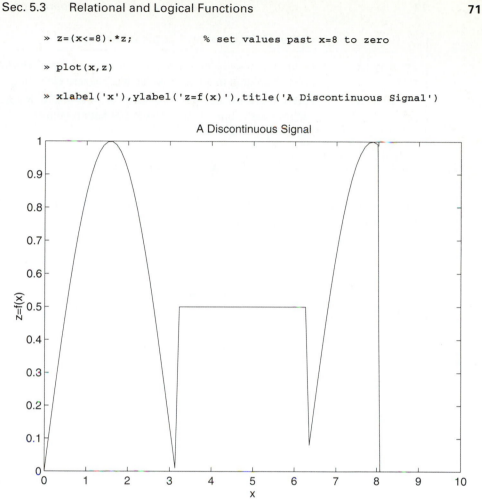

5.3 RELATIONAL AND LOGICAL FUNCTIONS

In addition to the above basic relational and logical operators, MATLAB provides a number of additional relational and logical functions including:

Other Relational and Logical Functions

`xor(x,y)`   Exclusive OR operation. Return ones where either **x** or **y** is nonzero (True). Return zeros where both **x** and **y** are zero (False) or both are nonzero (True).

Other Relational and Logical Functions

`any(x)`	Return one if *any* element in a vector **x** is nonzero. Return one for each column in a matrix **x** that has nonzero elements.
`all(x)`	Return one if *all* elements in a vector **x** are nonzero. Return one for each column in a matrix **x** that has all nonzero elements.

In addition to these functions, MATLAB provides numerous functions that test for the existence of specific values or conditions and return logical results.

Test Functions

`finite`	Return True where elements are finite
`isempty`	Return True if argument is empty
`isglobal`	Return True if argument is a global variable
`ishold`	Return True if current plot hold state is ON
`isieee`	Return True if computer performs IEEE arithmetic
`isinf`	Return True where elements are infinite
`isletter`	Return True where elements are letters of the alphabet
`isnan`	Return True where elements are NANs
`isreal`	Return True if argument has no imaginary part
`isspace`	Return True where elements are whitespace characters
`issparse`	Return True if argument is a sparse matrix
`isstr`	Return True if argument is a character string
`isstudent`	Return True if Student Edition of MATLAB
`isunix`	Return True if computer is UNIX
`isvms`	Return True if computer is VMS

5.4 NaNS AND EMPTY MATRICES

NaNs (Not-a-Numbers) and empty matrices (`[]`) require special treatment in MATLAB, especially when used in logical or relational expressions. According to IEEE mathematical standards, almost all operations on **NaNs** result in **NaNs**. For example,

```
» a=[1 2 nan inf nan] % note, in use, NaN is lowercase
a =
     1      2    NaN     ∞    NaN

» b=2*a
b =
     2      4    NaN     ∞    NaN

» c=sqrt(a)
c =
    1.0000      1.4142      NaN       ∞        NaN

» d=(a==nan)
d =
     0      0      0      0      0

» f=(a~=nan)
f =
     1      1      1      1      1
```

The first two computations above give **NaN** results for **NaN** inputs. However, the final two relational computations produce somewhat surprising results. **(a==nan)** produces all zeros or False results even when **NaN** is compared to **NaN**. At the same time **(a~=nan)** produces all ones or True results. Thus, individual **NaNs** are not equal to each other. As a result of this property of **NaNs**, MATLAB has a built-in logical function for finding **NaNs**.

```
» g=isnan(a)
g =
     0      0      1      0      1
```

This function makes it possible to find the indices of **NaNs** using the `find` command. For example,

```
» i=find(isnan(a)) % find indices of NaNs
i =
     3      5

» a(i)=zeros(size(i)) % changes NaNs in a to zeros
a =
     1      2      0      ∞      0
```

While **NaNs** are well-defined mathematically by IEEE standards, empty matrices are defined by the creators of MATLAB and have their own interesting properties. Empty matrices are simply that. They are MATLAB variables having zero size.

```
» size([])
ans =
        0      0
```

In MATLAB many functions return empty matrices when no other result is appropriate. Perhaps the most common example is the **find** function

```
» x=(1:5)-3 % new data
x =
      -2     -1      0      1      2

» y=find(x>2)
y =
      []
```

In this example, **x** contains no values greater than 2, so there are no indices to return. To test for empty results, MATLAB provides the logical function **isempty**.

```
» isempty(y)
ans =
        1
```

In MATLAB the empty matrix is not equal to any nonzero matrix (or scalar). This fact leads to the following example:

```
» y=[];
» a=(y==0)
a =
        0
```

which shows that an empty matrix is not equal to a scalar. So

```
» find(y==0)
ans =
      []
```

says there are no indices to return. Likewise,

```
» b=(y~=0)
b =
        1
```

Yet again, an empty matrix is not equal to a scalar. But,

```
» j=find(y~=0)
j =
        1
```

now there is an index even though **y** has zero size! This last example is an undocumented change in MATLAB since version 3.5. Prior to version 4.0, comparing an empty matrix to a nonempty matrix returned an empty matrix. **This new interpretation usually leads to problems, since y(find(y~=0)) does not exist.** For example,

```
» y(find(y~=0))
??? Index exceeds matrix dimensions.
```

Here MATLAB reports an error because the index is outside the dimensions of the empty **y**.

The properties of **NaNs** and empty matrices are summarized in the following table.

NaNs **and Empty Matrices**	
Data	a = [1 2 nan inf nan]
Expression	Result
2*a	[2 4 NaN ∞ NaN]
(a==nan)	[0 0 0 0 0]
(a~=nan)	[1 1 1 1 1]
isnan(a)	[0 0 1 0 1]
y=find(a==0)	y=[]
isempty(y)	1
(y==0)	0
find(y==0)	[]
(y~=0)	1
j=find(y~=0)	j=1
y(j)	Error! y(j) does not exist.

6

Text

MATLAB's true power is in its ability to crunch numbers. However, there are times when it is desirable to manipulate text, such as when putting labels and titles on plots. In MATLAB, text is referred to as character strings or simply strings.

6.1 CHARACTER STRINGS

Character strings in MATLAB are simply numerical arrays of ASCII values that are displayed as their character string representation. For example,

```
» t='How about this character string?'
t =
How about this character string?
```

```
» size(t)
ans =
    1 32
```

```
» whos
```

Name	Size	Elements	Bytes	Density	Complex
ans	1 by 2	2	16	Full	No
t	1 by 32	32	256	Full	No

A character string is simply text surrounded by single quote marks. Each character in a string is one element in the array, and storage of character strings requires 8 bytes per character, just as it does for other MATLAB variables. Since ASCII characters only require one byte, this storage requirement is wasteful since 7/8 of the storage allocated is not used. However, maintaining the same data structure for character strings simplifies the internal data structure of MATLAB. Given that character string manipulation is not intended to be a primary feature of MATLAB, this representation is convenient and acceptable.

To see the underlying ASCII representation of a character string, you need only perform some arithmetic operation on the string. The simplest and most computationally efficient approach is to take the absolute value of the array. For example,

```
» u=abs(t)
u =
  Columns 1 through 12
    72    111    119     32     97     98    111    117    116     32    116    104
  Columns 13 through 24
   105    115     32     99    104     97    114     97     99    116    101    114
  Columns 25 through 32
    32    115    116    114    105    110    103     63
```

```
» u=t+0
u =
  Columns 1 through 12
    72    111    119     32     97     98    111    117    116     32    116    104
  Columns 13 through 24
   105    115     32     99    104     97    114     97     99    116    101    114
  Columns 25 through 32
    32    115    116    114    105    110    103     63
```

In the latter example above, adding zero to the character string also changes its representation to ASCII. The function **setstr** provides the reverse transformation.

```
» v=setstr(u)
v =
How about this character string?
```

Since strings are numerical arrays, they can be manipulated with all the array manipulation tools available in MATLAB. For example,

```
» u=t(16:24)
u =
character
```

Strings are addressed just like arrays. Here elements 16 through 24 contain the word *character*.

```
» u=t(24:-1:16)
u =
retcarahc
```

This is the word *character* spelled backward.

```
» u=t(16:24)'
u =
c
h
a
r
a
c
t
e
r
```

Using the transpose operator changes the word *character* into a column.

```
» v='I can''t find the manual!'
v =
I can't find the manual!
```

Single quotes within a character string are symbolized by two consecutive single quotes. String catenation follows directly from array catenation.

```
» u='If a woodchuck could chuck wood,';
» v=' how much wood would a woodchuck chuck?';

» w=[u v]
w =
If a woodchuck could chuck wood, how much wood would a woodchuck chuck?
```

The function **disp** allows you to display a string without printing its variable name. For example,

```
» disp(u)
If a woodchuck could chuck wood,
```

Note that the **u** = statement is suppressed. This is useful for displaying helpful text within a script file.

As with matrices, character strings can have multiple rows, but each row must have an equal number of columns. Therefore, blanks are explicitly required to make all rows the same length. For example,

```
» v=['Character strings having more than'
     'one row must have the same number '
     'of columns just like matrices!    ']
v =
Character strings having more than
one row must have the same number
of columns just like matrices!
```

Consider the following example for converting a string to upper case. First the **find** function is used to find the indices of lower case characters. Then the difference between lower and upper case A is subtracted from the lower case elements only. Finally, the resulting array is converted to its string representation using **setstr**.

```
» disp(u)
If a woodchuck could chuck wood,

» i=find(u>='a' & u<= 'z'); % find is a very powerful function!

» u(i)=setstr(u(i)-('a'-'A'))
u =
IF A WOODCHUCK COULD CHUCK WOOD,
```

In fact, since matrices can be identified by a single index as described in Section 3.7, rather than by row and column indices, the above also works for the string matrix **v**:

```
» disp(v)
Character strings having more than
one row must have the same number
of columns just like matrices!

» i=find(v>='a' & v<= 'z'); % here i is a single index vector into v,

» v(i)=setstr(v(i)-('a'-'A')) % and matrix keeps the same orientation.
v =
CHARACTER STRINGS HAVING MORE THAN
ONE ROW MUST HAVE THE SAME NUMBER
OF COLUMNS JUST LIKE MATRICES!
```

Finally, strings are useful when using the function **input** described earlier in the section on script files.

```
» t='Enter number of rolls of tape > ';
» tape=input(t)
```

```
Enter number of rolls of tape > 5
tape =
    5
```

In addition, the function **input** can input a character string:

```
» x=input('Enter anything > ','s')

Enter anything > anything can be entered
x =
anything can be entered
```

Here the additional argument **'s'** in the function **input** tells MATLAB to simply pass the user input to the output variable as a character string. No quote marks are needed. In fact, if they are included, they become part of the returned string.

6.2 STRING CONVERSIONS

In addition to the conversion discussed above between the string and its ASCII representation, MATLAB offers a number of other useful string conversion functions. They include

---

String Conversions

abs	String to ASCII
dec2hex	Decimal number to hexadecimal string
fprintf	Write formatted text to file or screen
hex2dec	Hex string to decimal number
hex2num	Hexadecimal string to IEEE floating point number
int2str	Integer to string
lower	String to lower case
num2str	Number to string
setstr	ASCII to string
sprintf	Convert number to string with format control
sscanf	Convert string to number under format control
str2mat	Strings to a text matrix
str2num	String to number
upper	String to upper case

In many situations it is desirable to embed a numerical result within a string. Several string conversions perform this task.

```
» rad=2.5;area=pi*rad^2;

» t=['A circle of radius ' num2str(rad) ' has an area of ' num2str(area) '.'];

» disp(t)
A circle of radius 2.5 has an area of 19.63.
```

Here the function **num2str** has been used to convert numbers into strings, and string catenation was used to embed the converted numbers in a character-string sentence. In a similar manner, **int2str** converts integers into strings. Both **num2str** and **int2str** call the function **sprintf**, which uses C-like syntax for converting numbers into strings.

The function **fprintf** is often a useful replacement for the function **disp**, since it offers much more control over the result. While intended for writing formatted data to a file, by default it displays results in the *Command* window. For example,

```
» fprintf('See what this does')
See what this does»

» fprintf('See what this does\n')
See what this does


»
```

In the first example above, **fprintf** displays the string, then immediately gives the MATLAB prompt. On the other hand, in the second example the **\n** inserts a *newline character* creating a new line before the MATLAB prompt is issued.

Both **fprintf** and its cousin **sprintf** process input arguments in the same way, but **fprintf** sends its output to the screen or file, whereas **sprintf** returns its output to a character string. For example, the above example using **num2str** can be rewritten as

```
» t=sprintf('A circle of radius %.4g has an area of %.4g.',rad,area);

» disp(t)
A circle of radius 2.5 has an area of 19.63.

» fprintf('A circle of radius %.4g has an area of %.4g.\n',rad,area)
A circle of radius 2.5 has an area of 19.63.
```

Here the **%.4g** is the number formatting specification used in the function **num2str**. **%.4g** says to use exponential or fixed point notation, whichever is shorter, and to display up to 4 digits total digits. In addition to the **g** format, **e** (exponential) and **f** (fixed point)

conversions can be used. The table below shows how `pi` is displayed under a variety of conversion specifications.

Number Format Conversion Examples

Command	Result
`fprintf('%.0e\n',pi)`	3e+00
`fprintf('%.1e\n',pi)`	3.1e+00
`fprintf('%.3e\n',pi)`	3.142e+00
`fprintf('%.5e\n',pi)`	3.14159e+00
`fprintf('%.10e\n',pi)`	3.1415926536e+00
`fprintf('%.0f\n',pi)`	3
`fprintf('%.1f\n',pi)`	3.1
`fprintf('%.3f\n',pi)`	3.142
`fprintf('%.5f\n',pi)`	3.14159
`fprintf('%.10f\n',pi)`	3.1415926536
`fprintf('%.0g\n',pi)`	3
`fprintf('%.1g\n',pi)`	3
`fprintf('%.3g\n',pi)`	3.14
`fprintf('%.5g\n',pi)`	3.1416
`fprintf('%.10g\n',pi)`	3.141592654
`fprintf('%8.0g\n',pi)`	3
`fprintf('%8.1g\n',pi)`	3
`fprintf('%8.3g\n',pi)`	3.14
`fprintf('%8.5g\n',pi)`	3.1416
`fprintf('%8.10g\n',pi)`	3.141592654

Note that for the **e** and **f** formats, the number to the right of the decimal point says how many digits to the right of the decimal point to display. On the other hand, in the **g** format, the number to the right of the decimal specifies the total number of digits to display. In addition, note in the last five entries, a width of 8 characters is specified for the result, and the result is right justified. In the very last case, the 8 is ignored because more than 8 digits were specified.

To summarize, **fprintf** and **sprintf** are useful when you want more control than the default functions **disp**, **num2str**, and **int2str** provide.

The function **str2mat** converts a list of several strings into a string matrix. For example,

```
» a='one';b='two';c='three';

» disp(str2mat(a,b,c,'four'))
one
two
three
four
```

Although it is not apparent from the above, each of the above rows has the same number of elements. The shorter rows have been padded with blanks so that the result forms a valid matrix.

Sometimes it is convenient to convert in the reverse direction.

```
» s='[1 2;pi 4]' % a string of a MATLAB matrix
s =
[1 2;pi 4]
» str2num(s)

ans =
    1.0000    2.0000
    3.1416    4.0000

» s='123e+5' % a string containing a simple number
s =
123e+5

»str2num(s)
ans =
    12300000
```

The function **str2num** cannot accept user-defined variables nor can it perform arithmetic operations as part of the conversion process. For further information, see on-line help.

6.3 STRING FUNCTIONS

MATLAB offers a number of string functions, including those listed in the table below.

String Functions	
`eval(string)`	Evaluate string as a MATLAB command
`eval(try,catch)`	
`blanks(n)`	Return a string of n blanks or spaces
`deblank`	Remove trailing blanks from a string
`feval`	Evaluate function given by string

String Functions

`findstr`	Find one string within another
`isletter`	Return true where alphabet characters exist
`isspace`	Return true where whitespace characters exist
`isstr`	Return true if input is a string
`lasterr`	Return string of last MATLAB error issued
`strcmp`	Return true if strings are identical
`strrep`	Replace one string with another
`strtok`	Find first token in a string

The first function listed above **eval** provides MATLAB with **macro** capability. Among other things, this function provides the ability to pass the name of user-created functions to other functions for evaluation. Examples of its use include

```
» a=eval('sqrt(2)')
a =
       1.4142

» eval('a=sqrt(2)')
a =
       1.4142
```

While the above examples demonstrate the **eval** function, they are clearly not the simplest way to compute the square root of 2. The true utility of **eval** is when the string to be evaluated is composed by catenating substrings together or when the string is passed to a function for evaluation. An example illustrating this usage is covered later in this text.

In the event that the string passed to **eval** cannot be deciphered, MATLAB offers the following syntax:

```
» eval('a=sqrtt(2)','a=[]')
a =
     []
```

Here the second argument is executed since the first has an error, namely **sqrtt** is not a valid MATLAB function. This form is often described as **eval(try,catch)**.

The function **feval** is similar to eval, but much more restricted in usage. **feval('fun',x)** evaluates the function given by the string **'fun'** with its input argument being the variable **x**. That is, **feval('fun',x)** is equivalent to evaluating **fun(x)**. For example,

```
» a=feval('sqrt',2)
a =
      1.4142
```

As with the function **eval**, **feval** finds its primary use within user-created functions. In general, **feval** can evaluate functions having numerous input arguments, e.g., **feval('fun',x,y,z)** is equivalent to evaluating **fun(x,y,z)**.

A number of the string functions listed in the above table provide basic string parsing capabilities. For example, **findstr** returns the starting indices of one string within another.

```
» b='Peter Piper picked a peck of pickled peppers';

» findstr(b,' ') % find spaces
ans =
      6     12     19     21     26     29     37

» findstr(b,'p') % find the letter p
ans =
      9     13     22     30     38     40     41

» find(b=='p') % for single character searches the find command works too
ans =
      9     13     22     30     38     40     41

» findstr(b,'cow') % find the word cow
ans =
      []

» findstr(b,'pick') % find the string pick
ans =
     13     30
```

Note that this function is case sensitive and returns the empty matrix when no match is found. **findstr** does not work on string matrices.

```
» strrep(b,'p','P') % capitalize all p's
ans =
Peter PiPer Picked a Peck of Pickled PePPers

» strrep(b,'Peter','Pamela') % change Peter to Pamela
ans =
Pamela Piper picked a peck of pickled peppers
```

As shown above, **strrep** performs simple string replacement. **strrep** does not work on string matrices.

The function **strtok** finds tokens within a string delimited by specified characters, with whitespace being the default delimiting characters. For example,

```
» disp(b)
Peter Piper picked a peck of pickled peppers

» strtok(b) % find first token in above string separated by whitespace
ans =
Peter

» [c,r]=strtok(b) % return the remainder of the string array in r
c =
Peter
r =
Piper picked a peck of pickled peppers

» [d,s]=strtok(r) % find the next token by using the previous remainder
d =
Piper
s =
picked a peck of pickled peppers
```

By using whitespace as delimiters, **strtok** picks out words in an array. **strtok** does not work on string matrices.

```
» [d,s]=strtok(b,'pP') % let delimiter be lower or uppercase P
d =
eter
s =
Piper picked a peck of pickled peppers
```

If an optional character string is supplied, its characters are the delimiters. Note that the delimiters are not returned in the token, but all characters up to the next delimiter are. That is, there is a space at the end of the string **d = eter** above.

Decision Making: Control Flow

Computer programming languages and programmable calculators offer features that allow you to control the flow of command execution based on decision making structures. If you have used these features before, this section will be very familiar to you. On the other hand, if control flow is new to you, this material may seem complicated the first time through. If this is the case, take it slowly.

Control flow is extremely powerful since it lets past computations influence future operations. MATLAB offers three decision making or control flow structures. They are: For Loops, While Loops, and If-Else-End structures. Because these structures often encompass numerous MATLAB commands, they often appear in M-files, rather than being typed directly at the MATLAB prompt.

7.1 FOR LOOPS

For Loops allow a group of commands to be repeated a fixed, predetermined number of times. The general form of a For Loop is

```
for x = array
    {commands}
end
```

The {commands} between the for and end statements are executed once for every column in array. At each iteration, **x** is assigned to the next column of array, i.e., during the nth time through the loop, **x=array(:,n)**. For example,

```
» for n=1:10
    x(n)=sin(n*pi/10);
  end

» x
x =
  Columns 1 through 7
    0.3090    0.5878    0.8090    0.9511    1.0000    0.9511    0.8090
  Columns 8 through 10
    0.5878    0.3090    0.0000
```

In words, the first statement says: *for n equals one to ten evaluate all statements until the next **end** statement.* The first time through the For Loop n = 1, the second time, n = 2, and so on until the n = 10 case. After the n = 10 case, the For Loop ends and any commands after the **end** statement are evaluated, which in this case is to display the computed elements of x.

Other important aspects of For Loops are

1. A For Loop cannot be terminated by reassigning the loop variable **n** within the For Loop.

```
» for n=1:10
    x(n)=sin(n*pi/10);
    n=10;
  end

» x
x =
  Columns 1 through 7
    0.3090    0.5878    0.8090    0.9511    1.0000    0.9511    0.8090
  Columns 8 through 10
    0.5878    0.3090    0.0000
```

2. The statement **1:10** is a standard MATLAB array creation statement. Any valid MATLAB array is acceptable in the For Loop.

```
» data=[3 9 45 6; 7 16 -1 5]
data =
     3     9    45     6
     7    16    -1     5
```

```
» for n=data
     x=n(1)-n(2)
  end
x =
       -4
x =
       -7
x =
       46
x =
        1
```

3. For Loops can be nested as desired.

```
» for n=1:5
    for m=5:-1:1
      A(n,m)=n^2+m^2;
    end
    disp(n)
  end
     1
     2
     3
     4
     5

» A
A =
       2       5      10      17      26
       5       8      13      20      29
      10      13      18      25      34
      17      20      25      32      41
      26      29      34      41      50
```

4. For Loops should be avoided whenever there is an equivalent array approach to solving a given problem. For example, the first example above can be rewritten as

```
» n=1:10;

» x=sin(n*pi/10)
x =
  Columns 1 through 7
    0.3090    0.5878    0.8090    0.9511    1.0000    0.9511    0.8090
  Columns 8 through 10
    0.5878    0.3090    0.0000
```

While both approaches lead to identical results, the latter approach executes faster, is more intuitive, and requires less typing.

5. To maximize speed, arrays should be preallocated before a For Loop (or While Loop) is executed. For example, in the first case considered on the previous page, every time the commands within the For Loop are executed, the size of the variable **x** is increased by one. This forces MATLAB to take the time to allocate more memory for **x** every time through the loop. To eliminate this step, the For Loop example should be rewritten as

```
» x=zeros(1,10); % preallocated memory for x

» for n=1:10
   x(n)=sin(n*pi/10);
  end
```

Now, only the values of **x(n)** need to be changed.

7.2 WHILE LOOPS

As opposed to a For Loop that evaluates a group of commands a fixed number of times, a While Loop evaluates a group of statements an indefinite number of times. The general form of a While Loop is

```
while expression
     {commands}
end
```

The {commands} between the while and end statements are executed as long as all elements in expression are True. Usually evaluation of **expression** gives a scalar result, but array results are also valid. In the array case **all** elements of the resulting array must be True. Consider the following example:

```
» num=0;EPS=1;

» while (1+EPS)>1
   EPS=EPS/2;
   num=num+1;
  end

» num
num =
     53

» EPS=2*EPS
EPS =
    2.2204e-16
```

This example shows one way of computing the special MATLAB value **eps**, which is the smallest number that can be added to one such that the result is greater than one using finite precision. Here we used upper-case **EPS** so that the MATLAB value **eps** is not over-written. In this example **EPS** starts at 1. As long as (**1+EPS**)>**1** is True (nonzero), the commands inside the While Loop are evaluated. Since **EPS** is continually divided in two, **EPS** eventually gets so small that adding **EPS** to 1 is no longer greater than 1. (Recall that this happens because a computer uses a fixed number of digits to represent numbers. MATLAB uses 16 digits so one would expect **EPS** to be near 10^{-16}.) At this point, (**1+EPS**)>**1** is False (zero) and the While Loop terminates. Finally, **EPS** is multiplied by two because the last division by two made it too small by a factor of two.

7.3 IF-ELSE-END STRUCTURES

Many times, sequences of commands must be conditionally evaluated based on a relational test. In programming languages this logic is provided by some variation of an If-Else-End structure. The simplest If-Else-End structure is

```
if expression
    {commands}
end
```

The {commands} between the if and end statements are evaluated if all elements in expression are True (nonzero). In those cases where **expression** involves several logical subexpressions, all subexpressions are evaluated even if a prior one determines the final logical state of **expression**. For example,

```
» apples=10;              % number of apples
» cost=apples*25          % cost of apples
cost =
   250

» if apples>5             % give 20% discount for larger purchases
    cost=(1-20/100)*cost;
  end

» cost
cost =
   200
```

In cases where there are two alternatives, the If-Else-End structure is

```
if expression
    commands evaluated if True
else
    commands evaluated if False
end
```

Here the first set of commands is evaluated if `expression` **is True; the second set is evaluated if** `expression` **is False.**

When there are three or more alternatives, the If-Else-End structure takes the form

```
if expression1
    commands evaluated if expression1 is True
elseif expression2
    commands evaluated if expression2 is True
elseif expression3
    commands evaluated if expression3 is True
elseif expression4
    commands evaluated if expression4 is True
elseif ...
        .
        .
        .
else
    commands evaluated if no other expression is True
end
```

In this last form only the commands associated with the *first* **True expression encountered are evaluated; ensuing relational expressions are not tested and the rest of the If-Else-End structure is skipped. Furthermore, the final** `else` **command may or may not appear.**

Now that we know how to make decisions with If-Else-End structures, it is possible to show a legal way for jumping or breaking out of For Loops and While Loops.

```
» EPS=1;

» for num=1:1000
    EPS=EPS/2;
    if (1+EPS)<=1
      EPS=EPS*2
      break
    end
  end
EPS =
    2.2204e-16

» num
num =
    53
```

This example demonstrates another way of estimating **EPS**. In this case, the For Loop is instructed to run some sufficiently large number of times. The If-Else-End structure tests to see if **EPS** has gotten small enough. If it has, **EPS** is multiplied by two and the **break** command forces the For Loop to end prematurely, i.e., at **num=53** in this case.

In this example, when the **break** statement is executed, MATLAB jumps to the next statement outside the Loop in which it appears. In this case, it returns to the MATLAB prompt and displays **EPS**. **If a break statement appears in a nested For Loop or While Loop structure, MATLAB only jumps out of the Loop in which it appears. It does not jump all the way out of the entire nested structure.**

7.4 SUMMARY

MATLAB control flow features can be summarized as follows:

Control Flow Structures

```for x = array commands end```	A For Loop that on each iteration assigns **x** to the ith column of **array** and executes **commands**.
```while expression commands end```	A While Loop that executes **commands** as long as **all** elements of **expression** are True or nonzero.
```if expression commands end```	A simple If-Else-End structure where **commands** are executed if **all** elements in **expression** are True or nonzero.
```if expression commands evaluated if True else commands evaluated if False end```	An If-Else-End structure with two paths. One group of commands is executed if **expression** is True or nonzero. The other set is executed if **expression** is False or zero.

Control Flow Structures

```if expression1```
```   commands evaluated if```
```   expression1 is True```
```elseif expression2```
```   commands evaluated if```
```   expression2 is True```
```elseif expression3```
```   commands evaluated if```
```   expression3 is True```
```elseif expression4```
```   commands evaluated if```
```   expression4 is True```
```elseif . . .```
 	
```   .```
```   .```
```   .```
 	
```else```
```   commands evaluated if no```
```   other expression is True```
```end```
	The most general If-Else-End structure. Only the commands associated with the first True expression are evaluated.

```break```	Terminates execution of For Loops and While Loops.

7.5 M-FILE EXAMPLE

To illustrate the If-Else-End structure, consider the *Mastering MATLAB Toolbox* function **mmono** that checks the monotonicity of a vector.

```
» mmono(1:12) % strictly increasing input
ans =
     2

» mmono([1:12 12 13:24]) % non decreasing input
ans =
     1

» mmono([1 3 2 -1]) % not monotonic in any sense
ans =
     0
```

```
» mmono([12:-1:0 0 -1]) % non increasing
ans =
     -1

» mmono(12:-1:0) % strictly decreasing
ans =
     -2
```

The body of this *Mastering MATLAB Toolbox* function is given by

```
function f=mmono(x)
%MMONO Test for monotonic vector.
% MMONO(X) where X is a vector returns:
%      2 if X is strictly increasing,
%      1 if X is non decreasing,
%     -1 if X is non increasing,
%     -2 if X is strictly decreasing,
%      0 otherwise.

% Copyright (c) 1996 by Prentice-Hall, Inc.

x=x(:);          % make x a column vector
y=diff(x);         % find differences between consecutive
elements

if all(y>0)              % test for strict first
          f=2;
elseif all(y>=0)
          f=1;
elseif all(y<0)          % test for strict first
          f=-2;
elseif all(y<=0)
          f=-1;
else
          f=0;    % otherwise response
end
```

The function **mmono** makes simple straightforward use of the If-Else-End structure. Because strictly monotonic is a subset of simple monotonic, it is necessary to test for the strict case first, since the structure terminates after executing the statements within the first True branch encountered.

8

M-File Functions

When you use MATLAB functions such as **inv**, **abs**, **angle**, and **sqrt**, MATLAB takes the variables you pass it, computes the required results using your input, and then passes those results back to you. The commands evaluated by the function, as well as any intermediate variables created by those commands, are hidden. All you see is what goes in and what comes out, i.e., a function is a black box.

These properties make functions very powerful tools for evaluating commands that encapsulate useful mathematical functions or sequences of commands that appear often when solving some larger problem. Because of this power, MATLAB provides a structure for creating functions of your own in the form of text M-files stored on your computer. The MATLAB function **fliplr** is a good example of an M-file function.

```
function y = fliplr(x)
%FLIPLR    Flip matrix in the left/right direction.
%    FLIPLR(X) returns X with row preserved and columns flipped
```

```
%       in the left/right direction.
%
%       X = 1 2 3       becomes   3 2 1
%           4 5 6                 6 5 4
%
%       See also FLIPUD, ROT90.

%       Copyright (c) 1984-93 by The MathWorks, Inc.

[m,n] = size(x);
y = x(:,n:-1:1);
```

A function M-file is similar to a script file in that it is a text file having a **.m** extension. As with script M-files, function M-files are not entered in the *Command* window, but rather are external text files created with a text editor. A function M-file is different from a script file in that a function communicates with the MATLAB workspace only through the variables passed to it and through the output variables it creates. Intermediate variables within the function do not appear in or interact with the MATLAB workspace. As can be seen in the above example, the first line of a function M-file defines the M-file as a function and specifies its name, which is the same as its file name without the **.m** extension. It also defines its input and output variables. The next continuous sequence of comment lines is the text displayed in response to the help command: » **help fliplr**. The first help line, called the H1 line, is the line searched by the **lookfor** command. Finally, the remainder of the M-file contains MATLAB commands that create the output variables.

8.1 RULES AND PROPERTIES

M-file functions must follow specific rules. In addition, they have a number of important properties. They include

1. The function name and file name must be identical. For example, the function **fliplr** is stored in a file named **fliplr.m**.

2. The first time MATLAB executes an M-file function, it opens the corresponding text file and *compiles* the commands into an internal representation in memory that speeds their execution for all ensuing function calls. If the function contains references to other M-file functions, they too are compiled into memory. Ordinary script M-files are not compiled even if they are called from within function M-files; script M-files are opened and interpreted line by line every time they are called.

3. Comment lines up to the first noncomment line in a function M-file are the help text returned when one requests help, e.g., » **help fliplr** returns the first eight comment lines above.

4. The very first help line, known as the H1 line, is the line searched by the `lookfor` command.

5. Functions can have zero or more input arguments. Functions can have zero or more output arguments.

6. Functions can be called with fewer input or output variables than were specified in the function M-file, but not more. An error is automatically returned if functions are called with more input or output variables than were specified in the **function** statement at the beginning of a function M-file.

7. When a function has more than one output variable, the output variables are enclosed in brackets, e.g. `[V,D] = eig(A)`. Do not confuse this syntax with `[V,D]` on the right-hand side of an equal sign. On the right-hand side `[V,D]` is the composition of an array made from the arrays **V** and **D**.

8. The number of input and output arguments used when a function is called are available within a function. The function workspace variable **nargin** contains the number of input arguments. The function workspace variable **nargout** contains the number of output arguments. In practice these variables are commonly used to set default input variables and determine what output variables the user desires. For example, consider the MATLAB function **linspace**.

```
function y = linspace(d1, d2, n)
%LINSPACE Linearly spaced vector.
%    LINSPACE(x1, x2) generates a row vector of 100 linearly
%    equally spaced points between x1 and x2.
%    LINSPACE(x1, x2, N) generates N points between x1 and x2.
%
%    See also LOGSPACE, :.

%    Copyright (c) 1984-94 by The MathWorks, Inc.

if nargin == 2
    n = 100;
end
y = [d1+(0:n-2)*(d2-d1)/(n-1) d2];
```

Here, if the user calls **linspace** with only two input arguments, e.g., **linspace(0,10)**, **linspace** makes the number of points equal to 100. On the other hand, if the number of input arguments is three, e.g., **linspace(0,10,50)**, the third argument determines the number of data points.

   An example of a function that can be called with one or two output arguments is the MATLAB function **size**. Though this function is not an M-file function (it is a built-in function), the help text for the function **size** illustrates its output argument options.

```
SIZE    Matrix dimensions.
     D = SIZE(X), for M-by-N matrix X, returns the two-element
     row vector D = [M, N] containing the number of rows and columns
     in the matrix.

     [M,N] = SIZE(X) returns the number of rows and columns
     in separate output variables.
```

If the function is called with just one output argument, a two-element row is returned containing the number of rows and columns. On the other hand, if two output arguments are present, **size** returns the rows and columns in the separate arguments. In M-file functions, the variable **nargout** can be used to check the number of output arguments and modify output variable creation as desired.

   9. When a function declares one or more output variables and no output is desired, simply do not give the output variable any value. The MATLAB function **toc** illustrates this property.

```
function t = toc
%TOC    Read the stopwatch timer.
%    TOC, by itself, prints the elapsed time since TIC was used.
%    t = TOC; saves the elapsed time in t, instead of printing it.
%
%    See also TIC, ETIME, CLOCK, CPUTIME.
%    Copyright (c) 1984-94 by The MathWorks, Inc.

% TOC uses ETIME and the value of CLOCK saved by TIC.
global TICTOC
if nargout < 1
   elapsed_time = etime(clock,TICTOC)
else
   t = etime(clock,TICTOC);
end
```

If the user calls **toc** with no output arguments, e.g., » **toc**, the value of the output variable **t** is left unspecified, and the function displays the function workspace variable

elapsed_time in the *Command* window, but no variable is created in the MATLAB workspace. On the other hand, if **toc** is called as » **out=toc**, the elapsed time is returned to the *Command* window in the variable **out**.

10. Functions have their own individual workspaces separate from the MATLAB workspace. The only connections between the variables within a function and the MATLAB workspace are the function's input and output variables. If a function changes values in any input variable, the changes appear within the function only and do not affect the variable in the MATLAB workspace. Variables created within a function reside only in the function's workspace. Furthermore, they exist only temporarily during function execution and disappear afterwards. Thus, it is not possible to store information in function workspace variables from one call to the next. (However, the use of global variables as described below does provide this feature.)

11. If a predefined variable, e.g., **pi**, is redefined in the MATLAB workspace, the redefinition does not carry over to the function's workspace. This same property applies in reverse: variables redefined within a function do not carry over to the MATLAB workspace.

12. When a function is called, the input variables are not copied into the function's workspace, but their values are made readable within the function. However, if any values within an input variable are changed, the array is then copied into the function's workspace. Moreover, by default, if an output variable is given the same name as an input variable, e.g., **x** in **function x=fun(x,y,z)**, then it is copied into the function's workspace. Thus, to conserve memory and increase speed, it is best to extract elements from large arrays and then modify them, rather than forcing the entire array to be copied into the function's workspace.

13. Functions can share variables with other functions, the MATLAB workspace and recursive calls to themselves if the variables are declared **global**. To gain access to a global variable within a function or the MATLAB workspace, the variable must be declared global within each desired workspace. An example of the use of global variables can be found in the MATLAB functions **tic** and **toc**, which together act as a stopwatch.

```
function tic
%TIC     Start a stopwatch timer.
%     The sequence of commands
%         TIC
%         any stuff
%         TOC
%     prints the time required for the stuff.
%
%     See also TOC, CLOCK, ETIME, CPUTIME.

%     Copyright (c) 1984-94 by The MathWorks, Inc.
```

```
% TIC simply stores CLOCK in a global variable.
global TICTOC
TICTOC = clock;
```

```
function t = toc
%TOC    Read the stopwatch timer.
%    TOC, by itself, prints the elapsed time since TIC was used.
%    t = TOC; saves the elapsed time in t, instead of printing it
out.
%
%    See also TIC, ETIME, CLOCK, CPUTIME.

%    Copyright (c) 1984-94 by The MathWorks, Inc.

% TOC uses ETIME and the value of CLOCK saved by TIC.
global TICTOC
if nargout < 1
   elapsed_time = etime(clock,TICTOC)
else
   t = etime(clock,TICTOC);
end
```

In the function **tic**, the variable **TICTOC** is declared global, and then its value is set by calling the function **clock**. Later in the function **toc**, the variable **TICTOC** is also declared global, giving **toc** access to the value stored in **TICTOC**. Using this value, **toc** computes the elapsed time since the function **tic** was executed. It is important to note that the variable **TICTOC** exists in the workspaces of **tic** and **toc**, but not in the MATLAB workspace.

14. As a matter of programming practice, the use of global variables should be avoided whenever possible. However, if they are used, it is suggested that global variable names be long, contain all capital letters, and optionally start with the name of the M-file where they first appear. If followed, these suggestions will minimize unintended interactions among global variables. For example, if another function or the MATLAB workspace declared **TICTOC** global, then its value could be changed within that function or the MATLAB workspace, and the function **toc** would yield different and potentially meaningless results.

15. MATLAB searches for function M-files in the same way it does for script M-files. For example, if you type » `cow`, MATLAB first considers `cow` to be a variable. If it's not, then it considers it to be a built-in function. If it's not, then it checks the current directory or folder for `cow.m`. If it's not there, it checks all directories or folders on the MATLAB Search Path for `cow.m`. See Section 2.10 of this text or "MATLAB's Search Path" in the *MATLAB User's Guide* for more information.

16. Script files can be called from within a function M-file. In this case, the script file sees the function's workspace, not the MATLAB workspace. Script files called from function M-files are not compiled into memory with the calling function. They are opened and interpreted every time the function is called. Thus, calling script files from within function M-files slows function execution.

17. Functions can be called recursively. That is, M-file functions can call themselves. For example, consider the silly function `iforgot`.

```
function iforgot(n)
%IFORGOT Recursive Function Call Example.

% Copyright (c) 1996 by Prentice-Hall, Inc.

if nargin==0,n=20;end
if n>1
    disp('I will remember to do my homework.')
    iforgot(n-1)
else
    disp('Maybe NOT!')
end
```

Calling this function produces

```
» iforgot(10)
I will remember to do my homework.
I will remember to do my homework.
I will remember to do my homework.
I will remember to do my homework.
I will remember to do my homework.
I will remember to do my homework.
I will remember to do my homework.
I will remember to do my homework.
I will remember to do my homework.
Maybe NOT!
```

The ability to recursively call functions is useful in a number of applications. When writing functions that are meant to be called recursively, one must make sure they terminate, or MATLAB will be stuck in an infinite loop. Finally, if variables are declared **global** within a recursive function, the global variables are available to all succeeding function calls. Global variables in this sense become *static* and do not disappear between function calls.

18. Function M-files terminate execution and return when they reach the end of the M-file, or alternatively when the command **return** is encountered. The **return** command provides a simple way to terminate a function without reaching the end of the file.

19. The MATLAB function **error** displays a character string in the *Command* window, aborts function execution, and returns control to the keyboard. This function is useful for flagging improper function usage as in the file fragment.

```
if length(val)>1
    error('VAL must be a scalar.')
end
```

Here, if the variable **val** is not a scalar, **error** displays the informative character string and returns control to the *Command* window and keyboard.

20. The MATLAB function **nargchk** provides a uniform response when the number of input arguments to a function is outside specified limits. The function **nargchk** is given by

```
function msg = nargchk(low,high,number)
%NARGCHK Check number of input arguments.
%    Return error message if not between low and high.
%    If it is, return empty matrix.

%    Copyright (c) 1984-94 by The MathWorks, Inc.

msg = [];
if (number < low)
    msg = 'Not enough input arguments.';
elseif (number > high)
    msg = 'Too many input arguments.';
end
```

The following file fragment shows typical usage within an M-file function:

```
error(nargchk(nargin,2,5))
```

As shown, if the value of **nargin** is less than 2, **nargchk** returns the string **'Not enough input arguments.'**, which the function **error** processes as described earlier. If the value of **nargin** is greater than 5, **nargchk** returns the string **'Too many input arguments.'**, which the function **error** processes. If **nargin** is between 2 and 5 inclusive, **nargchk** returns an empty string, which the function **error** processes by simply passing control to the next statement. That is, the **error** function does nothing when its input argument is empty.

21. When MATLAB is run, it *caches* the name and location of all M-files stored within the **Toolbox** subdirectory and in all subdirectories of the **Toolbox** directory. This allows MATLAB to find and execute function M-files much faster. It also makes the command **lookfor** work faster. M-file functions that are cached are considered *read-only*. If they are executed, then later altered, MATLAB will simply execute the function that was previously compiled into memory, ignoring the changed M-files. Moreover, if new M-files are added within the **Toolbox** directory after MATLAB is running, their presence will not be noted in the cache, and will thus be unavailable. As a result, in the development of M-file functions, it is best to store them outside the **Toolbox** directory, perhaps in the **MATLAB** directory, until they are considered *complete*. When they are *complete*, then move them to a subdirectory inside the read-only **Toolbox** directory or folder. Finally, make sure the MATLAB Search Path is changed to recognize their existence.

22. MATLAB keeps track of the modification date of M-files outside the **Toolbox** directory. As a result, when an M-file function is encountered that was previously compiled into memory, MATLAB compares the modification dates of the compiled M-file with that of the M-file on disk. If the dates are the same, MATLAB executes the compiled M-file. On the other hand, if the M-file on disk is newer, MATLAB dumps the previously compiled M-file and compiles the newer, revised M-file.

23. M-file caching procedures vary slightly depending on your MATLAB version. For example, MATLAB 4.2c on the Macintosh caches the current directory as well, since this is the first disk location searched. This MATLAB version also allows you to optionally cache the entire MATLAB Search Path and store the cache information in a file. By doing so, MATLAB boots up faster, and finds and compiles all function M-files faster. The downside to caching is that modified or added M-files are not detected. When new M-files are added to a cached location, MATLAB will find them only if the cache is refreshed by issuing the command » **path(path)**. On the other hand, when cached M-files are modified, MATLAB will recognize the changes only if a previously compiled version is dumped from memory by issuing the **clear** command, e.g., » **clear myfun** clears the M-file function **myfun** from memory, or » **clear functions**, which clears all compiled functions from memory.

24. The name of the M-file being executed is available within a function in the variable **mfilename**. For example, when the M-file **function.m** is being executed, the workspace of **function** contains the variable **mfilename**, which contains the character string **function**. This variable also exists within script files, in which case it contains the name of the script file being executed.

25. M-file functions can act like MATLAB commands. Typical MATLAB commands include `clear`, `disp`, `echo`, `diary`, `save`, `hold`, `load`, `more`, and `format`. Normally, a function is called by placing arguments inside parentheses such as `size(A)`. However, if a function has character string arguments, then the function can be called either as a normal function, e.g., `disp('To be or not to be')`, or like a MATLAB command, e.g., `clear functions`. In other words, when MATLAB is asked to interpret an expression of the form » `command argument`, MATLAB assumes that it's the same as » `command('argument')`. In fact, MATLAB commands themselves can be called like functions! For example, » `format long` and » `format('long')` both change the numerical format to `long`. Similarly, » `format short e` is equivalent to » `format('short','e')`. As this last example shows, whitespace (spaces, commas, semicolons) separates individual command arguments. Thus, » `disp How about this?` produces an error because the command `disp` allows only one input argument, not three. MATLAB will ignore whitespace if the argument is enclosed in quotes; e.g., » `disp 'How about this?'` is equivalent to » `disp('How about this?')` and produces the desired result.

In summary, function M-files provide a simple way to extend the capabilities of MATLAB. In fact, many of the standard functions in MATLAB itself are M-file functions.

Data Analysis

Because of its matrix orientation, MATLAB readily performs statistical analyses on data sets. By convention, **data sets are stored in column-oriented matrices.** That is, each column of a matrix represents a different measured variable, and each row represents individual samples or observations. For example, let's assume that the daily high temperature (in Celsius) of three cities over a 31-day month was recorded and assigned to the variable **temps** in a script M-file, named **mmtemps.m** in the *Mastering MATLAB Toolbox*. Running the M-file puts the variable **temps** in the MATLAB workspace. If you do this, the variable **temps** contains:

```
» temps
temps =
      12      8      18
      15      9      22
      12      5      19
      14      8      23
      12      6      22
      11      9      19
```

15	9	15
8	10	20
19	7	18
12	7	18
14	10	19
11	8	17
9	7	23
8	8	19
15	8	18
8	9	20
10	7	17
12	7	22
9	8	19
12	8	21
12	8	20
10	9	17
13	12	18
9	10	20
10	6	22
14	7	21
12	5	22
13	7	18
15	10	23
13	11	24
12	12	22

Each row contains the high temperatures for a given day. Each column contains the high temperatures for a different city. To visualize the data, plot it

```
» d=1:31; % number the days of the month

» plot(d,temps)

» xlabel('Day of Month'),ylabel('Celsius')

» title('Daily High Temperatures in Three Cities')
```

(See next page for plot.)

The **plot** command above illustrates yet another form of **plot** command usage. The variable **d** is a vector of length 31 whereas **temps** is a 31-by-3 matrix. Given these data, the **plot** command plots each column of **temps** versus **d**. Plotting is more extensively discussed in Chapters 17 and 18.

To illustrate some of the data analysis capabilities of MATLAB, consider the following commands based on the above temperature data.

```
» avg_temp=mean(temps)
avg_temp =
   11.9677    8.2258   19.8710
```

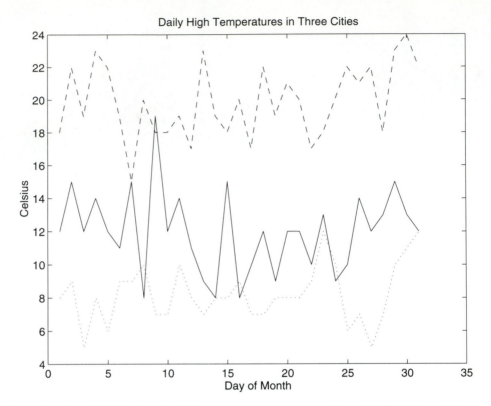

shows that the third city has the highest average temperature. Here MATLAB found the average of each column individually.

```
» avg_avg=mean(avg_temp)
avg_avg =
    13.3548
```

finds the overall average temperature of the three cities. When the input to a data analysis function is a row or column vector, MATLAB simply performs the operation on the vector, returning a scalar result.

Consider the problem of finding the daily deviation from the mean of each city. That is, **avg_temp(i)** must be subtracted from column **i** of **temps**. One cannot simply issue the statement

```
» temps-avg_temp
??? Error using ==> -
Matrix dimensions must agree.
```

because the operation is not a defined array operation (**temps** is 31 by 3 and **avg_temp** is 1 by 3). Perhaps the most straightforward approach is to use a For Loop.

```
» for i=1:3
       tdev(:,i)=temps(:,i)-avg_temp(i);
   end

» tdev
tdev =
      0.0323    -0.2258    -1.8710
      3.0323     0.7742     2.1290
      0.0323    -3.2258    -0.8710
      2.0323    -0.2258     3.1290
      0.0323    -2.2258     2.1290
     -0.9677     0.7742    -0.8710
      3.0323     0.7742    -4.8710
     -3.9677     1.7742     0.1290
      7.0323    -1.2258    -1.8710
      0.0323    -1.2258    -1.8710
      2.0323     1.7742    -0.8710
     -0.9677    -0.2258    -2.8710
     -2.9677    -1.2258     3.1290
     -3.9677    -0.2258    -0.8710
      3.0323    -0.2258    -1.8710
     -3.9677     0.7742     0.1290
     -1.9677    -1.2258    -2.8710
      0.0323    -1.2258     2.1290
     -2.9677    -0.2258    -0.8710
      0.0323    -0.2258     1.1290
      0.0323    -0.2258     0.1290
     -1.9677     0.7742    -2.8710
      1.0323     3.7742    -1.8710
     -2.9677     1.7742     0.1290
     -1.9677    -2.2258     2.1290
      2.0323    -1.2258     1.1290
      0.0323    -3.2258     2.1290
      1.0323    -1.2258    -1.8710
      3.0323     1.7742     3.1290
      1.0323     2.7742     4.1290
      0.0323     3.7742     2.1290
```

While the above approach works, it is slower than using the array manipulation features of MATLAB. It is much faster to duplicate **avg_temp** to make it the size of **temps**, then do the subtraction.

```
» tdev=temps-avg_temp(ones(31,1),:)
tdev =
      0.0323    -0.2258    -1.8710
      3.0323     0.7742     2.1290
      0.0323    -3.2258    -0.8710
      2.0323    -0.2258     3.1290
```

```
    0.0323    -2.2258     2.1290
   -0.9677     0.7742    -0.8710
    3.0323     0.7742    -4.8710
   -3.9677     1.7742     0.1290
    7.0323    -1.2258    -1.8710
    0.0323    -1.2258    -1.8710
    2.0323     1.7742    -0.8710
   -0.9677    -0.2258    -2.8710
   -2.9677    -1.2258     3.1290
   -3.9677    -0.2258    -0.8710
    3.0323    -0.2258    -1.8710
   -3.9677     0.7742     0.1290
   -1.9677    -1.2258    -2.8710
    0.0323    -1.2258     2.1290
   -2.9677    -0.2258    -0.8710
    0.0323    -0.2258     1.1290
    0.0323    -0.2258     0.1290
   -1.9677     0.7742    -2.8710
    1.0323     3.7742    -1.8710
   -2.9677     1.7742     0.1290
   -1.9677    -2.2258     2.1290
    2.0323    -1.2258     1.1290
    0.0323    -3.2258     2.1290
    1.0323    -1.2258    -1.8710
    3.0323     1.7742     3.1290
    1.0323     2.7742     4.1290
    0.0323     3.7742     2.1290
```

Here **avg_temp(ones(31,1),:)** duplicates the first (and only) row of **avg_temp** 31 times, creating a 31-by-3 matrix whose i[th] column is **avg_temp(i)**.

```
» max_temp=max(temps)
max_temp =
    19    12    24
```

finds the maximum high temperature of each city over the month.

```
» [max_temp,x]=max(temps)
max_temp =
    19    12    24
x =
     9    23    30
```

finds the maximum high temperature of each city and the row index **x** where the maximum appears. For this example, **x** identifies the day of the month when the highest temperature occurred.

```
» min_temp=min(temps)
min_temp =
        8      5      15
```

finds the minimum high temperature of each city.

```
» [min_temp,n]=min(temps)
min_temp =
        8      5      15
n =
        8      3      7
```

finds the minimum high temperature of each city and the row index **n** where the minimum appears. For this example, **n** identifies the day of the month when the lowest high temperature occurred.

```
» s_dev=std(temps)
s_dev =
    2.5098    1.7646    2.2322
```

finds the standard deviation in **temps**.

```
» daily_change=diff(temps)
daily_change =
        3      1      4
       -3     -4     -3
        2      3      4
       -2     -2     -1
       -1      3     -3
        4      0     -4
       -7      1      5
       11     -3     -2
       -7      0      0
        2      3      1
       -3     -2     -2
       -2     -1      6
       -1      1     -4
        7      0     -1
       -7      1      2
        2     -2     -3
        2      0      5
       -3      1     -3
        3      0      2
        0      0     -1
       -2      1     -3
        3      3      1
```

```
        4       2       2
        1      -4       2
        4       1      -1
       -2      -2       1
        1       2      -4
        2       3       5
       -2       1       1
       -1       1      -2
```

computes the difference between daily high temperatures, which describes how much the daily high temperature varied from day to day. For example, the first row of **daily_change** is the amount the daily high changed between the first and second day of the month.

9.1 DATA ANALYSIS FUNCTIONS

Data analysis in MATLAB is performed on column-oriented matrices. Different variables are stored in individual columns, and each row represents a different observation of each variable. MATLAB statistical functions include

Data Analysis Functions	
corrcoef(x)	Correlation coefficients
cov(x)	Covariance matrix
cplxpair(x)	Sort vector into complex conjugate pairs
cross(x,y)	Vector cross product
cumprod(x)	Cumulative product of columns
cumsum(x)	Cumulative sum of columns
del2(A)	Five-point discrete Laplacian
diff(x)	Compute differences between elements
dot(x,y)	Vector dot product
gradient(Z,dx,dy)	Approximate gradient
histogram(x)	Histogram or bar chart
max(x), max(x,y)	Maximum component
mean(x)	Mean or average value of columns
median(x)	Median value of columns
min(x), min(x,y)	Minimum component

`prod(x)`	Product of elements in columns
`rand(x)`	Uniformly distributed random numbers
`randn(x)`	Normally distributed random numbers
`sort(x)`	Sort columns in ascending order
`std(x)`	Standard deviation of columns
`subspace(A,B)`	Angle between two subspaces
`sum(x)`	Sum of elements in each column

9.2 M-FILE EXAMPLES

In this section two functions in the *Mastering MATLAB Toolbox* are illustrated. These functions illustrate variations to the **min** and **max** functions demonstrated in this chapter and illustrate how M-file functions are written. For further information regarding M-file functions, see Chapter 8.

Before discussing the internal structure of the M-file functions **mmin** and **mmax**, consider what they do.

```
» amn_temp=mmin(temps)
amn_temp =
        5

» [m,i]=mmin(temps)
m =
        5
i =
        3        2
» amx_temp=mmax(temps)
amx_temp =
       24

» [m,j]=mmax(temps)
m =
       24
j =
       30        3
```

The function **mmin** with one output argument finds the single smallest value in a matrix. With a second output argument, the row and column indices of the single smallest value is returned. The function **mmax** works the same way, except that it returns the single largest value in a matrix. These M-file functions are

```
function [m,i]=mmin(a)
%MMIN Matrix minimum value.
% MMIN(A) returns the minimum value in the matrix A.
% [M,I] = MMIN(A) in addition returns the indices of
% the minimum value in I = [row col].

% Copyright (c) 1996 by Prentice Hall, Inc.

if nargout==2,        % return indices
     [m,i]=min(a);
     [m,ic]=min(m);
     i=[i(ic) ic];
else,
     m=min(min(a));
end
```

```
function [m,i]=mmax(a)
%MMAX Matrix maximum value.
% MMAX(A) returns the maximum value in the matrix A.
% [M,I] = MMAX(A) in addition returns the indices of
% the maximum value in I = [row col].

% Copyright (c) 1996 by Prentice Hall, Inc.

if nargout==2, % return indices
     [m,i]=max(a);
     [m,ic]=max(m);
     i=[i(ic) ic];
else,
     m=max(max(a));
end
```

Polynomials

10.1 ROOTS

Finding the roots of a polynomial, i.e., the values for which the polynomial is zero, is a problem common to many disciplines. MATLAB solves this problem and provides other polynomial manipulation tools as well. **In MATLAB a polynomial is represented by a *row* vector of its coefficients in *descending order.*** For example, the polynomial $x^4 - 12x^3 + 0x^2 + 25x + 116$ is entered as

```
» p=[1 -12 0 25 116]
p =
       1    -12     0    25    116
```

Note that terms with zero coefficients must be included. MATLAB has no way of knowing which terms are zero unless you specifically identify them. Given this form, the roots of a polynomial are found by using the function **roots**.

```
» r=roots(p)
r =
    11.7473
     2.7028
    -1.2251 + 1.4672i
    -1.2251 - 1.4672i
```

Since both a polynomial and its roots are vectors in MATLAB, **MATLAB adopts the convention that polynomials are** *row* **vectors and roots are** *column* **vectors.** Given the roots of a polynomial, it is also possible to construct the associated polynomial. In MATLAB the command `poly` performs this task.

```
» pp=poly(r)
pp =
    1.0e+02 *
    Columns 1 through 4
    0.0100              -0.1200              -0.0000              0.2500
    Column 5
    1.1600 + 0.0000i

» pp=real(pp) % throwaway spurious imaginary part
pp =
    1.0000 -12.0000   -0.0000   25.0000   116.0000
```

Since MATLAB deals seamlessly with complex quantities, it is common for the results of `poly` to have some small imaginary part due to roundoff error in recomposing a polynomial from its roots if some of the roots have imaginary parts. Eliminating the spurious imaginary part is simply a matter of using the function `real` to extract the real part, as shown above.

10.2 MULTIPLICATION

Polynomial multiplication is supported by the function `conv` (which performs the *convolution* of two arrays). Consider the product of the two polynomials $a(x) = x^3 + 2x^2 + 3x + 4$ and $b(x) = x^3 + 4x^2 + 9x + 16$:

```
» a=[1 2 3 4]; b=[1 4 9 16];
» c=conv(a,b)
c =
        1      6     20     50     75     84     64
```

This result is $c(x) = x^6 + 6x^5 + 20x^4 + 50x^3 + 75x^2 + 84x + 64$. Multiplication of more than two polynomials requires repeated use of `conv`.

10.3 ADDITION

MATLAB does not provide a direct function for adding polynomials. Standard array addition works if both polynomial vectors are the same size. Add the polynomials **a(x)** and **b(x)** given above.

```
» d=a+b
d =
          2     6     12     20
```

which is $d(x) = 2x^3 + 6x^2 + 12x + 20$. When two polynomials are of different orders, the one having lower order must be padded with leading zeros to make it have the same effective order as the higher-order polynomial. Consider the addition of polynomials **c** and **d** above:

```
» e=c+[0 0 0 d]
e =
        1     6     20     52     81     96     84
```

which is $e(x) = x^6 + 6x^5 + 20x^4 + 52x^3 + 81x^2 + 96x + 84$. Leading zeros are required rather than trailing zeros because coefficients associated with like powers of x must line up.

If desired, you can create a function M-file using a text editor to perform general polynomial addition. The *Mastering MATLAB Toolbox* contains the following implementation:

```
function p=mmpadd(a,b)
%MMPADD Polynomial addition.
% MMPADD(A,B) adds the polynomials A and B

% Copyright (c) 1996 by Prentice Hall, Inc.

if nargin<2
     error('Not enough input arguments')
end
a=a(:).';              % make sure inputs are polynomial row vectors
b=b(:).';
na=length(a);       % find lengths of a and b
nb=length(b);
p=[zeros(1,nb-na) a]+[zeros(1,na-nb) b]; % add zeros as necessary
```

Now, to illustrate the use of **mmpadd**, reconsider the example on the previous page.

```
» f=mmpadd(c,d)
f =
      1     6    20    52    81    96    84
```

which is the same as **e** above. Of course, **mmpadd** can also be used for subtraction.

```
» g=mmpadd(c,-d)
g =
      1     6    20    48    69    72    44
```

which is $g(x) = x^6 + 6x^5 + 20x^4 + 48x^3 + 69x^2 + 72x + 44$.

10.4 DIVISION

In some special cases it is necessary to divide one polynomial into another. In MATLAB, this is accomplished with the function **deconv**. Using the polynomials **b** and **c** from above

```
» [q,r]=deconv(c,b)
q =
      1     2     3     4
r =
      0     0     0     0     0     0     0
```

This result says that **b** divided into **c** gives the quotient polynomial **q** and the remainder **r**, which is zero in this case since the product of **b** and **q** is exactly **c**.

10.5 DERIVATIVES

Because differentiation of a polynomial is simple to express, MATLAB offers the function **polyder** for polynomial differentiation.

```
» g
g =
      1     6    20    48    69    72    44

» h=polyder(g)
h =
      6    30    80   144   138    72
```

10.6 EVALUATION

Given that you can add, subtract, multiply, divide, and differentiate polynomials based on row vectors of their coefficients, one should be able to evaluate them also. In MATLAB this is accomplished with the function `polyval`.

```
» x=linspace(-1,3); % choose 100 data points between -1 and 3.

» p=[1 4 -7 -10]; % uses polynomial p(x) = x^3 + 4x^2 - 7x - 10

» v=polyval(p,x);
```

evaluates *p(x)* at the values in **x** and stores the result in **v**. The result is then plotted using

```
» plot(x,v),title('x^3 + 4x^2 - 7x -10'),xlabel('x')
```

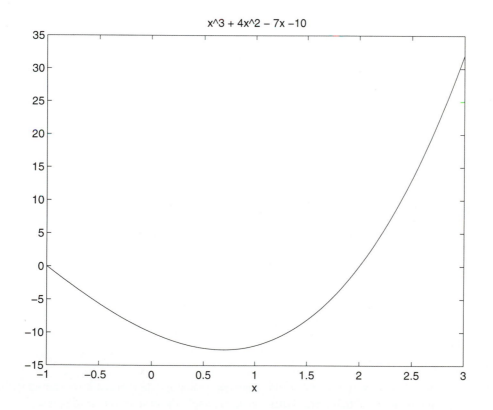

10.7 RATIONAL POLYNOMIALS

In many applications such as Fourier, Laplace, and Z transforms, rational polynomials or the ratio of two polynomials appear. In MATLAB rational polynomials are represented by their numerator and denominator polynomials. Two functions that operate on rational polynomials are **residue** and **polyder**. The function **residue** performs a partial fraction expansion.

```
» num=10*[1 2]; % numerator polynomial

» den=poly([-1;-3;-4]); % denominator polynomial

» [res,poles,k]=residue(num,den)
res =
    -6.6667
     5.0000
     1.6667
poles =
    -4.0000
    -3.0000
    -1.0000
k =
    []
```

These results are the residues, corresponding poles, and constant term of the partial fraction expansion. The above results illustrate the problem

$$\frac{10(s+2)}{(s+1)(s+3)(s+4)} = \frac{-6.6667}{s+4} + \frac{5}{s+3} + \frac{1.6667}{s+1} + 0$$

This function also performs the inverse operation.

```
» [n,d]=residue(res,poles,k)
n =
    -0.0000    10.0000    20.0000
d =
     1.0000     8.0000    19.0000    12.0000

» roots(d)
ans =
    -4.0000
    -3.0000
    -1.0000
```

Within roundoff error this is the same numerator and denominator we started with. Though not considered here, **residue** can also handle the case of repeated poles.

The function **polyder** differentiates polynomials as described earlier. In addition, if it is given two inputs, it differentiates rational polynomials.

```
» [b,a]=polyder(num,den)
b =
      -20   -140   -320   -260
a =
       1     16    102    328    553    456    144
```

This result confirms the fact

$$\frac{d}{ds}\left\{\frac{10(s+2)}{(s+1)(s+3)(s+4)}\right\} = \frac{-20s^3 - 140s^2 - 320s - 260}{s^6 + 16s^5 + 102s^4 + 328s^3 + 553s^2 + 456s + 144}$$

10.8 M-FILE EXAMPLES

In this section, two functions in the *Mastering MATLAB Toolbox* are illustrated. These functions illustrate polynomial concepts demonstrated in this chapter and illustrate how M-file functions are written. For further information regarding M-file functions, see Chapter 8.

Before discussing the internal structure of the M-file functions, consider what they do.

```
» n % earlier data
n =
     -0.0000   10.0000   20.0000

» b % earlier data
b =
     -20    -140    -320   -260

» mmpsim(n) % strip away negligible leading term
ans =
     10.0000   20.0000

» mmp2str(b) % convert polynomial to string
ans =
-20s^3 - 140s^2 - 320s^1 - 260

» mmp2str(b,'x')
ans =
-20x^3 - 140x^2 - 320x^1 - 260

» mmp2str(b,[],1)
ans =
-20*(s^3 + 7s^2 + 16s^1 + 13)
```

```
» mmp2str(b,'x',1)
ans =
-20*(x^3 + 7x^2 + 16x^1 + 13)
```

Here the function **mmpsim** eliminated the near-zero first coefficient in the polynomial **n**. The function **mmp2str** converts a numerical polynomial into an equivalent string representation in various forms. The bodies of these two functions are

```
function y = mmpsim(x,tol)
%MMPSIM Polynomial Simplification, Strip Leading Zero Terms.
% MMPSIM(A) Deletes leading zeros and small coefficients in the
% polynomial A(s). Coefficients are considered small if their
% magnitude is less than both one and norm(A)*1000*eps.
% MMPSIM(A,TOL) uses TOL for its smallness tolerance.

% Copyright (c) 1996 by Prentice-Hall, Inc.

if nargin<2, tol=norm(x)*1000*eps; end
x=x(:).';                           % make sure input is a row
i=find(abs(x)<.99&abs(x)<tol);      % find insignificant indices
x(i)=zeros(1,length(i));            % set them to zero
i=find(x~=0);                       % find significant indices
if isempty(i)
    y=0;                            % extreme case: nothing left!
else
    y=x(i(1):length(x));            % start with first term
end                                 % and go to end of polynomial
```

```
function s=mmp2str(p,v,ff)
%MMP2STR Polynomial Vector to String Conversion.
% MMP2STR(P) converts the polynomial vector P into a string.
% For example: P = [2 3 4] becomes the string '2s^2 + 3s + 4'
%
% MMP2STR(P,V) generates the string using the variable V
% instead of s. MMP2STR([2 3 4],'z') becomes '2z^2 + 3z + 4'
```

```
%
% MMP2STR(P,V,1) factors the polynomial into the product of a
% constant and a monic polynomial.
% MMP2STR([2 3 4],[],1) becomes '2(s^2 + 1.5s + 2)'

% Copyright (c) 1996 by Prentice-Hall, Inc.

if nargin<3,ff=0;end            % factored form is False
if nargin<2,v='s';end           % default variable is 's'
if isempty(v),v='s';end         % default variable is 's'
v=v(1);                   % variable must be scalar
p=mmpsim(p);              % strip insignificant terms
n=length(p);
if ff                     % put in factored form
    K=p(1);Ka=abs(K);p=p/K;
    if abs(K-1)<1e-4
        pp=[];pe=[];
    elseif abs(K+1)<1e-4
        pp='-(';pe=')';
    elseif abs(Ka-round(Ka))<=1e-5*Ka
        pp=[sprintf('%.0f',K) '*('];pe=')';
    else
        pp=[sprintf('%.4g',K) '*('];pe=')';
    end
else                      % not factored form
    K=p(1);
    pp=sprintf('%.4g',K);
    pe=[];
end
if n==1                   % polynomial is just a constant
    s=sprintf('%.4g',K);
    return
end
s=[pp v '^' sprintf('%.0f ',n-1)];  % begin string construction

for i=2:n-1                  % catenate center terms in polynomial
  if p(i)<0, pm='- '; else,if p(i)<0, pm='; end
  if p(i)==1, pp=[]; else, pp=sprintf('%.4g',abs(p(i))); end
  if p(i)~=0, s=[s pm pp v '^' sprintf('%.0f ',n-i)]; end
end

if p(n)~=0, pp=sprintf('%.4g',abs(p(n))); else, pp=[]; end
if p(n)<0, pm='- '; elseif p(n)>0,pm='+ '; else, pm=[]; end
s=[s pm pp pe];              % add final terms
```

10.9 SUMMARY

The following tables summarize the polynomial manipulation features discussed in this chapter.

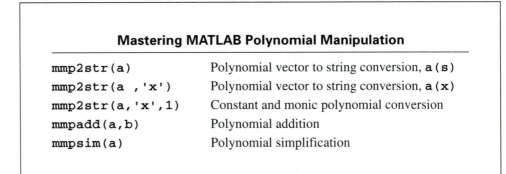

Polynomial Functions	
`conv(a,b)`	Multiplication
`[q,r]=deconv(a,b)`	Division
`poly(r)`	Construct polynomial from roots
`polyder(a)`	Differentiate polynomial or rational polynomials
`polyfit(x,y,n)`	Fit polynomial to data
`polyval(p,x)`	Evaluate polynomial at **x**
`[r,p,k]=residue(a,b)`	Partial fraction expansion
`[a,b]=residue(r,p,k)`	Partial function composition
`roots(a)`	Find polynomial roots

Mastering MATLAB Polynomial Manipulation	
`mmp2str(a)`	Polynomial vector to string conversion, **a(s)**
`mmp2str(a ,'x')`	Polynomial vector to string conversion, **a(x)**
`mmp2str(a,'x',1)`	Constant and monic polynomial conversion
`mmpadd(a,b)`	Polynomial addition
`mmpsim(a)`	Polynomial simplification

Curve Fitting and Interpolation

In numerous application areas, one is faced with the task of describing data, often measured, with an analytic function. There are two approaches to this problem. In *interpolation,* the data are assumed to be correct and what is desired is some way to describe what happens between the data points. This approach is discussed in the next section. In the method to be discussed here, *curve fitting* or *regression,* one seeks to find some smooth curve that "best fits" the data, but does not necessarily pass through any data points. The figure shown on the next page illustrates these two approaches. The 'o' marks are the data points; the solid lines connecting them depict linear interpolation, and the dashed curve is a "best fit" to the data.

11.1 CURVE FITTING

Curve fitting involves answering two fundamental questions: What is meant by **best fit?** And, what kind of a curve should be used? *Best fit* can be defined in many different ways and there are an infinite number of curves. So where do we go from here? As it turns out, when *best fit* is interpreted as minimizing the sum of the squared error at the data points and

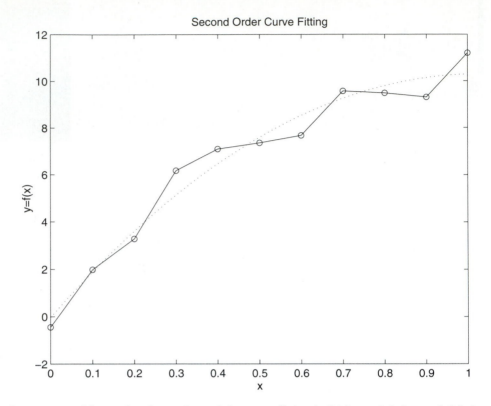

the curve used is restricted to polynomials, curve fitting is fairly straightforward. Mathematically, this is called least squares curve fitting to a polynomial. If this description is confusing to you, study the above figure again. The vertical distance between the dashed curve and a marked data point is the error at that point. Squaring this distance at each data point and adding the squared distances together is the **sum of the squared error.** The dashed curve is the curve that makes this sum of squared error as small as it can be, i.e., it is a *best fit*. The term **least squares** is just an abbreviated way of saying **minimizing the sum of the squared error.**

In MATLAB, the function **polyfit** solves the least-squares curve-fitting problem. To illustrate the use of this function, let's start with the data in the above plot.

```
» x=[0 .1 .2 .3 .4 .5 .6 .7 .8 .9 1];

» y=[-.447 1.978 3.28 6.16 7.08 7.34 7.66 9.56 9.48 9.30 11.2];
```

To use **polyfit**, we must give it the above data and the order or degree of the polynomial we wish to best fit to the data. If we choose **n=1** as the order, the best straight line approximation will be found. This is often called linear regression. On the other hand, if we choose **n=2** as the order, a quadratic polynomial will be found. For now, let's choose a quadratic polynomial.

```
» n=2;   % polynomial order

» p=polyfit(x,y,n)
p =
   -9.8108 20.1293 -0.0317
```

The output of **polyfit** is a row vector of the polynomial coefficients. Here the solution is $y = -9.8108x^2 + 20.1293x - 0.0317$. To compare the curve-fit solution to the data points, let's plot both.

```
» xi=linspace(0,1,100); % x-axis data for plotting
» z=polyval(p,xi);
```

calls the MATLAB function **polyval** to evaluate the polynomial **p** at the data points in **xi**.

```
» plot(x,y,'o',x,y,xi,z,':')
```

plots the original data **x** and **y** marking the data points with **'o'**, plots the original data again drawing straight lines between the data points, and plots the polynomial data **xi** and **z** using a dotted line **':'**.

```
» xlabel('x'),ylabel('y=f(x)'),title('Second Order Curve Fitting')
```

labels the plot. The result of these steps is shown in the preceding plot.

The choice of polynomial order is somewhat arbitrary. It takes two points to define a straight line or first order polynomial. It takes three points to define a quadratic or second order polynomial. Following this progression, it takes n + 1 data points to uniquely specify an nth order polynomial. Thus, in the above case where there are 11 data points, we could choose up to a tenth order polynomial. However, given the poor numerical properties of higher order polynomials, one should not choose a polynomial order any higher than necessary. In addition, as the polynomial order increases, the approximation becomes less smooth since higher order polynomials can be differentiated more times before they become zero. For example, choosing a tenth order polynomial

```
» pp=polyfit(x,y,10);
» format short e % change display format

» pp.' % display polynomial coefficients as a column
ans =
  -4.6436e+05
   2.2965e+06
  -4.8773e+06
   5.8233e+06
  -4.2948e+06
```

```
       2.0211e+06
      -6.0322e+05
       1.0896e+05
      -1.0626e+04
       4.3599e+02
      -4.4700e-01
```

Note the size of the polynomial coefficients in this case as compared to those of the earlier quadratic fit. Note also the seven orders of magnitude difference between the smallest $(-4.4700e - 01)$ and largest $(5.8233e + 06)$ coefficients. How about plotting this solution and comparing it to the original data and quadratic curve fit.

```
» zz=polyval(pp,xi); % evaluate 10th order polynomial

» plot(x,y,'o',xi,z,':',xi,zz) % plot data

» xlabel('x'),ylabel('y=f(x)'),title('2nd and 10th Order Curve Fitting')
```

In the plot below, the original data are marked with **'o'**, the quadratic curve fit is dotted, and the 10th order fit is solid. Note the wave-like ripples that appear between the data points

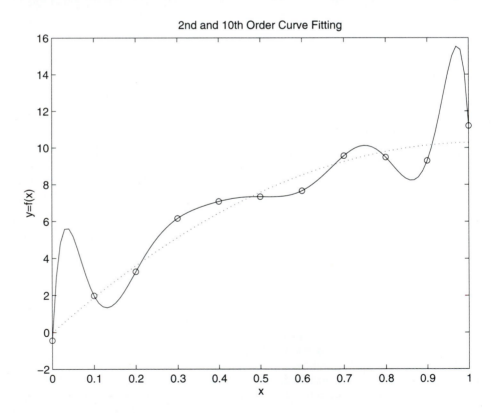

at the left and right extremes in the 10th order fit. This ripple phenomenon commonly occurs whenever a high order curve fit is attempted. Based on the above plot it is clear that a "more is better" philosophy does not necessarily apply here.

11.2 ONE-DIMENSIONAL INTERPOLATION

As described in the previous section on curve fitting, interpolation is defined as a way of estimating values of a function between those given by some set of data points. Interpolation is a valuable tool when one cannot quickly evaluate the function at the desired intermediate points. For example, this is true when the data points are the result of some experimental measurements or lengthy computational procedure.

Perhaps the simplest example of interpolation is MATLAB plots. By default, MATLAB draws straight lines connecting the data points used to make a plot. **This *linear* interpolation guesses that intermediate values fall on a straight line between the data points.** Certainly as the number of data points increases and the distance between them decreases, linear interpolation becomes more accurate. For example,

```
» x1=linspace(0,2*pi,60);

» x2=linspace(0,2*pi,6);

» plot(x1,sin(x1),x2,sin(x2),'-')

» xlabel('x'),ylabel('sin(x)'),title('Linear Interpolation')
```

(See next page for plot.)

Of the two plots of the sine function, the one using 60 points is much smoother and more accurate between the data points than the one using only 6 points.

As with curve fitting, there are decisions to be made. There are multiple approaches to interpolation depending on the assumptions made. Moreover, it is possible to interpolate in more than one dimension. That is, if you have data reflecting a function of two variables, $z=f(x,y)$, you can interpolate between values of both x and y to find intermediate values of z. MATLAB provides a number of interpolation options in the one-dimensional function `interp1` and in the two-dimensional function `interp2`. Each of these functions will be illustrated in the following.

To illustrate one-dimensional interpolation consider the following problem. The outside temperature is measured once an hour for twelve hours. The data are stored in two MATLAB variables.

```
» hours=1:12; % index for hour data was recorded

» temps=[5 8 9 15 25 29 31 30 22 25 27 24]; % recorded temperatures

» plot(hours,temps,hours,temps,'+') % view temperatures
```

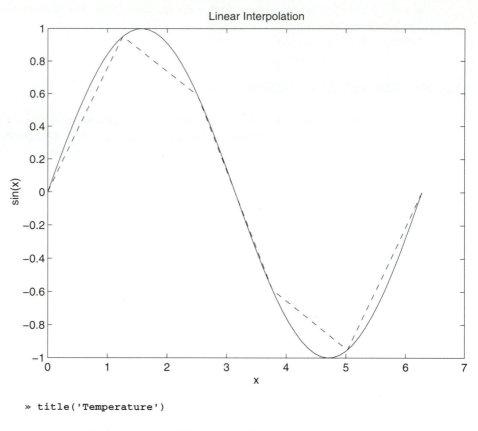

```
» title('Temperature')

» xlabel('Hour'),ylabel('Degrees Celsius')
```

As shown in the plot on the next page, MATLAB draws lines that linearly interpolate the data points. To estimate the temperature at any given time, one could try to interpret the plot visually. Alternatively, the function **interp1** could be used.

```
» t=interp1(hours,temps,9.3) % estimate temperature at hour=9.3
t =
    22.9000

» t=interp1(hours,temps,4.7) % estimate temperature at hour=4.7
t =
    22

» t=interp1(hours,temps,[3.2 6.5 7.1 11.7]) % find temp at many points!
t =
    10.2000
    30.0000
    30.9000
    24.9000
```

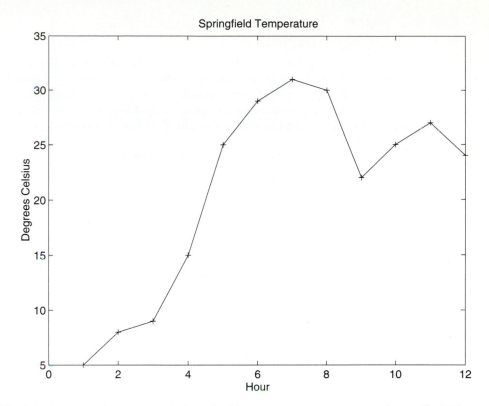

This default usage of `interp1` is described by `interp1(x,y,xo)` where **x** is the independent variable (abscissa), **y** is the dependent variable (ordinate), and **xo** is an array of the values to interpolate. In addition, this default usage assumes *linear* interpolation.

Rather than assume that a straight line connects the data points, we can assume that some *smoother* curve fits the data points. The most common assumption is that a third order polynomial, i.e., a cubic polynomial, is used to model each segment between consecutive data points and that the first two derivatives of each cubic polynomial match at the data points. This type of interpolation is called ***cubic splines*** or just ***splines.*** The function **interp1** also performs cubic spline interpolation.

```
» t=interp1(hours,temps,9.3,'spline') % estimate temperature at hour=9.3
t =
   21.8577

» t=interp1(hours,temps,4.7,'spline') % estimate temperature at hour=4.7
t =
   22.3143

» t=interp1(hours,temps,[3.2 6.5 7.1 11.7],'spline')
t =
    9.6734
   30.0427
```

```
31.1755
25.3820
```

Note that the answers for spline interpolation are different from the linear interpolation results shown above. Since interpolation is a process of estimating or guessing values, it
makes sense that applying different estimation rules leads to different results.

One of the most common uses of spline interpolation is to smooth data. That is, given
a set of data, use spline interpolation to evaluate the data at a finer interval. For example,

```
» h=1:0.1:12; % estimate temperature every 1/10 hour

» t=interp1(hours,temps,h,'spline');

» plot(hours,temps,'—',hours,temps,'+',h,t) % plot comparative results

» title('Temperature')

» xlabel('Hour'),ylabel('Degrees Celsius')
```

In the plot below, the dashed line is linear interpolation, the solid line is the smoothed spline
interpolation, and the original data are marked with **'+'**. By asking for a finer resolution

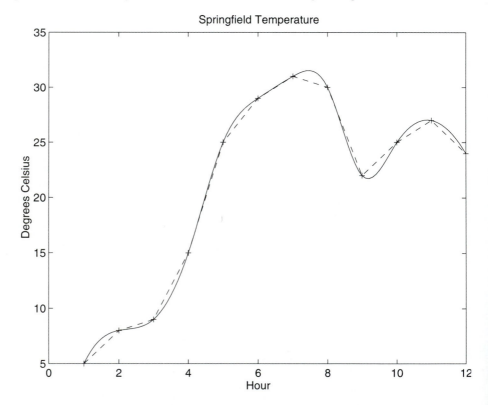

on the **hours** axis and using spline interpolation, we have a smoother, but not necessarily more accurate, estimate of the temperature. In particular, note how the slope of the spline solution does not change abruptly at the data points. In return for this smoother interpolation, cubic spline interpolation requires a great deal more computational effort since a cubic polynomial must be found to describe the behavior of the data between the given data. More detailed information regarding splines can be found in the next chapter.

Before discussing two-dimensional interpolation, it is important to recognize the two major restrictions enforced by **interp1**. First, one cannot ask for results outside the range of the independent variable, e.g., **interp1(hours,temps,13.5)** leads to an error since **hours** varies between 1 and 12. Second, the independent variable must be **monotonic.** That is, the independent variable must always increase or must always decrease in value. In our example, **hours** is monotonic. However, if we had defined the independent variable to be the actual time of day,

```
» time_of_day=[7:12 1:6] % start at 7AM, end at 6PM
time_of_day =
      7   8   9  10  11  12   1   2   3   4   5   6
```

the independent variable would not be monotonic since **time_of_day** increases until 12, then drops to 1, and then increases again. If **time_of_day** were used instead of **hours** in **interp1**, an error would be returned. By the same reasoning, one cannot interpolate **temps** to find the hour when some temperature occurred, because **temps** is not monotonic.

11.3 TWO-DIMENSIONAL INTERPOLATION

Two-dimensional interpolation is based on the same underlying ideas as one-dimensional interpolation. However, as the name implies, two-dimensional interpolation interpolates functions of two variables, $z = f(x,y)$. To illustrate this added dimension, consider a problem where one is interested in estimating the temperature distribution on a flat plate given temperature values taken at equally distributed grid points on the surface of the plate.

The following data are collected:

```
» width=1:5; % index for width of plate (i.e., the x-dimension)

» depth=1:3; % index for depth of plate (i.e., the y-dimension)

» temps=[82 81 80 82 84;79 63 61 65 81;84 84 82 85 86] % temperature data
temps =
     82    81    80    82    84
     79    63    61    65    81
     84    84    82    85    86
```

The matrix **temps** shows the temperature distribution over the plate as measured at the points indexed. The columns of **temps** are associated with the index **depth** or y-dimension, and the rows are associated with the index **width** or x-dimension. To estimate the temperature at intermediate points, we must identify them.

```
» wi=1:0.2:5;   % estimate across width of plate

» d=2;          % at a depth of 2

» zlinear=interp2(width,depth,temps,wi,d);   % linear interpolation

» zcubic=interp2(width,depth,temps,wi,d,'cubic'); % cubic interpolation

» plot(wi,zlinear,'—',wi,zcubic)   % plot results

» xlabel('Width of Plate'), ylabel('Degrees Celsius')

» title(['Temperature at Depth = ' num2str(d)])
```

Alternatively, we can interpolate in both directions. First plot the raw data in 3-D to see the roughness of the data.

```
» mesh(width,depth,temps) % use mesh plot

» xlabel('Width of Plate'), ylabel('Depth of Plate')

» zlabel('Degrees Celsius'), axis('ij'), grid
```

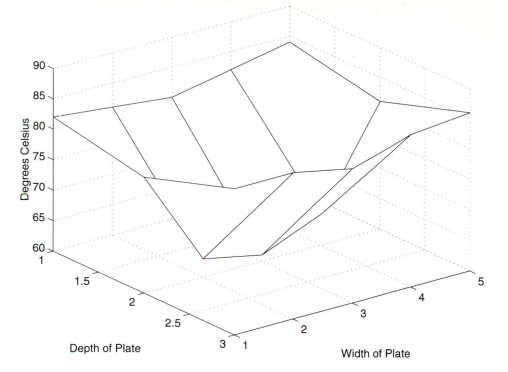

Then interpolate in both directions to smooth the data.

```
» di=1:0.2:3; % choose higher resolution for depth

» wi=1:.2:5; % choose higher resolution for width

» zcubic=interp2(width,depth,temps,wi,di,'cubic'); % cubic

» mesh(wi,di,zcubic)

» xlabel('Width of Plate'), ylabel('Depth of Plate')

» zlabel('Degrees Celsius'), axis('ij'), grid
```

The above example clearly demonstrates that two-dimensional interpolation is more complicated simply because there is more to keep track of. The basic form of **interp2** is **interp2(x,y,z,xi,yi,method)**. Here **x** and **y** are the two independent variables and **z** is a matrix of the dependent variable. The association of **x** and **y** to **z** is

$$z(i,:) = f(x,y(i)) \text{ and } z(:,j) = f(x(j),y).$$

That is, the ith row of **z** is associated with i^{th} element of **y**, **y(i)**, as **x** varies, and the jth column of **z** is associated with j^{th} element of **x**, **x(j)**, as **y** varies. **xi** is an array of the values to interpolate along the x-axis; **yi** is an array of values to interpolate along the

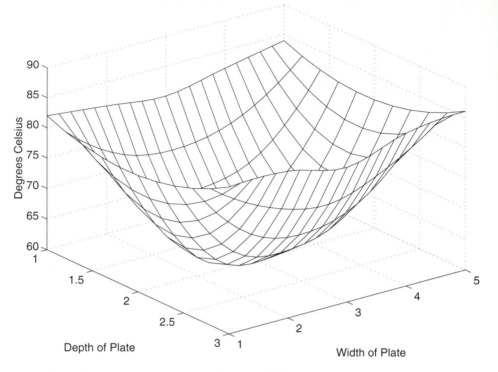

y-axis. The optional parameter **method** can be **'linear'**, **'cubic'**, or **'nearest'**. In this case, **cubic** does not mean cubic splines, but rather another algorithm using cubic polynomials. The method **linear** is linear interpolation just as that used to connect data points on plots. The **nearest** method simply chooses the rough data point closest to each estimation point. In all cases, it is assumed that the independent variables **x** and **y** are linearly spaced and monotonic. For more information regarding these methods, ask for on-line help, e.g., » **help interp2**, or see the *MATLAB Reference Guide*.

11.4 M-FILE EXAMPLES

While the functions **interp1** and **interp2** are useful for many applications, they are limited to interpolating over a monotonic vector. In some situations this limitation is overly restrictive. For example, consider interpolating the following:

```
» x=linspace(0,5);

» y=1-exp(-x).*sin(2*pi*x);

» plot(x,y)
```

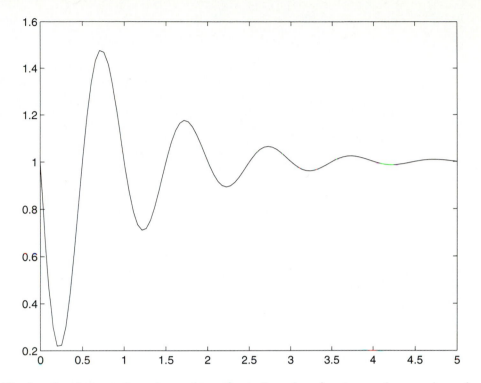

The function **interp1** can be used to estimate the value of **y** at any value or values of **x**.

```
» yi=interp1(x,y,1.8)
yi =
      1.1556
```

However, **interp1** cannot find the values of **x** corresponding to some values of **y**. For example, consider finding the values of **x** where **y=1.1** as shown in the plot

```
» plot(x,y,[0 5],[1.1 1.1])
```

(See next page for plot.)
From the plot we see that there are four crossings. Using **interp1** we get

```
» xi=interp1(y,x,1.1)
??? Error using ==> table1
First column of the table must be monotonic.
```

The function **interp1** fails because **y** is not monotonic.

The *Mastering MATLAB Toolbox* M-file example illustrated in this section eliminates the monotonic requirement.

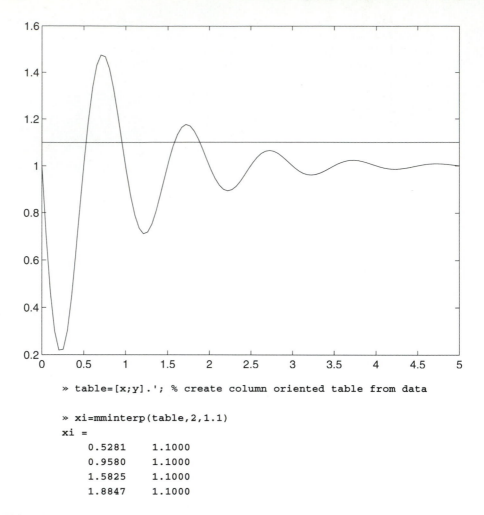

```
» table=[x;y].'; % create column oriented table from data

» xi=mminterp(table,2,1.1)
xi =
    0.5281      1.1000
    0.9580      1.1000
    1.5825      1.1000
    1.8847      1.1000
```

Using linear interpolation, the function **mminterp** estimates the four points where **y=1.1**. Because of the general nature of the function **mminterp**, the data to be interpolated are given by a column-oriented matrix, called **table** in the above example. The second input argument is the column of the matrix **table** to be searched, and the third argument is the value to be found.

The body of this *Mastering MATLAB Toolbox* function is given below:

```
function y=mminterp(tab,col,val)
%MMINTERP 1-D Table Search by Linear Interpolation.
% Y=MMINTERP(TAB,COL,VAL) linearly interpolates the table
% TAB searching for the scalar value VAL in the column COL.
```

```
% All crossings are found and TAB(:,COL) need not be monotonic.
% Each crossing is returned as a separate row in Y and Y has as
% many columns as TAB. Naturally, the column COL of Y contains
% the value VAL. If VAL is not found in the table, Y=[].

% Copyright (c) 1996 by Prentice-Hall, Inc.

[rt,ct]=size(tab);
if length(val)>1, error('VAL must be a scalar.'), end
if col>ct|col<1, error('Chosen column outside table width.'), end
if rt<2, error('Table too small or not oriented in columns.'), end

above=tab(:,col)>val; % True where > VAL
below=tab(:,col)<val; % True where < VAL
equal=tab(:,col)==val; % True where = VAL

if all(above==0)|all(below==0), % handle simplest case
    y=tab(find(equal),:); return
end
pslope=find(below(1:rt-1)&above(2:rt)); % indices where slope is +
nslope=find(below(2:rt)&above(1:rt-1)); % indices where slope is -

ib=sort([pslope;nslope+1]); % put indices below in order
ia=sort([nslope;pslope+1]); % put indices above in order
ie=find(equal);             % indices where equal to val

[tmp,ix]=sort([ib;ie]);     % find where equals fit in result
ieq=ix>length(ib);          % True where equals values fit
ry=length(tmp);             % # of rows in result y

y=zeros(ry,ct); % poke data into a zero matrix

alpha=(val-tab(ib,col))./(tab(ia,col)-tab(ib,col));
alpha=alpha(:,ones(1,ct)); % duplicate for all columns
y(~ieq,:)=alpha.*tab(ia,:)+(1-alpha).*tab(ib,:);%interpolated values

y(ieq,:)=tab(ie,:); % equal values
y(:,col)=val*ones(ry,1); % remove roundoff error
```

As you can see, **mminterp** makes use of the **find** and **sort** functions, logical arrays, and array-manipulation techniques. **There are no For loops or While loops**. An implementation using either of them would run much slower, especially for large tables. Note that **mminterp** works with tables containing any number of columns greater than or equal to two, just as the function **interp1** does. In this case, however, the interpolating variable can be any column of the table. For example,

```
» z=sin(pi*x); % add more data to table

» table=[x;y;z].';

» t=mminterp(table,2,1.1) % same interpolation as earlier
t =
    0.5281    1.1000    0.9930
    0.9580    1.1000    0.1314
    1.5825    1.1000   -0.9639
    1.8847    1.1000   -0.3533

» t=mminterp(table,3,-.5) % search third column now
t =
    1.1669    0.7316   -0.5000
    1.8329    1.1377   -0.5000
    3.1671    0.9639   -0.5000
    3.8331    1.0187   -0.5000
```

These last results estimate the values of **x** and **y** where **z=-.5**.

While it would be very informative to go through the function **mminterp** line-by-line explaining how it works, doing so would require more space and time than available here. The easiest way to decipher what **mminterp** does is to create a small table, then call the function after deleting the semicolons at the ends of important statements. By doing so, the intermediate results will assist you in understanding how the function finds data values straddling the desired value and performs the interpolation.

Earlier the use of **interp1** was illustrated. When used for linear interpolation **interp1** works fine as long as the number of requested interpolation points is small. In those cases where many interpolated points are requested, **interp1** is slow because of the algorithm used. To eliminate this problem, the *Mastering MATLAB Toolbox* includes the function **mmtable** whose help text is

```
» help table


MMTABLE 1-D Monotonic Table Search by Linear Interpolation.
  YI=MMTABLE(TAB,COL,VALS) linearly interpolates the table TAB
  searching for values VALS in the column COL.


  TAB(:,COL) must be monotonic, but need NOT be equally spaced.
  YI has as many rows as VALS and as many columns as TAB
  NaNs are returned where VALS are outside the range of TAB(:,COL).
```

```
YI=MMTABLE(TAB,VALS) interpolates using COL=1 and does not return
TAB(:,1) in Y. This matches the usage of TABLE1(TAB,X0).

YI=MMTABLE(X,Y,XI) interpolates the vector X to find YI associated
with XI. This matches the usage of INTERP1(X,Y,XI)

This routine is 10X faster than TABLE1 which is called by INTERP1.
```

As described previously, **mmtable** can be called in several ways. In addition, the column or vector interpolated does not need to be linearly spaced. For this reason, **mmtable** is more general than the **ilinear** function which will be used by **interp1** to perform linear interpolation in MATLAB version 5.

11.5 SUMMARY

The table below summarizes the curve fitting and interpolation functions found in MATLAB.

Curve Fitting and Interpolation Functions	
`polyfit(x,y,n)`	Least-squares curve-fitting of data describing y=f(x) to an nth order polynomial
`interp1(x,y,xo)`	1-D linear interpolation
`interp1(x,y,xo,'spline')`	1-D cubic spline interpolation
`interp1(x,y,xo,'cubic')`	1-D cubic interpolation
`interp2(x,y,Z,xi,yi)`	2-D linear interpolation
`interp2(x,y,Z,xi,yi,'cubic' )`	2-D cubic interpolation
`interp2(x,y,Z,xi,yi,'neares t')`	2-D nearest neighbor interpolation

12

Cubic Splines

It is well-known that interpolation using high order polynomials often produces ill-behaved results. There are numerous approaches to eliminating this poor behavior. Of these approaches, cubic splines are very popular. In MATLAB, basic cubic spline interpolation is accomplished by the functions `spline`, `ppval`, `mkpp`, and `unmkpp`. Of these only `spline` is documented in the *MATLAB Reference Guide*. In the following sections, the basic features of cubic splines as implemented in these M-file functions will be demonstrated.

12.1 BASIC FEATURES

In cubic splines, cubic polynomials are found to approximate the curve between each pair of data points. In the language of splines, these data points are called the ***break points.*** Since a straight line is uniquely defined by two points, an infinite number of cubic polynomials can be used to approximate a curve between two points. Therefore, in cubic splines additional constraints are placed on the cubic polynomials to make the result unique. By constraining the first and second derivatives of each cubic polynomial to match at the break

142

points, all internal cubic polynomials are well-defined. Moreover, both the slope and curvature of the approximating polynomials are continuous across the break points. However, the first and last cubic polynomials do not have adjoining cubic polynomials beyond the first and last break points. As a result, the remaining constraints must be determined by some other means. The most common approach, which is adopted by the function **spline**, is to adopt a ***not-a-knot*** condition. This condition forces the third derivative of the first and second cubic polynomials to be identical, and likewise for the last and second-to-the-last cubic polynomials.

Based on the above description, one could guess that finding cubic spline polynomials requires solving a large set of linear equations. In fact, given N break points, there are N-1 cubic polynomials to be found, each having 4 unknown coefficients. Thus, the set of equations to be solved involves 4*(N-1) unknowns. By writing each cubic polynomial in a special form and by applying the constraints, the cubic polynomials can be found by solving a set of N equations in N unknowns. Thus, if there are 50 break points, there are 50 equations in 50 unknowns. Luckily, the equations can be concisely written and solved using sparse matrices, which are what the function **spline** uses to compute the unknown coefficients.

12.2 PIECEWISE POLYNOMIALS

In its simplest use, **spline** takes data **x** and **y** and desired values **xi**, finds the cubic spline interpolation polynomials that fit **x** and **y**, and then evaluates the polynomials to find the corresponding **yi** values for each **xi** value. For example,

```
» x=0:12;

» y=tan(pi*x/25);

» xi=linspace(0,12);

» yi=spline(x,y,xi);

» plot(x,y,'o',xi,yi), title('Spline fit')
```

(See next page for plot.)
This approach is appropriate if only one set of interpolated values is required. However, if another set of interpolated values is needed from the same set of data, it doesn't usually make sense to compute the cubic spline coefficients a second time. In this situation, one can call **spline** with only the first two arguments.

```
» pp=spline(x,y)
pp =
  Columns 1 through 7
    10.0000     1.0000    12.0000          0    1.0000    2.0000    3.0000
  Columns 8 through 14
     4.0000     5.0000     6.0000     7.0000    8.0000    9.0000   10.0000
```

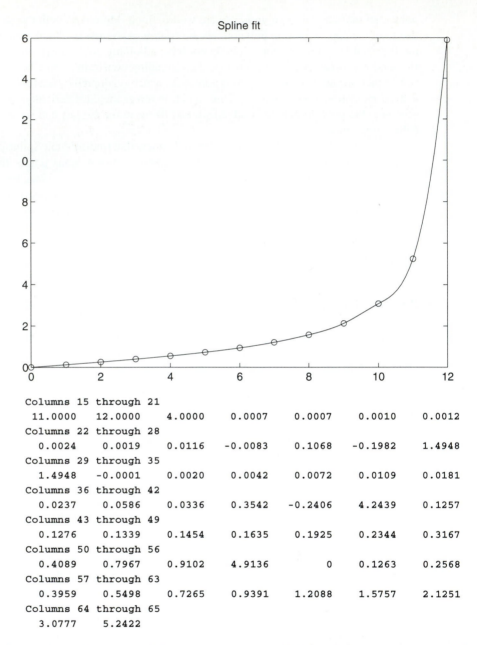

```
Columns 15 through 21
  11.0000    12.0000     4.0000     0.0007     0.0007     0.0010     0.0012
Columns 22 through 28
   0.0024     0.0019     0.0116    -0.0083     0.1068    -0.1982     1.4948
Columns 29 through 35
   1.4948    -0.0001     0.0020     0.0042     0.0072     0.0109     0.0181
Columns 36 through 42
   0.0237     0.0586     0.0336     0.3542    -0.2406     4.2439     0.1257
Columns 43 through 49
   0.1276     0.1339     0.1454     0.1635     0.1925     0.2344     0.3167
Columns 50 through 56
   0.4089     0.7967     0.9102     4.9136          0     0.1263     0.2568
Columns 57 through 63
   0.3959     0.5498     0.7265     0.9391     1.2088     1.5757     2.1251
Columns 64 through 65
   3.0777     5.2422
```

When called in this way, **spline** returns an array called the **pp-form** or **piecewise polynomial form** of the cubic splines. This array contains all the information necessary to evaluate the cubic splines for any set of desired interpolation values. Given the pp-form, the function **ppval** evaluates the cubic splines. For example,

```
» yi=ppval(pp,xi);
```

computes the same `yi` values computed earlier. Similarly,

> ```
> » xi2=linspace(10,12);
> ```

> ```
> » yi2=ppval(pp,xi2);
> ```

uses the pp-form again to evaluate the cubic splines over a finer spacing restricted to the region between 10 and 12.

> ```
> » xi3=10:15;
> ```

> ```
> » yi3=ppval(pp,xi3)
> yi3 =
>       3.0777      5.2422      15.8945      44.0038      98.5389      188.4689
> ```

shows that cubic splines can be evaluated outside the region over which the cubic polynomials were computed. When data appear beyond the last or before the first break point, the last and first cubic polynomials respectively are used to find interpolated values.

The cubic splines pp-form given above stores the break points and polynomial coefficients, as well as other information regarding the cubic splines representation. This form is a convenient data structure in MATLAB, since all information is stored in a single vector. When a cubic spline representation is evaluated, the pp-form must be broken into its representative pieces. In MATLAB this process is performed by the function **unmkpp**. Using this function on the above pp-form gives

```
» [breaks,coefs,npolys,ncoefs]=unmkpp(pp)
breaks =
  Columns 1 through 12
     0      1      2      3      4      5      6      7      8      9      10      11
  Column 13
     12
coefs =
      0.0007    -0.0001     0.1257          0
      0.0007     0.0020     0.1276     0.1263
      0.0010     0.0042     0.1339     0.2568
      0.0012     0.0072     0.1454     0.3959
      0.0024     0.0109     0.1635     0.5498
      0.0019     0.0181     0.1925     0.7265
      0.0116     0.0237     0.2344     0.9391
     -0.0083     0.0586     0.3167     1.2088
      0.1068     0.0336     0.4089     1.5757
     -0.1982     0.3542     0.7967     2.1251
      1.4948    -0.2406     0.9102     3.0777
      1.4948     4.2439     4.9136     5.2422
npolys =
     12
ncoefs =
     4
```

Here **breaks** is the break points, **coefs** is a matrix whose i^th row is the ith cubic polynomial, **npolys** is the number of polynomials, and **ncoefs** is the number of coefficients per polynomial. Note that this form is sufficiently general in that the spline polynomials need not be cubic. This fact is useful when the spline is integrated or differentiated.

Given the broken-apart form above, the function **mkpp** restores the pp-form.

```
» pp=mkpp(breaks,coefs)
pp =
  Columns 1 through 7
   10.0000      1.0000    12.0000           0      1.0000      2.0000      3.0000
  Columns 8 through 14
    4.0000      5.0000     6.0000      7.0000      8.0000      9.0000     10.0000
  Columns 15 through 21
   11.0000     12.0000     4.0000      0.0007      0.0007      0.0010      0.0012
  Columns 22 through 28
    0.0024      0.0019     0.0116     -0.0083      0.1068     -0.1982      1.4948
  Columns 29 through 35
    1.4948     -0.0001     0.0020      0.0042      0.0072      0.0109      0.0181
  Columns 36 through 42
    0.0237      0.0586     0.0336      0.3542     -0.2406      4.2439      0.1257
  Columns 43 through 49
    0.1276      0.1339     0.1454      0.1635      0.1925      0.2344      0.3167
  Columns 50 through 56
    0.4089      0.7967     0.9102      4.9136           0      0.1263      0.2568
  Columns 57 through 63
    0.3959      0.5498     0.7265      0.9391      1.2088      1.5757      2.1251
  Columns 64 through 65
    3.0777      5.2422
```

Since the size of the matrix **coefs** determines **npolys** and **ncoefs**, **npolys** and **ncoefs** are not needed by **mkpp** to reconstruct the pp-form. The pp-form data structure is given simply in **mkpp** as **pp=[10 1 npolys breaks(:)' ncoefs coefs(:)']**. The first two elements appear in all pp-forms as a means to identify a vector as a pp-form.

12.3 INTEGRATION

In many situations it is desirable to know the area under a function described by a cubic spline as a function of the independent variable x. That is, if the function is denoted $y = s(x)$, we are interested in computing

$$S(x) = \int_{x_1}^{x} s(x)\,dx, \qquad \text{with } S(x_1) = 0$$

where x_1 is the first spline break point. Since $s(x)$ is composed of connected cubic polynomials, with the kth cubic polynomial being

$$s_k(x) = a_k(x - x_k)^3 + b_k(x - x_k)^2 + c_k(x - x_k) + d_k, \qquad x_k \leq x \leq x_{k+1}$$

and whose area over the region $x_k \leq x \leq x_{k+1}$ is

$$S_k(x) = \int_{x_1}^{x} s_k(x)dx = \frac{a_k}{4}(x - x_k)^4 + \frac{b_k}{3}(x - x_k)^3 + \frac{c_k}{2}(x - x_k)^2 x_k + d_k(x - x_k)$$

the area under a cubic spline is easily computed as

$$S(x) = \sum_{i=1}^{k-1} \int_{x_i}^{x_{i+1}} s_i(x)\,dx + \int_{x_k}^{x} s_k(x)\,dx, \qquad \text{where } x_k \leq x \leq x_{k+1}$$

or

$$S(x) = \sum_{i=1}^{k-1} S_i(x_{i+1}) + S_k(x), \qquad \text{where } x_k \leq x \leq x_{k+1}$$

The summation term is the cumulative sum of the areas under all preceding cubic polynomials. As such, it is readily computed and forms the constant term in the polynomial describing $S(x)$ since $S_k(x)$ is a polynomial. With this understanding, the integral itself can be written as a spline. In this case it is a quartic spline since the individual polynomials are of order four.

Because the pp-form used in MATLAB can support splines of any order, the above splines integration is embodied in the *Mastering MATLAB Toolbox* function `spintgrl`. The body of this function is given by

```
function z=spintgrl(x,y,xi)
%SPINTGRL Cubic Spline Integral Interpolation
% YI=SPINTRGL(X,Y,XI) uses cubic spline interpolation to fit the
% data in X and Y, integrates the spline and returns
% values of the integral evaluated at the points in XI.
%
% PPI=SPINTGRL(PP) returns the piecewise polynomial vector PPI
% describing the integral of the cubic spline described by
% the piecewise polynomial in PP. PP is returned by the function
% SPLINE and is a data vector containing all information to
% evaluate and manipulate a spline.
%
% YI=SPINTGRL(PP,XI) integrates the cubic spline given by
% the piecewise polynomial PP, and returns the values of the
% integral evaluated at the points in XI.
%
% See also SPLINE, PPVAL, MKPP, UNMKPP, SPDERIV

% Copyright (c) 1996 by Prentice-Hall, Inc.
```

```
if nargin==3
pp=spline(x,y);
else
     pp=x;
end
[br,co,npy,nco]=unmkpp(pp); % take apart pp
if pp(1)~=10
     error('Spline data does not have the correct form.')
end
sf=nco:-1:1; % scale factors for integration
ico=[co./sf(ones(npy,1),:) zeros(npy,1)]; % integral coefficients
nco=nco+1; % spline order increases
for k=2:npy % find constant terms
     ico(k,nco)=polyval(ico(k-1,:),br(k)-br(k-1));
end
ppi=mkpp(br,ico); % build pp form for integral
if nargin==1
     z=ppi;
elseif nargin==2
     z=ppval(ppi,y);
else
     z=ppval(ppi,xi);
end
```

Consider the following example using `spintgrl`:

```
» x=(0:.1:1)*2*pi;

» y=sin(x); % create rough data

» pp=spline(x,y); % pp-form fitting rough data

» ppi=spintgrl(pp); % pp-form of integral

» xi=linspace(0,2*pi); % finer points for interpolation

» yi=ppval(pp,xi); % evaluate curve

» yyi=ppval(ppi,xi); % evaluate integral

» plot(x,y,'o',xi,yi,xi,yyi,'-') % plot results
```

As shown on the next page, note that this qualitatively shows the identity

$$\int_{x_1}^{x} \sin(x)\,dx = 1 - \cos(x)$$

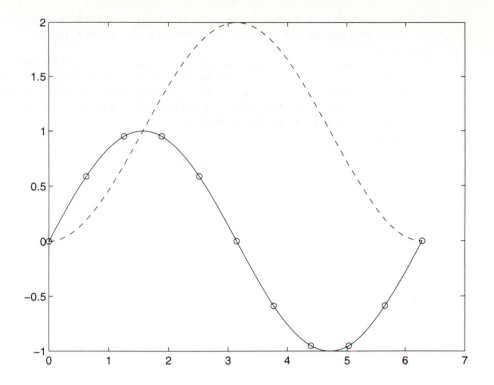

12.4 DIFFERENTIATION

Just as one may be interested in spline integration, the derivative or slope of a function described by splines is also be useful. Given the kth cubic polynomial being

$$s_k(x) = a_k(x - x_k)^3 + b_k(x - x_k)^2 + c_k(x - x_k) + d_k, \, x_k \le x \le x_{k+1}$$

the derivative is easily written as

$$\frac{ds_k(x)}{dx} = 3a_k(x - x_k)^2 + 2b_k(x - x_k) + c_k, \, x_k \le x \le x_{k+1}$$

As with integration, the derivative is also a spline. However, in this case it is a quadratic spline, since the order of the polynomial is two.

Based on the above expression, the *Mastering MATLAB Toolbox* function **spderiv** performs spline differentiation. The body of **spderiv** is

```
function z=spderiv(x,y,xi)
%SPDERIV Cubic Spline Derivative Interpolation
% YI=SPDERIV(X,Y,XI) uses cubic spline interpolation to fit the
```

```
% data in X and Y, differentiates the spline and returns values
% of the spline derivatives evaluated at the points in XI.
%
% PPD=SPDERIV(PP) returns the piecewise polynomial vector PPD
% describing the cubic spline derivative of the curve described by
% the piecewise polynomial in PP. PP is returned by the function
% SPLINE and is a data vector containing all information to
% evaluate and manipulate a spline.
%
% YI=SPDERIV(PP,XI) differentiates the cubic spline given by
% the piecewise polynomial PP, and returns the values of the
% spline derivatives evaluated at the points in XI.
%
% See also SPLINE, PPVAL, MKPP, UNMKPP, SPINTGRL

% Copyright (c) 1996 by Prentice-Hall, Inc.

if nargin==3
     pp=spline(x,y);
else
     pp=x;
end
[br,co,npy,nco]=unmkpp(pp); % take apart pp
if nco==1|pp(1)~=10
     error('Spline data does not have the correct PP form.')
end
sf=nco-1:-1:1; % scale factors for differentiation
dco=sf(ones(npy,1),:).*co(:,1:nco-1); % derivative coefficients
ppd=mkpp(br,dco); % build pp form for derivative
if nargin==1
     z=ppd;
elseif nargin==2
     z=ppval(ppd,y);
else
     z=ppval(ppd,xi);
end
```

To demonstrate the use of **spderiv**, consider the following example:

```
» x=(0:.1:1)*2*pi;      % same data as earlier

» y=sin(x);
```

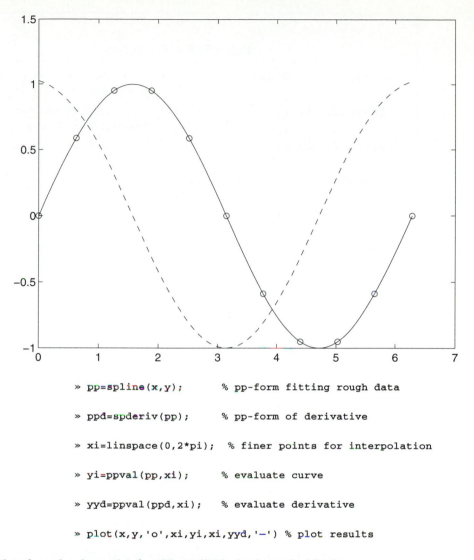

```
» pp=spline(x,y);         % pp-form fitting rough data

» ppd=spderiv(pp);        % pp-form of derivative

» xi=linspace(0,2*pi);    % finer points for interpolation

» yi=ppval(pp,xi);        % evaluate curve

» yyd=ppval(ppd,xi);      % evaluate derivative

» plot(x,y,'o',xi,yi,xi,yyd,'-') % plot results
```

Note from the above plot that this qualitatively shows the identity

$$\frac{d}{dx} \sin(x) = \cos(x)$$

12.5 SUMMARY

The tables on the next page summarize the spline functions discussed in this chapter.

Cubic Spline Functions

`yi=spline(x,y,xi)`	Cubic spline interpolation of y=f(x) at points in **xi**
`pp=spline(x,y)`	Return piecewise polynomial representation of y=f(x)
`yi=ppval(pp,xi)`	Evaluate piecewise polynomial at points in **xi**
`[breaks,coefs,npolys,ncoefs] =unmkpp(pp)`	Unmake piecewise polynomial representation
`pp=mkpp(breaks,coefs)`	Make piecewise polynomial representation

Mastering MATLAB Spline Functions

`yi=spintgrl(x,y,xi)`	Cubic spline interpolation of integral of y=f(x) at points in **xi**
`ppi=spintgrl(pp)`	Return piecewise polynomial representation of integral of y=f(x), given piecewise polynomial representation of y=f(x)
`yi=spintgrl(pp,xi)`	Find piecewise polynomial representation of integral of y = f(x), given piecewise polynomial representation of y = f(x) and evaluate at points in **xi**
`yi=spderiv(x,y,xi)`	Cubic spline interpolation of derivative of y = f(x) at points in **xi**
`ppi=spderiv(pp)`	Return piecewise polynomial representation of derivative of y = f(x), given piecewise polynomial representation of y = f(x)
`yi=spderiv(pp,xi)`	Find piecewise polynomial representation of derivative of y = f(x), given piecewise polynomial representation of y = f(x) and evaluate at points in **xi**

13

Numerical Analysis

Whenever it is difficult to integrate, differentiate, or determine some specific value of a function analytically, a computer can be called upon to numerically approximate the desired solution. This area of computer science and mathematics is known as numerical analysis. As you may have guessed by now, MATLAB provides tools to solve these problems. In this chapter, the use of these tools will be illustrated.

13.1 PLOTTING

Up to this point, plots of a function have been generated by simply evaluating the function over some range and plotting the resulting vectors. For many cases this is sufficient. However, sometimes a function is flat and unexciting over some range and then acts wildly over another. Using the traditional plotting approach in this case can lead to a plot that misrepresents the true nature of a function. As a result MATLAB provides a smart plotting function called **fplot**. This function carefully evaluates the function to be plotted and makes sure that all its oddities are represented in the output plot. As its input, this function needs to know the name of the function as a character string and the plotting range as a two-element array. For example,

```
» fplot('humps',[0 2])
```

```
» title('FPLOT of humps')
```

evaluates the function **humps** between 0 and **2** and displays the plot.

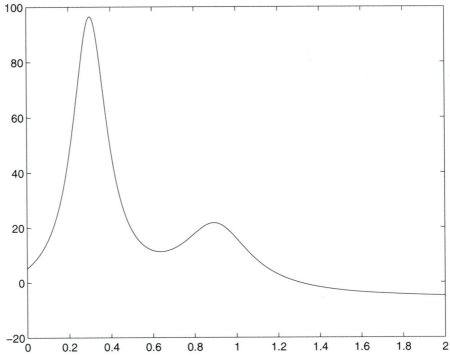

In this example, **'humps'** is the MATLAB M-file function.

```
function y = humps(x)
%HUMPS    A function used by QUADDEMO, ZERODEMO and FPLOTDEMO.
%    HUMPS(X) is a function with strong maxima near x = .3 and
x = .9.
%    See QUADDEMO, ZERODEMO and FPLOTDEMO.
```

```
%    Copyright (c) 1984-93 by The MathWorks, Inc.

y = 1 ./ ((x-.3).^2 + .01) + 1 ./ ((x-.9).^2 + .04) - 6;
```

fplot works for any function M-file with one vector input and one vector output. That is, as in **humps**, the output variable **y** returns an array the same size as the input **x**, with **y** and **x** associated in the array-to-array sense. The most common mistake made in using **fplot** (as well as other numerical analysis functions) is forgetting to put the name of the function in quotes. That is, **fplot** needs to know the name of the function as a character string. If **fplot(humps, [0 2])** is typed, MATLAB thinks **humps** is a variable in the workspace rather than the name of a function. Note that this problem can be avoided by defining the variable **humps** to be the desired character string.

```
» humps='humps';

» fplot(humps,[0 2])
```

Now MATLAB evaluates the variable **humps** to find the string **'humps'**.

For simple functions that can be expressed as a single character string, such as y = $2e^{-x}\sin(x)$, **fplot** can plot the function without creating an M-file by simply writing the function to be plotted as a complete character string using x as the independent variable.

```
» f='2*exp(-x).*sin(x)';
```

Here, the function f(x) = $2e^{-x}\sin x$ is defined using array multiplication.

```
» fplot(f,[0 8])

» title(f),xlabel('x')
```

plots the function over the range $0 \leq x \leq 8$, producing the plot.

Beyond these basic features, the function **fplot** has many more powerful capabilities. For more information see the *MATLAB Reference Guide* or on-line help.

13.2 MINIMIZATION

In addition to the visual information provided by plotting, it is often necessary to determine other more specific attributes of a function. Of particular interest in many applications are

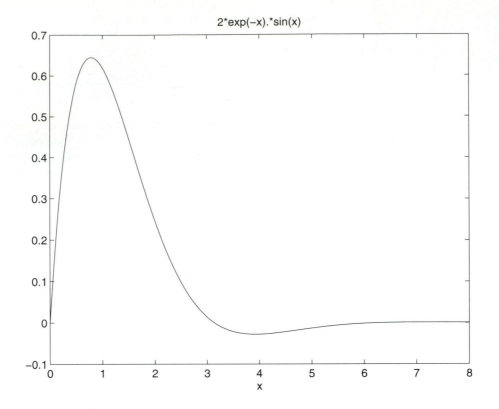

2*exp(−x).*sin(x)

function extremes, i.e., the ***maxima*** (peaks) and the ***minima*** (valleys). Mathematically, these extremes are found analytically by determining where the derivative (slope) of a function is zero. This fact can be readily understood by inspecting the slope of the **humps** plot at its peaks and valleys. Clearly, when a function is simply defined, this process often works. However, even for many simple functions that can be differentiated readily, it is often not possible to find where the derivative is zero. In these cases and in cases where it is difficult or impossible to find the derivative analytically, it is necessary to search for function extremes numerically. MATLAB provides two functions that perform this task, **fmin** and **fmins**. These two functions find minima of one-dimensional and n-dimensional functions respectively. Only **fmin** will be discussed here. Further information regarding **fmins** can be found in the *MATLAB Reference Guide*. Since a maximum of f(x) is equal to a minimum of $-f(x)$, **fmin** and **fmins** can be used to find both minima and maxima. If this fact is not clear, visualize the preceding plot flipped upside down. In the upside-down state, the peaks become valleys and the valleys become peaks.

     To illustrate one-dimensional minimization and maximization, consider the preceding example once again. From the figure, there is a maximum near $x_{max} = 0.7$ and a minimum near $x_{min} = 4$. Analytically, these points can be shown to be $x_{max} = \pi/4 \approx .785$ and $x_{min} = 5\pi/4 \approx 3.93$. Writing a script M-file using a text editor for convenience and using **fmin** to find the points numerically gives

```
% ex_fmin.m

fn='2*exp(-x)*sin(x)';   % define function for min
xmin=fmin(fn,2,5)        % search over range 2<x<5

emin=5*pi/4-xmin         % find error
x=xmin;                  % need x since fn has x as its variable
ymin=eval(fn)            % evaluate at xmin

fx='-2*exp(-x)*sin(x)';  % define for max: note minus sign
xmax=fmin(fx,0,3)        % search over range 0<x<3

emax=pi/4-xmax           % find error

x=xmax;                  % need x since fn has x as its variable
ymax=eval(fn)            % evaluate at xmax
```

Running this M-file results in the following:

```
» ex_fmin
xmin =
      3.9270
emin =
      1.4523e-06
ymin =
     -0.0279
xmax =
      0.7854
emax =
     -1.3781e-05
ymax =
      0.6448
```

These results agree well with the preceding plot. Note that **fmin** works a lot like **fplot**. The function to be evaluated can be expressed in a function M-file or just given as a character string with x being the independent variable. The latter was done here. This example also uses the function **eval**, which takes a character string and interprets it as if the string were typed at the MATLAB prompt. Since the function to be evaluated was given as a character string with an independent variable **x**, setting **x** equal to **xmin** and **xmax** allows **eval** to evaluate the function to find **ymin** and **ymax**.

Finally, it is important to note that numerical minimization involves searching for a minimum; **fmin** evaluates the function over and over looking for a minimum. This searching can take a significant amount of time if evaluating the function requires many computations or if the function has more than one minimum within the search range. In some cases the searching process does not find a solution at all! When **fmin** cannot find a minimum, it stops and provides an explanation.

Like the function **fmin**, the function **fmins** searches for a minimum. However, **fmins** searches for the minimum of a scalar function of a vector variable. That is, **fmins** seeks

$$\min_{\mathbf{x}} f(\mathbf{x})$$

where **x** is the vector argument of the function f(·) that returns a scalar value. The function **fmins** uses the Simplex method, which does not require explicit gradient computation. More extensive optimization algorithms are available in the optional *Optimization Toolbox*.

13.3 ZERO FINDING

Just as one is interested in finding function extremes, it is sometimes important to find out where a function crosses zero or some other constant value. Trying to find this point analytically is often difficult and many times impossible. In the preceding **humps** function plot, on the next page, the function crosses zero near x = 1.2.

Once again MATLAB provides a numerical solution to this problem. The function **fzero** searches for the zero of a one-dimensional function. To illustrate the use of this function, let's use the **humps** example again.

```
» xzero=fzero('humps',1.2) % look for a zero near 1.2
xzero =
    1.2995

» yzero=humps(xzero) % evaluate at xzero
yzero =
    3.5527e-15
```

So the zero actually occurs close to 1.3. As before, the zero searching process may not find a solution. If **fzero** does not find one, it will stop and provide an explanation.

The function **fzero** must be given the name of a function when it's called. For some reason, it was never given the capability to accept a function described by a character string

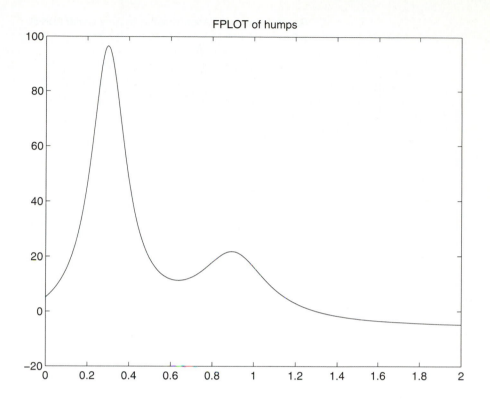
FPLOT of humps

using x as the independent variable. Thus, even though this feature is available in both **fplot** and **fmin**, it does not work with **fzero**.

   While **fzero** finds where a function is zero, it can also be used to find where a function is equal to any constant. All that's required is a simple redefinition. For example, to find where the function f(x) equals the constant c, define the function g(x) as g(x) = f(x) − c. Then using g(x) in **fzero** will find the value of x where g(x) is zero, which occurs when f(x) = c.

13.4 INTEGRATION

The integral or the area under a function is yet another useful attribute. MATLAB provides three functions for numerically computing the area under a function over a finite interval: **trapz**, **quad**, and **quad8**. The function **trapz** approximates the integral under a function by summing the area of trapezoids formed from the data points as shown on the next page using the function **humps**.

   As is apparent from the figure, the area of individual trapezoids underestimates the true area in some segments and overestimates it in others. As with linear interpolation, this

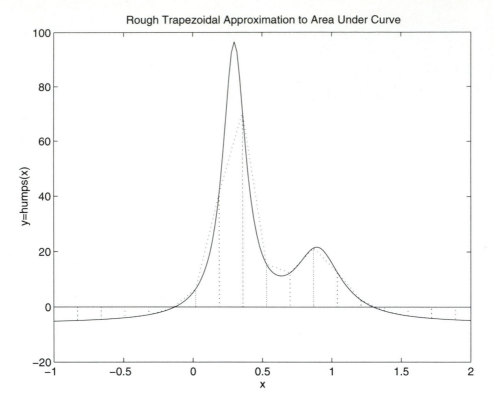

Rough Trapezoidal Approximation to Area Under Curve

approximation gets better as the number of trapezoids increases. For example, if we roughly double the number of trapezoids used in the above figure, we get a much better approximation as shown on the following page.

To compute the area under y = humps(x) over the range -1 < x < 2 using **trapz** for each of the two plots shown above:

```
» x=-1:.17:2;          % rough approximation

» y=humps(x);

» area=trapz(x,y)      % call trapz just like the plot command
area =
   25.9174

» x=-1:.07:2;          % better approximation

» y=humps(x);

» area=trapz(x,y)
area =
   26.6243
```

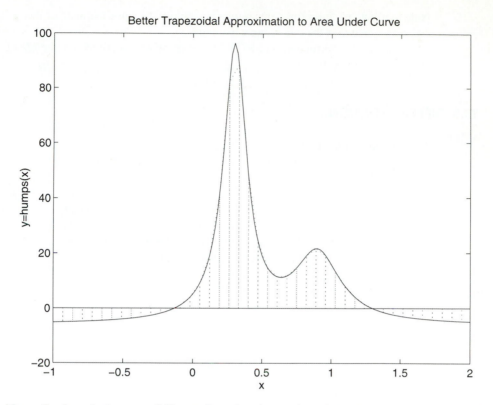

Naturally the solutions are different. Based on inspection of the plots, the rough approximation probably underestimates the area. Nothing certain can be said about the better approximation except that it is likely to be much more accurate. Clearly, if one can somehow change individual trapezoid widths to match the characteristics of the function, i.e., make them narrower where the function changes more rapidly, much greater accuracy can be achieved.

The MATLAB functions **quad** and **quad8**, which are based on the mathematical concept of quadrature, take this approach. These integration functions operate in the same way. Both evaluate the function to be integrated at whatever intervals are necessary to achieve accurate results. Moreover, both functions make higher order approximations than a simple trapezoid, with **quad8** being more rigorous than **quad**. These functions are called in the same way that **fzero** is

```
» area=quad('humps',-1,2) % find area between -1 and 2
area =
   26.3450

» area=quad8('humps',-1,2)
area =
   26.3450
```

Note that both of these functions return essentially the same estimate of the area, and that estimate is between the two **trapz** estimates.

For more information on MATLAB's integration functions, see the *MATLAB Reference Guide* and on-line help.

13.5 DIFFERENTIATION

As opposed to integration, numerical differentiation is much more difficult. Integration describes an overall, or macroscopic, property of a function, whereas differentiation describes the slope of a function at a point, which is a microscopic property of a function. As a result, integration is not sensitive to minor changes in the shape of a function, whereas differentiation is. Any small change in a function can easily create large changes in its slope in the neighborhood of the change.

Because of this inherent difficulty with differentiation, numerical differentiation is avoided whenever possible, especially if the data to be differentiated are obtained experimentally. In this case, it is best to perform a least-squares curve fit to the data, then differentiate the resulting polynomial. Or alternatively, one could fit cubic splines to the data, then find the spline representation of the derivative as discussed in Chapter 11. For example, reconsider the curve fitting example from Chapter 11.

```
» x=[0 .1 .2 .3 .4 .5 .6 .7 .8 .9 1];

» y=[-.447 1.978 3.28 6.16 7.08 7.34 7.66 9.56 9.48 9.30 11.2]; % data

» n=2; % order of fit

» p=polyfit(x,y,n) % find polynomial coefficients
p =
    -9.8108    20.1293    -0.0317

» xi=linspace(0,1,100);

» z=polyval(p,xi); % evaluate polynomial

» plot(x,y,'o',x,y,xi,z,':')

» xlabel('x'),ylabel('y=f(x)'),title('Second Order Curve Fitting')
```

The derivative in this case is found by using the polynomial derivative function **polyder**.

```
» pd=polyder(p)
pd =
    -19.6217    20.1293
```

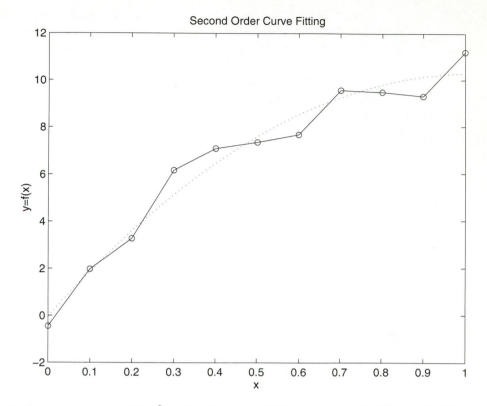

The derivative of $y = -9.8108x^2 + 20.1293x = 0.0317$ is $dy/dx = -19.6217x + 20.1293$.
Since the derivative of a polynomial is yet another polynomial of the next lower order, the
derivative can also be evaluated and plotted.

```
» z=polyval(pd,xi); % evaluate derivative
```

```
» plot(xi,z)
```

```
» xlabel('x'),ylabel('dy/dx'),title('Derivative of a Curve Fit Polynomial')
```

(See next page for plot.)
In this case, the polynomial fit was second order, making the resulting derivative first order. As a result, the derivative is a straight line, meaning that it changes linearly with x.

MATLAB provides one function for computing a very rough derivative given the data
describing some function. This function, named **diff**, computes the difference between
elements in an array. Since differentiation is defined as

$$\frac{dy}{dx} = \lim_{h\to 0} \frac{f(x + h) - f(x)}{(x + h) - (x)}$$

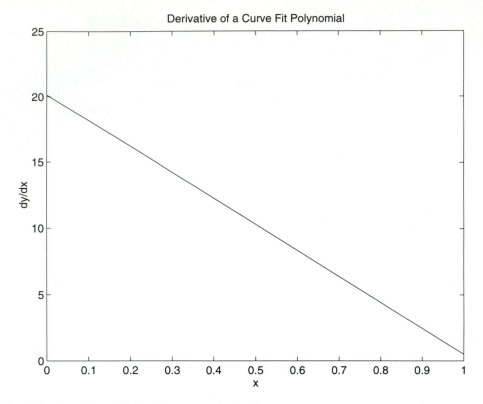

the derivative of y = f(x) can be approximated by

$$\frac{dy}{dx} \approx \frac{f(x + h) - f(x)}{(x + h) - (x)} \quad \text{where } h>0$$

which is the finite difference of y divided by the finite difference in x. Since `diff` computes differences between array elements, differentiation can be approximated in MATLAB. Continuing with the prior example,

```
» dy=diff(y)./diff(x); % compute differences and use array division

» xd=x(1:length(x)-1); % create new x axis since dy is shorter than y

» plot(xd,dy)

» title('Approximate Derivative Using DIFF')

» ylabel('dy/dx'),xlabel('x')
```

Since `diff` computes the difference between elements of an array, the resulting output contains one less element than the original array. Thus, to plot the derivative, one element of

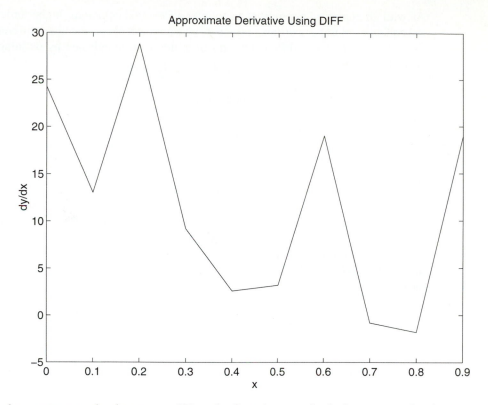

the x array must be thrown out. When the first element of x is thrown out, the above procedure gives a backward difference approximation, whereas throwing out the last element gives a forward difference approximation. Comparing the last two plots, it is overwhelmingly apparent that approximating the derivative by finite differences can lead to poor results, especially if the data are corrupted with noise.

13.6 DIFFERENTIAL EQUATIONS

Ordinary differential equations describe how the rate of change of variables within a system is influenced by variables within the system and by external stimuli, i.e., inputs. When ordinary differential equations can be solved analytically, features in MATLAB's *Symbolic Toolbox* can be used to find exact solutions. Some features of this toolbox are described later in this text.

In those cases where the equations cannot be readily solved analytically, it is convenient to solve them numerically. For illustration purposes, consider the classic Van der Pol differential equation which describes an oscillator.

$$\frac{d^2x}{dt^2} - \mu(1 - x^2)\frac{dx}{dt} + x = 0$$

As with all numerical approaches to solving differential equations, higher order differential equations must be rewritten in terms of an equivalent set of first order differential equations. For the previous differential equation, this is accomplished by defining two new variables.

$$\text{let } y_1 = x, \text{ and } y_2 = \frac{dx}{dt}$$

$$\text{then } \frac{dy_1}{dt} = y_2$$

$$\frac{dy_2}{dt} = \mu(1 - y_2) - y_1$$

From this set of equations, the MATLAB functions **ode23** and **ode45** can be used to find the motion of this system as time evolves. Doing so requires that we write a function M-file that returns the above derivatives given the present time and the present values of y_1 and y_2. In MATLAB the derivatives are given by a column vector, called **yprime** in this case. Similarly, y_1 and y_2 are written as a column vector **y**. The resulting function M-file is

```
function yprime=vdpol(t,y);
%VDPOL(t,y) returns derivatives of the Van der Pol equation:
%
%    x'' - mu*(1-x^2)*x' + x = 0    (' = d/dx, '' = d^2/dx^2)
%
%    let y(1) = x    and y(2) = x'
%
%    then y(1)' = y(2)
%         y(2)' = MU*(1-y(1)^2)*y(2) -y(1)

global MU % choose 0< MU < 10 in Command workspace

yprime=[y(2)
        MU*(1-y(1)^2)*y(2)-y(1)]; % output must be a column
```

Given this function that completely describes the differential equation, the solution is computed as:

```
» global MU % define MU as a global variable in the Command Workspace

» MU=2;  % set global parameter to desired value
```

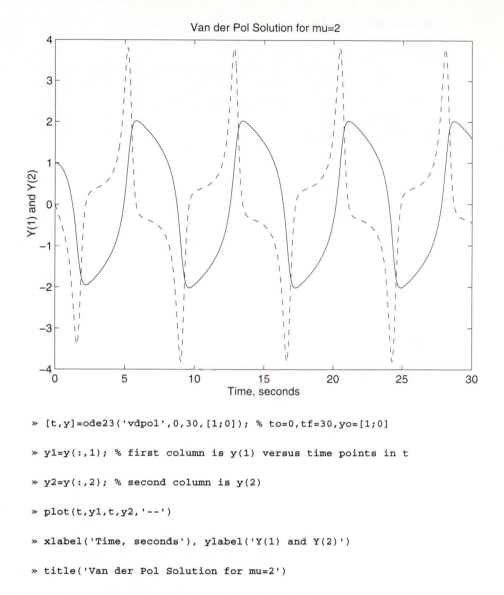

```
» [t,y]=ode23('vdpol',0,30,[1;0]); % to=0,tf=30,yo=[1;0]

» y1=y(:,1); % first column is y(1) versus time points in t

» y2=y(:,2); % second column is y(2)

» plot(t,y1,t,y2,'--')

» xlabel('Time, seconds'), ylabel('Y(1) and Y(2)')

» title('Van der Pol Solution for mu=2')
```

In this plot, y_2 (dashed) is the derivative of y_1 (solid). The parameters passed to **ode23** are described by **ode23(f_name,to,tf,yo,tol)** where **f_name** is the character string name of the M-file function that computes the derivatives, **to** is the initial time, **tf** is the final time, **yo** is the vector of initial conditions, and the optional parameter **tol** (**tol=1e-3** is default) is the desired relative accuracy. In the above example, the solution starts at time equal to zero, ends at time equal to 30 seconds, and begins with the initial condition **y=[1;0]**. The two output parameters are a column vector **t**, containing the time points where the response was estimated, and the matrix **y** that has as many columns as there are differential equations (two in this case) and as many rows as there are in **t**. The time points in **t** are not linearly spaced, since the integration algorithm varies the step size to maintain the desired relative accuracy.

The function **ode45** is used exactly the same way as **ode23**. The difference between the two functions has to do with the internal algorithm used. Both functions use variations of the basic Runge-Kutta numerical integration approach. **ode23** uses a combination second/third-order Runge-Kutta-Fehlberg algorithm, whereas **ode45** uses a combination fourth/fifth-order Runge-Kutta-Fehlberg algorithm. In general, **ode45** can take larger time steps, and therefore fewer time steps, between **to** and **tf** while maintaining the same relative error that **ode23** does. However, at the same time, **ode23** calls the function **f_name** a minimum of three times per time step, whereas **ode45** calls **f_name** at least six times per time step.

Just as interpolation using higher order polynomials does not always lead to the best solution, **ode45** is not always better than **ode23**. If **ode45** produces results that are spaced too far apart for plotting, additional time may be required to interpolate the data over a finer time scale, e.g., with the function **interp1**. This additional time may make **ode23** more efficient. As a general rule, lower order algorithms are more efficient when there are repetitive discontinuities in the computed derivatives that cause a higher order algorithm to reduce its time step size to maintain accuracy. It is for this reason that electric circuit analysis programs by default use a first order algorithm, and at most offer a second order algorithm for solving transient time-response problems. In addition, requesting higher accuracy by setting **tol** smaller does not necessarily make the absolute error smaller. **tol** sets the relative accuracy per time step, which does not necessarily cause the absolute error to decrease.

To sum things up: **Don't use a numerical method blindly**. Try various alternatives for a given problem before deciding which is best. For further information regarding the numerical solution of differential equations, consult a numerical analysis text. Some provide very practical information regarding algorithm selection and dealing with *stiff* equations, i.e., those with widely varying time constants.

13.7 M-FILE EXAMPLES

The *Mastering MATLAB Toolbox* M-files illustrated here approximate the integrals and derivatives of functions given by their sample values. It is assumed that the functions themselves are not available and that the independent variable may not be linearly spaced. For example, the data may be the result of experimental measurements that have been loaded into MATLAB for analysis.

Lacking a functional description of the data involved, there are a number of approaches to integration and differentiation. As described earlier, one could find a least-squares polynomial fit to the data and then operate on the polynomial description. Alternatively, one could find a cubic spline representation of the data and then use the *Mastering MATLAB Toolbox* functions **spintgrl** and **spderiv** to find spline representations of the integral and derivative respectively. The approaches illustrated here offer simpler alternatives. The integral is computed using the trapezoidal rule, and the derivative is computed using weighted central differences. In addition, the functions are designed to work on matrices whose columns represent dependent variables each associated with the independent variable.

As described earlier in this chapter, the MATLAB function **trapz** performs trapezoidal integration over some finite region. Here we seek the integral as a function of the independent variable x, i.e., if y = f(x), we seek

$$S(x) = \int_{x_1}^{x} f(x)\,dx$$

where x_1 is the first element of the vector x. Using the trapezoidal rule, this integral is approximated by

$$S(x_k) = \sum_{i=1}^{k} 0.5(y_i + y_{i-1})(x_i - x_{i-1}) \text{ with } S(x_1) = 0$$

Thus, the integral at the kth data point is the cumulative sum of the preceding trapezoidal areas. The function **mmintgrl** implements this algorithm as follows:

```
function z=mmintgrl(x,y)
%MMINTGRL Compute Integral using Trapezoidal Rule.
% MMINTGRL(X,Y) computes the integral of the function y=f(x) given the
% data in X and Y. X must be a vector, but Y may be a column oriented
% data matrix. The length of X must equal the length of Y if Y is a
% vector, or it must equal the number of rows in Y if Y is a matrix.
%
% X need not be equally spaced. The trapezoidal algorithm is used.
%
% See also mmderiv

% Copyright (c) 1996 by Prentice-Hall, Inc.

flag=0;                     % flag is True if y is a row
x=x(:);nx=length(x);        % make x a column
[ry,cy]=size(y);
if ry==1&cy==nx,y=y.';ry=cy;cy=1;flag=1;end
if nx~=ry, error('X and Y not the right size'),end
```

```
dx=x(2:nx)-x(1:nx-1);              % width of each trapezoid

dx=dx(:,ones(1,cy));               % duplicate for each column in y

yave=(y(2:ry,:)+y(1:ry-1,:))/2;    % average of heights

z=[zeros(1,cy); cumsum(dx.*yave)]; % Use cumsum to find area

if flag,z=z';end                   % if y was a row, return a row
```

Before illustrating the use of the above function, consider differentiation. In this case, one is interested in approximating slopes at data points given just the data points. Rather than use a simple forward or backward difference, a weighted central difference is considered here as illustrated below:

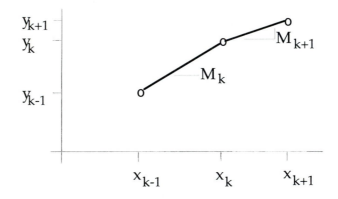

Based on the above figure, the approximate derivative at the kth data point is

$$D(x_k) = (1 - \alpha_k)M_k + (\alpha_k)M_{k+1}$$

where

$$\alpha_k = \frac{x_k - x_{k-1}}{x_{k+1} - x_{k-1}}$$

and M_k is the slope of the line connecting y_{k-1} to y_k. Thus, the derivative at the kth point is a weighted average of the slopes between adjacent points, with the closer point getting the higher weight. At the first and last data points, the above approach doesn't work simply because there are no adjoining segments. For these data points a different approach is required. The approach taken here is to fit a quadratic polynomial to the first three (last three) data points and compute the derivative of the polynomial at the first (last) point. The function **mmderiv** implements this algorithm as follows:

```
function z=mmderiv(x,y)
%MMDERIV Compute Derivative Using Weighted Central Differences.
% MMDERIV(X,Y) computes the derivative of the function y=f(x) given the
% data in X and Y. X must be a vector, but Y may be a column oriented
% data matrix. The length of X must equal the length of Y if Y is a
% vector, or it must equal the number of rows in Y if Y is a matrix.
%
% X need not be equally spaced. Weighted central differences are used.
% Quadratic approximation is used at the endpoints.
%
% See also mmintgrl

% Copyright (c) 1996 by Prentice-Hall, Inc.

flag=0;                  % flag is True if y is a row
x=x(:);nx=length(x);     % make x a column
[ry,cy]=size(y);
if ry==1&cy==nx,y=y.';ry=cy;cy=1;flag=1;end
if nx~=ry, error('X and Y not the right size'),end
if nx<3, error('X and Y must have at least three elements'),end

dx=x(2:nx)-x(1:nx-1);       % first difference in x
dx=dx+(dx==0)*eps;          % make infinite slopes finite
dxx=x(3:nx)-x(1:nx-2);      % second difference in x
dxx=dxx+(dxx==0)*eps;       % make infinite slopes finite
alpha=dx(1:nx-2)./dxx;      % central difference weight
alpha=alpha(:,ones(1,cy)); % duplicate for each column in y

dy=y(2:ry,:)-y(1:ry-1,:);   % first difference in y
dx=dx(:,ones(1,cy));        % duplicate dx for each column in y

% now apply weighting to dy
z=alpha.*dy(2:ry-1,:)./dx(2:nx-1,:)+(1-alpha).*dy(1:ry-2,:)./dx(1:nx-2,:);

z1=zeros(1,cy)>=z1;
for i=1:cy % fit quadratic at endpoints of each column
  p1=polyfit(x(1:3),y(1:3,i),2);        % quadratic at first point
  z1(i)=2*p1(1)*x(1)+p1(2);             % evaluate poly derivative
  pn=polyfit(x(nx-2:nx),y(ry-2:ry,i),2); % quadratic at last point
  zn(i)=2*pn(1)*x(nx)+pn(2);            % evaluate poly derivative
end
z=[z1;z;zn];
if flag,z=z';en     % if y was a row, return a row
```

Finally, an example:

```
» x=linspace(0,2*pi,30);

» y=sin(x); % create data

» yi=mmintgrl(x,y); % find integral

» yd=mmderiv(x,y); % find derivative

» plot(x,y,x,yi,'—',x,yd,':') % plot results
```

Note that the integral qualitatively shows the identity

$$\int_0^x \sin(x)\,dx = 1 - \cos(x)$$

whereas the derivative qualitatively shows the identity

$$\frac{d}{dx}\sin(x) = \cos(x)$$

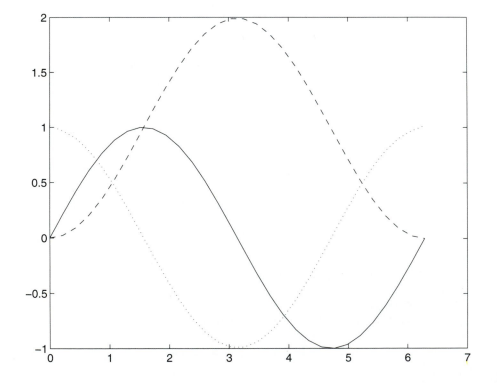

13.8 SUMMARY

The following table summarizes the functions discussed in this chapter.

Numerical Analysis Functions	
`fplot('fname',[lb ub])`	Plot function between lower and upper bounds
`fmin('fname',[lb ub])`	Find scalar minimum between lower and upper bounds
`fmins('fname',Xo)`	Find vector minimum near Xo
`fzero('fname',xo)`	Find zero in scalar function near xo
`trapz(x,y)`	Trapezoidal integration of area under $y = f(x)$ given data points in x and y
`diff(x)`	Difference between array elements
`[t,y]= ode23('fname',to,tf ,yo)`	Solution of a set of differential equations using a 2nd/3rd order Runge-Kutta algorithm
`[t,y]= ode45('fname',to,tf ,yo)`	Solution of a set of differential equations using a 4th/5th order Runge-Kutta algorithm

14

Fourier Analysis

Frequency domain tools such as Fourier Series, Fourier Transform, and their discrete time counterparts form a cornerstone in signal processing. To facilitate this analysis, MATLAB offers the functions `fft`, `ifft`, `fft2`, `ifft2`, and `fftshift`. This collection of functions performs the discrete Fourier transform and its inverse in one and two dimensions. These functions allow one to perform a variety of signal processing tasks. More extensive signal processing tools are available in the optional *Signal Processing Toolbox*.

Because signal processing encompasses such a diverse area, it is beyond the scope of this text to illustrate even a small sample of the type of problems that can be solved using the discrete Fourier transform functions in MATLAB. As a result, one example using the function `fft` to approximate the Fourier transform of a continuous time signal will be illustrated. In addition, a collection of *Mastering MATLAB Toolbox* functions for manipulating Fourier Series will be discussed.

14.1 FAST FOURIER TRANSFORM

In MATLAB, the function `fft` computes the discrete Fourier transform of a signal. In those cases where the length of the data is a power of two, or a product of prime factors, fast

Fourier transform (FFT) algorithms are utilized to compute the discrete Fourier transform. Because of the substantial increase in computational speed that occurs when data length is a power of two, whenever possible it is important to choose data lengths equal to a power of two, or to pad data with zeros to give the data a length equal to a power of two. A discussion of this issue can be found in the *MATLAB Reference Guide*.

The fast Fourier transform implemented in MATLAB follows that commonly used in engineering texts.

$$F(k) = FFT\{f(n)\}$$

$$F(k) = \sum_{n=0}^{N-1} f(n)\, e^{-j2\pi nk/N} \qquad k = 0,1,\ldots,N-1$$

Since MATLAB does not allow zero indices, the values are shifted by one index value.

$$F(k) = \sum_{n=1}^{N} f(n)\, e^{-j2\pi(n-1)(k-1)/N} \qquad k = 1,2,\ldots,N$$

The inverse transform follows accordingly.

$$f(n) = FFT^{-1}\{F(k)\}$$

$$f(n) = \frac{1}{N}\sum_{k=1}^{N} F(k)\, e^{j2\pi(n-1)(k-1)/N} \qquad n = 1,2,\ldots,N$$

To illustrate use of the FFT, consider the problem of estimating the continuous Fourier transform of the signal.

$$f(t) = \begin{cases} 12e^{-3t} & t \geq 0 \\ 0 & t < 0 \end{cases}$$

Analytically, this Fourier transform is given by

$$F(\omega) = \frac{12}{3 + j\omega}$$

Although using the FFT in this case has little real value since the analytic solution is known, this example illustrates an approach to estimating the Fourier transform of less-common signals, especially those whose Fourier transform is not readily found analytically. The following MATLAB statements estimate $F(\omega)$ using the FFT and graphically compare it to the analytic expression above:

```
» N=128; % choose a power of 2 for speed

» t=linspace(0,3,N); % time points for function evaluation

» f=2*exp(-3*t); % evaluate the function and minimize aliasing: f(3) ~ 0

» Ts=t(2)-t(1); % the sampling period
```

```
» Ws=2*pi/Ts; % the sampling frequency in rad/sec

» F=fft(f); % compute the fft

» Fp=F(1:N/2+1)*Ts;
```

extracts only the positive frequency components from **F** and multiplies them by sampling period to estimate F(ω).

```
» W=Ws*(0:N/2)/N;
```

creates the continuous frequency axis, which starts at zero and ends at the Nyquist frequency **Ws/2**,

```
» Fa=2./(3+j*W); % evaluate analytical Fourier transform

» plot(W,abs(Fa),W,abs(Fp),'+') % generate plot, '+' mark fft results

» xlabel('Frequency, Rad/s'),ylabel('|F(w)|')
```

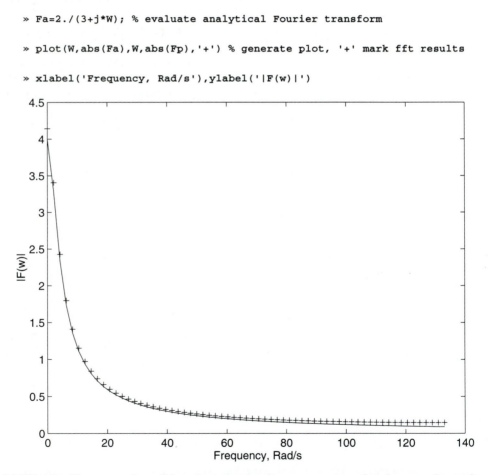

MATLAB offers a number of functions that implement common signal processing tasks. They include

Signal Processing Functions	
conv	Convolution
conv2	2-D convolution
fft	Fast Fourier Transform
fft2	2-D Fast Fourier Transform
ifft	Inverse Fast Fourier Transform
ifft2	2-D Inverse Fast Fourier Transform
filter	Discrete time filter
filter2	2-D discrete time filter
abs	Magnitude
angle	Four quadrant phase angle
unwrap	Remove phase angle jumps at 360 degree boundaries
fftshift	Shift FFT results so negative frequencies appear first
nextpow2	Next higher power of two

14.2 FOURIER SERIES

MATLAB itself offers no functions specifically tailored to Fourier series analysis and manipulation. However, they are easily added by creating M-file functions. In this section, Fourier series functions in the *Mastering MATLAB Toolbox* will be illustrated. Before doing so, let's define the Fourier series expressions for a real periodic signal f(t).

The **complex exponential form** of the Fourier series is given by

$$f(t) = \sum_{n=\infty}^{\infty} F_n e^{jn\omega_0 t}$$

where the Fourier series coefficients are

$$F_n = \frac{1}{T_0} \int_t^{t+T_0} f(t)e^{-jn\omega_0 t} \, dt$$

and the fundamental frequency is $\omega_0 = 2\pi/T_0$, where T_0 is such that $f(t + T_0) = f(t)$.

The **trigonometric form** of the Fourier series is given by

$$f(t) = A_0 + \sum_{n=1}^{\infty} \left\{ A_n \cos(n\omega_0 t) + B_n \sin(n\omega_0 t) \right\}$$

where the Fourier series coefficients are

$$A_0 = \frac{1}{T_o} \int_t^{t+T_0} f(t) \, dt$$

$$A_n = \frac{2}{T_o} \int_t^{t+T_0} f(t)\cos(n\omega_o t) \, dt$$

$$B_n = \frac{2}{T_o} \int_t^{t+T_0} f(t)\sin(n\omega_o t) \, dt$$

and the fundamental frequency is $\omega_o = 2\pi/T_o$, where T_o is such that $f(t + T_o) = f(t)$.

Of these two forms, the complex exponential Fourier series is generally easier to manipulate analytically, whereas the trigonometric form provides a more intuitive understanding because it is generally easier to visualize sine and cosine waveforms than the complex exponential $e^{jn\omega_o t}$. For this reason, the Fourier series functions in the *Mastering MATLAB Toolbox* generally assume the complex exponential form. However, the toolbox does have a function for converting between the two forms. The Fourier series functions in the *Mastering MATLAB Toolbox* are described in the following table:

Mastering MATLAB Fourier Series Functions

`fsderiv(Kn,Wo)`	Derivative of a Fourier series
`fseval(Kn,t,Wo)`	Evaluate Fourier series
`fsfind('fname',T,N)`	Find Fourier series coefficient vector for a time function
`[An,Bn,Ao]=fsform(Kn)`	Fourier series form conversion
`Kn=fsform(An,Bn,Ao)`	
`fsharm(Kn,i)`	Extract a particular Fourier series harmonic
`fsmsv(Kn)`	Compute mean square value of signal
`fsresize(Kn,N)`	Resize a Fourier series coefficient vector
`fsresp(Num,Den,Un,Wo)`	Fourier series response of a linear system to an input Fourier series, `Un`
`fsround(Kn)`	Set insignificant Fourier series coefficients to zero
`fswindow(N,'type')`	Generate a window function to minimize Gibb's phenomenon
`fswindow(Kn,'type')`	

Because there is an infinite number of harmonics or Fourier series coefficients, it is necessary to truncate the Fourier series by considering only a finite number of harmonics. As a result, if N harmonics are considered, a Fourier series is represented in MATLAB by a row vector of length 2N + 1 of its complex exponential Fourier series coefficients. The vector contains the Fourier series coefficients in increasing order, i.e.,

$$F = [F_{-N} \; F_{-N+1} \; \ldots \; F_{-1} \; F_0 F_1 \; \ldots \; F_{N-1} \; F_N]$$

To illustrate the use of these functions, consider finding the Fourier series representation of the sawtooth signal.

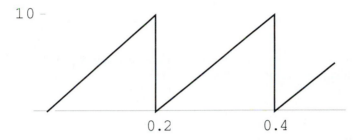

Although the Fourier series coefficients of this signal are readily found, the function **fsfind** can be used to approximate them.

```
function [Fn,nwo,f,t]=fsfind(fun,T,N,P)
%FSFIND Find Fourier Series Approximation.
% Fn=FSFIND(FUN,T,N) computes the Complex Exponential
% Fourier Series of a signal described by the function 'FUN'.
% FUN is the character string name of a user created M-file function.
% The function is called as f=FUN(t) where t is a vector over
% the range 0<=t<=T.
%
% The FFT is used. Choose sufficient harmonics to minimize aliasing.
%
% T is the period of the function. N is the number of harmonics.
% Fn is the vector of FS coefficients.
%
% [Fn,nWo,f,t]=FSFIND(FUN,T,N) returns the frequencies associated
% with Fn in nWo and returns values of the function FUN
% in f evaluated at the points in t over the range 0<=t<=T.
%
% FSFIND(FUN,T,N,P) passes the data in P to the function FUN as
% f=FUN(t,P). This allows parameters to be passed to FUN.
```

```
% Copyright (c) 1996 by Prentice-Hall, Inc.

n=2*N;
t=linspace(0,T,n+1);
if nargin==3
 f=feval(fun,t);
else
 f=feval(fun,t,P);
end
Fn=fft(f(1:n));
Fn=[0 conj(Fn(N:-1:2)) Fn(1:N) 0]/n;
nwo=2*pi/T*(-N:N);
```

From the above description, we must first create a function that computes values of the sawtooth over its period. Doing so we have

```
function f=sawtooth(t,T)
%SAWTOOTH Sawtooth Waveform Generation.
% SAWTOOTH(t,T) computes values of a sawtooth having a
% period T at the values in t.
%
% Used in Fourier series example.

f=10*rem(t,T)/T;
```

Now approximate the Fourier series coefficients.

```
» T=.2; % desired period

» N=25; % number of harmonics

» Fn=fsfind('sawtooth',T,N,T); % compute Fourier series coefficients
```

The resulting coefficients are not exact since the FFT was used to approximate the coefficients. However, if sufficient coefficients are requested, the error is small, especially if the time function or one or more derivatives are continuous (the sawtooth is not continuous). If more accuracy is required, an integration routine can be called to evaluate the integral relationship for each Fourier series coefficient. This latter approach requires much more time because the FFT approximates all the coefficients at once. Despite this error, the FFT approximation is usually sufficiently accurate for many applications.

Fourier series evaluation is accomplished using the function **fseval**.

```
function y=fseval(kn,t,wo)
%FSEVAL Fourier Series Function Evaluation.
% FSEVAL(Kn,t,Wo) computes values of a real valued function given
% its complex exponential Fourier series coefficients Kn, at the
% points given in t where the fundamental frequency is Wo rad/s.
% K contains the Fourier coefficients in ascending order:
% Kn = [k    k       ... k  ... k     k ]
%           -N  -N+1       0       N-1  N
% if Wo is not given, Wo=1 is assumed.
% Note: this function creates a matrix of size:
% rows = length(t) and columns = (length(K)-1)/2

% Copyright (c) 1996 by Prentice-Hall, Inc.

if nargin==2,wo=1;end
nk=length(kn);
if rem(nk-1,2)|(nk==1)
  error('Number of elements in K must be odd and greater than 1')
end
n=0.5*(nk-1);      % highest harmonic
nwo=wo*(1:n);      % harmonic frequencies
ko=kn(n+1);        % average value
kn=kn(n+2:nk).';   % positive frequency coefs
y=ko+2*(real(exp(j*t(:)*nwo)*kn))';
```

Following the required syntax, we get

```
» n=-N:N; % harmonic index

» Fna=j*10./((2*pi*n).^2).*(1+2*pi*n); % actual Fourier series coefficients

» Fna(N+1)=5; % poke in average value

» t=linspace(0,.4); % time points to evaluate functions

» wo=2*pi/T; % fundamental frequency

» f=fseval(Fn,t,wo); % evaluate approximated Fourier series
```

```
» fa=fseval(Fna,t,wo); % evaluate actual Fourier series
```

```
» plot(t,f,t,fa) % plot results for comparison
```

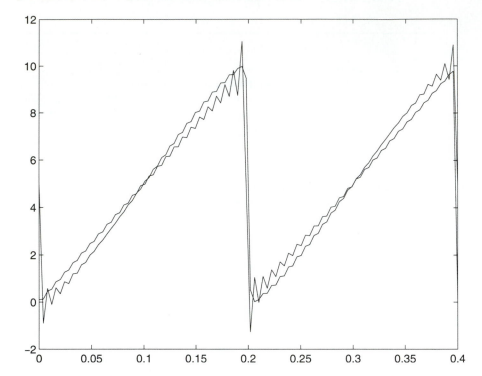

Note that both Fourier series give similar results. From the plot, *Gibb's phenomenon ripple* around the discontinuity in the sawtooth is apparent. This ripple can be reduced by applying a ***window*** to the Fourier series coefficients by using the function **fswindow**.

```
» help fswindow

  FSWINDOW Generate Window Functions.
    FSWINDOW(N,TYPE) creates a window vector of type TYPE having
    a length equal to the scalar N.
    FSWINDOW(X,TYPE) creates a window vector of type TYPE having
    a length and orientation the same as the vector X.
    FSWINDOW(X,TYPE,alpha) provides a parameter alpha as required
    for some window types.
    FSWINDOW with no input arguments returns a string matrix whose
    i-th row is the i-th TYPE given below.
    TYPE is a string designating the window type desired:
    'rec' = Rectangular or Boxcar
    'tri' = Triangular or Bartlet
    'han' = Hann or Hanning
```

```
'ham' = Hamming
'bla' = Blackman common coefs.
'blx' = Blackman exact coefs.
'rie' = Riemann {sin(x)/x}
'tuk' = Tukey, 0< alpha < 1; alpha = 0.5 is default
'poi' = Poisson, 0< alpha < inf; alpha = 1 is default
'cau' = Cauchy, 1< alpha < inf; alpha = 1 is default
'gau' = Gaussian, 1< alpha < inf; alpha = 1 is default
```

```
Reference: F.J. Harris,''On the Use of Windows for Harmonic Analysis with
the Discrete Fourier Transform,'' IEEE Proc, vol. 66, no. 1, Jan. 1978,
pp. 51-83.
```

```
» Fnh=Fn.*fswindow(Fn,'han'); % apply Hanning window
```

```
» f=fseval(Fnh,t,wo); % evaluate windowed FS
```

```
» plot(t,f) % plot results
```

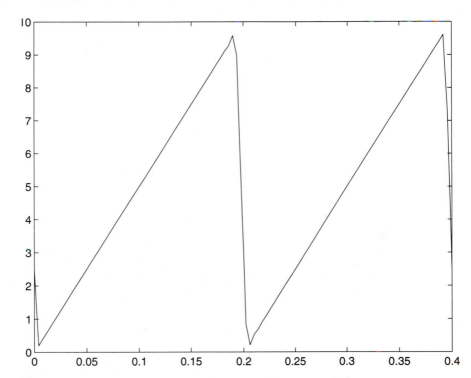

Now the plot looks better because we distorted the Fourier series coefficients by multiplying by a window function.

Next, let's apply the sawtooth waveform to a linear system having the transfer function.

$$H(s) = \frac{60}{s + 60}$$

The function **fsresp** solves this problem.

```
function y=fsresp(num,den,un,wo)
%FSRESP Fourier Series Linear System Response.
% FSRESP(N,D,Un,Wo) returns the complex exponential FS of the
% output of a linear system when the input is given by a FS.
% N and D are the numerator and denominator coefficients
% respectively of the system transfer function.
% Un is the complex exponential Fourier Series of the system input.
% Wo is the fundamental frequency associated with the input.

% Copyright (c) 1996 by Prentice-Hall, Inc.

if nargin<4,wo=1;end
N=(length(un)-1)/2; % highest harmonic
jnWo=sqrt(-1)*(-N:N)*wo;% frequencies of harmonics
y=(polyval(num,jnWo)./polyval(den,jnWo)).*un; % output is Yn=H(jnWo)*Un
```

Using **fsresp**, we get

```
» Yn=fsresp(60,[1 60],Fn,wo); % find response coefficients

» y=fseval(Yn,t,wo); % evaluate output

» plot(t,f,t,y) % plot system input and output
```

As a final example, convert the sawtooth Fourier series to trigonometric form using **fsform**.

```
function [a,b,ao]=fsform(c,d,co)
%FSFORM Fourier Series Format Conversion
% Kn=FSFORM(An,Bn,Ao) converts the trigonometric FS with
% An being the COSINE and Bn being the SINE coefficients to
% the complex exponential FS with coefficients Kn.
% Ao is the DC component and An, Bn and Ao are assumed to be real.
```

```
%
% [Kn,i]=FSFORM(An,Bn,Ao) returns the index vector i that
% identifies the harmonic number of each element of Kn.
%
% [An,Bn,Ao]=FSFORM(Kn) does the reverse format conversion.

% Copyright (c) 1996 by Prentice-Hall, Inc.

nc=length(c);
if nargin==1 % complex exp -> trig form
 if rem(nc-1,2)|(nc==1)
  error('Number of elements in K must be odd and greater than 1')
 end
 nn=(nc+3)/2;
 a=2*real(c(nn:nc));
 b=-2*imag(c(nn:nc));
 ao=real(c(nn-1));
elseif nargin==3    % trig form -> complex exp form
 nd=length(d);
 if nc~=nd,
  error('A and B must be the same length')
 end
 a=0.5*(c-sqrt(-1)*d);
 a=[conj(a(nc:-1:1)) co(1) a];
```

```
    b=-nc:nc;
   else
    error('Improper number of input arguments.')
   end
```

```
» [An,Bn,Ao]=fsform(Fn)
An =
  Columns 1 through 7
   -0.2000    -0.2000    -0.2000    -0.2000    -0.2000    -0.2000    -0.2000
  Columns 8 through 14
   -0.2000    -0.2000    -0.2000    -0.2000    -0.2000    -0.2000    -0.2000
  Columns 15 through 21
   -0.2000    -0.2000    -0.2000    -0.2000    -0.2000    -0.2000    -0.2000
  Columns 22 through 25
   -0.2000    -0.2000    -0.2000         0
Bn =
  Columns 1 through 7
   -3.1789    -1.5832    -1.0484    -0.7789    -0.6155    -0.5051    -0.4250
  Columns 8 through 14
   -0.3638    -0.3151    -0.2753    -0.2418    -0.2130    -0.1878    -0.1655
  Columns 15 through 21
   -0.1453    -0.1269    -0.1100    -0.0941    -0.0792    -0.0650    -0.0514
  Columns 22 through 25
   -0.0382    -0.0253    -0.0126         0
Ao =
    4.9000
```

Ignoring the average value, this sawtooth exhibits odd symmetry. While all **An** coefficients should be zero, clearly they are not. This occurred because we used the FFT to approximate the coefficients. In addition, the average value **Ao** is slightly in error as it should be 5.0.

14.3 SUMMARY

The following tables summarize the functions discussed in this chapter.

Signal Processing Functions	
conv	Convolution
conv2	2-D convolution
fft	Fast Fourier Transform

`fft2`	2-D Fast Fourier Transform
`ifft`	Inverse Fast Fourier Transform
`ifft2`	2-D Inverse Fast Fourier Transform
`filter`	Discrete time filter
`filter2`	2-D discrete time filter
`abs`	Magnitude
`angle`	Phase angle
`unwrap`	Remove phase angle jumps at 360 degree boundaries
`fftshift`	Shift FFT results so negative frequencies appear first
`nextpow2`	Next higher power of two

Mastering MATLAB Fourier Series Functions

`fsderiv(Kn,Wo)`	Derivative of a Fourier series
`fseval(Kn,t,Wo)`	Evaluate Fourier series
`fsfind('fname',T,N)`	Find Fourier series coefficient vector for a time function
`[An,Bn,Ao]=fsform(Kn)` `Kn= fsform(An,Bn,Ao)`	Fourier series form conversion
`fsharm(Kn,i)`	Extract a particular Fourier series harmonic
`fsmsv(Kn)`	Compute mean square value of signal
`fsresize(Kn,N)`	Resize a Fourier series coefficient vector
`fsresp(Num,Den,Un,Wo)`	Fourier series response of a linear system to an input Fourier series, `Un`
`fsround(Kn)`	Set insignificant Fourier series coefficients to zero
`fswindow(N,'type')` `fswindow(Kn,'type')`	Generate a window function to minimize Gibb's phenomenon

Low-Level File I/O

For most users, the MATLAB functions **load** and **save** provide sufficient means for loading and saving data. **load** and **save** assume data is stored in a platform independent binary data format with a file name ending in the **.mat** extension, or in simple ASCII format called a *flat* file. In those cases where the flat ASCII or **.mat** format is not sufficient, MATLAB provides low-level file I/O functions that are based on their C language counterparts. Using these low-level functions, MATLAB can read or write any file format, provided you know what that format is. For example, if you know the format used by a spreadsheet or database program, you could read those data files into MATLAB matrices. Similarly, you could create spreadsheet or database files. The basic low-level file I/O commands available in MATLAB are the following:

MATLAB Low-Level File I/O Functions

`fclose`	Close file
`feof`	Test for end-of-file
`ferror`	Inquire file I/O error status
`fgetl`	Read line from file, ignoring newline character
`fgets`	Read line from file, including new line character
`fopen`	Open file
`fprintf`	Write formatted data to file or screen
`fread`	Read binary data from file
`frewind`	Rewind to start of file
`fscanf`	Read formatted data from file
`fseek`	Set file position indicator
`ftell`	Get file position indicator
`fwrite`	Write binary data to file

In addition to these functions, your version of MATLAB may offer specific function M-files for reading and writing files for one or more popular software packages. For further information regarding these functions, use on-line help: » **help iofun**.

16

Debugging Tools

In the process of developing function M-files, it is inevitable that errors, i.e., *bugs*, appear. MATLAB provides a number of approaches and functions to assist in the *debugging* of functions.

Two types of errors appear in MATLAB expressions: syntax errors and run-time errors. Syntax errors are found when MATLAB evaluates an expression or when a function is compiled into memory. MATLAB flags these errors immediately and provides feedback about the type of error encountered and the line number in the M-file where it occurs. Given this feedback, these errors are usually easy to spot.

Run-time errors, on the other hand, are generally more difficult to find, even though MATLAB flags them also. When a run-time error is found, MATLAB returns control to the *Command* window and the MATLAB workspace. Access to the function workspace where the error occurred is lost, so one cannot interrogate the contents of the function workspace in an effort to isolate the problem.

Based on the authors' experience, the most common run-time errors occur when the result of some operation leads to empty matrices or **NaNs**. All operations on **NaNs** return

NaNs, so if **NaNs** are a possible result, it is good to use the logical function **isnan** to perform some default action when **NaNs** occur. Addressing matrices that are empty always leads to an error since empty matrices have zero dimension. The **find** function represents a common situation where an empty matrix may result. If the empty matrix output of the **find** function is used to index some other array, the result returned will also be empty. That is, empty matrices tend to propagate empty matrices. For example,

```
» x=pi*(1:4)      % example data
x =
     3.1416      6.2832      9.4248      12.5664

» i=find(x>20) % use find function
i =
     []

» y=2*x(i)        % propagate the empty matrix

y =
     []
```

Clearly when **y** is expected to have a finite dimension and value(s), a run-time error is likely to occur. When performing operations or using functions that can return empty results, the logical function **isempty** is useful to define a default result for the empty matrix case, thereby avoiding a run-time error.

There are several approaches to debugging function M-files. For simple problems, it is straightforward to use a combination of the following:

1. Remove semicolons from selected lines within the function, so that intermediate results are displayed in the *Command* window.

2. Add statements that display variables of interest within the function.

3. Place the **keyboard** command at selected places in the M-file to give temporary control to the keyboard. By doing so, the function workspace can be interrogated and values changed as necessary.

4. Change the function M-file into a script M-file by placing a **%** before the **function** statement at the beginning of the M-file. When executed as a script file, the workspace is the MATLAB workspace, and thus it can be interrogated after the error occurs.

When the M-file is large, the M-file is recursive, or when the M-file is highly nested—it calls other M-file functions, which call still other functions, etc.—it may be more convenient to use MATLAB debugging functions. As opposed to the approaches listed above, these functions do not require one to edit the M-files in question. These functions, given on the next page, are similar to those found in other high-level programming languages. For further information, and examples of their use, see the *MATLAB User's Guide*.

MATLAB Debugging Functions

`dbclear`	Remove breakpoint
`dbcont`	Resume execution after breakpoint
`dbdown`	Drop down on workspace level
`dbquit`	Quit debug mode
`dbstack`	List who called whom
`dbstatus`	List all breakpoints
`dbstep`	Execute one or more lines
`dbstop`	Set breakpoint
`dbtype`	List M-file with line numbers
`dbup`	Move up one workspace level

2-D Graphics

Throughout this text several of MATLAB's graphics features were introduced. In this and the next several chapters, the graphics features in MATLAB will be more rigorously illustrated.

17.1 THE plot FUNCTION

As you have seen in earlier examples, the most common command for plotting two-dimensional data is the **plot** command. This versatile command plots sets of data arrays on appropriate axes and connects the points with straight lines. For example,

```
» x=linspace(0,2*pi,30);

» y=sin(x);

» plot(x,y)
```

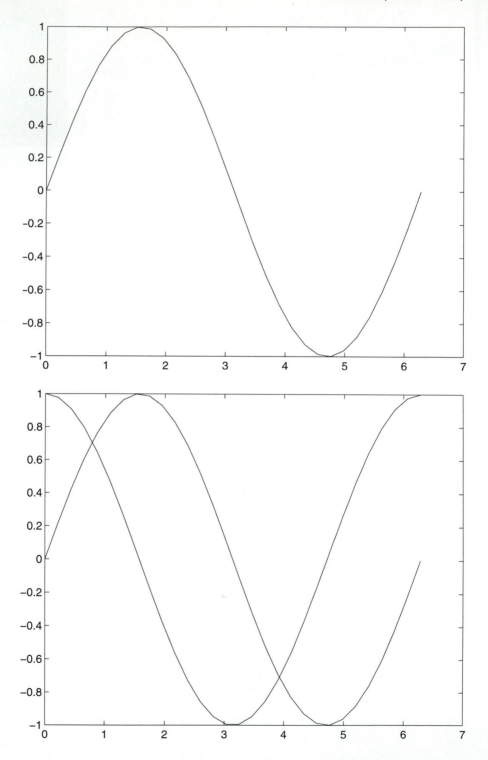

This example creates 30 data points in the range $0 \leq x \leq 2\pi$ to form the horizontal axis of the plot, and creates another vector **y** containing the sine of the data points in **x**. The `plot` command opens a window called a *Figure* window, scales the axes to fit the data, and plots the data by connecting the points with straight lines. It also adds a numerical scale and tic marks to both axes automatically. If a *Figure* window already exists, `plot` clears the current *Figure* window and draws a new plot.

Let's plot a sine and cosine on the same plot.

```
» z=cos(x);

» plot(x,y,x,z)
```

The example as depicted on the previous page shows that you can plot more than one set of arrays at the same time, just by giving `plot` another pair of arguments. This time `sin(x)` vs. **x**, and `cos(x)` vs. **x** were plotted on the same plot. `plot` automatically drew the second curve in a different color on the screen. Many curves may be plotted at one time by supplying additional pairs of arguments to `plot`.

If one of the arguments is a matrix and the other a vector, the `plot` command plots each column of the matrix versus the vector.

```
» W=[y;z]; % create a matrix of the sin and cosine

» plot(x,W) % plot the columns of W vs. x
```

If you change the order of the arguments, the plot rotates 90 degrees.

```
» plot(W,x) % plot x vs. the columns of W
```

(See plot on next page.)
When the `plot` command is called with only one input argument, e.g., `plot(Y)`, the plot function acts differently depending on the data contained in **Y**. If **Y** is a *complex*-valued vector, `plot(Y)` is interpreted as `plot(real(Y),imag(Y))`. In all other cases, the imaginary components of input vectors are ignored. On the other hand, if **Y** is *real*-valued, then `plot(Y)` is interpreted as `plot(1:length(Y),Y)`, i.e., **Y** is plotted versus an index of its values. When **Y** is a matrix, the above interpretations are applied to each column of **Y**.

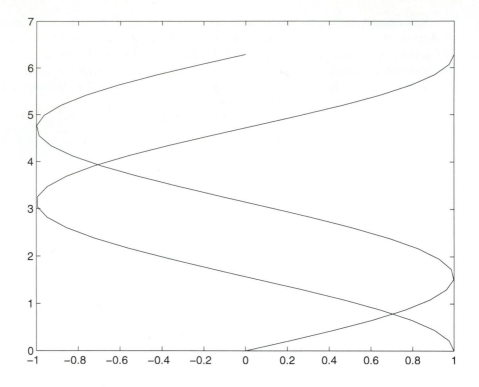

17.2 LINESTYLES, MARKERS, AND COLORS

In the previous examples, MATLAB chose the *solid* linestyle and the colors *yellow* and *magenta* for the plots. You can specify the colors and linestyles you want by giving **plot** an additional argument after each pair of data arrays. The optional additional argument is a character string consisting of 1, 2, or 3 characters from the table on page 197.

If you do not specify a color, MATLAB starts with *yellow* and cycles through the first six colors in the table for each additional line. The default linestyle is the solid line unless you specify a different linestyle. Use of the point, circle, x-mark, plus, or star symbol places the chosen symbol at each data point, but does not connect the data points with a straight line.

Here is an example using different linestyles, colors, and point markers:

```
» plot(x,y,'g:',x,z,'r-',x,y,'wo',x,z,'c+')
```

(See page 198 for plot.)
As with many of the plots in this chapter, your computer displays color but the figures printed here do not.

Basic Plot Line Types and Colors

Symbol	Color	Symbol	Linestyle
y	yellow	.	point
m	magenta	o	circle
c	cyan	x	x-mark
r	red	+	plus
g	green	*	star
b	blue	-	solid line
w	white	:	dotted line
k	black	-.	dash-dot line
		--	dashed line

17.3 ADDING GRIDS AND LABELS

The **grid on** command adds grid lines to the current plot at the tic marks. The **grid off** command removes the grid. **grid** with no arguments alternately turns them on and off, i.e., *toggles* them. Horizontal and vertical axes can be labeled with the **xlabel** and **ylabel** commands, respectively. The **title** command adds a line of text at the top of the plot. Let's use the sine and cosine plot again as an example.

```
» x=linspace(0,2*pi,30);

» y=sin(x); z=cos(x);

» plot(x,y,x,z)
```

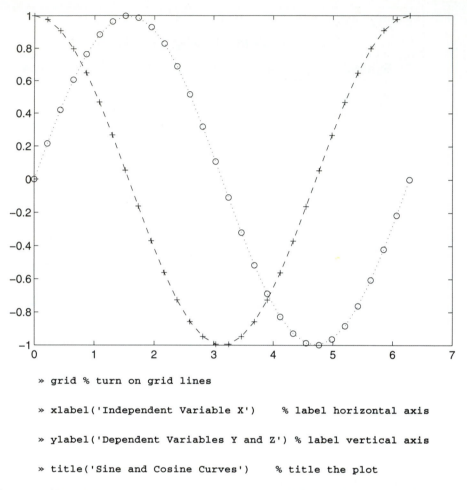

```
» grid % turn on grid lines

» xlabel('Independent Variable X')      % label horizontal axis

» ylabel('Dependent Variables Y and Z') % label vertical axis

» title('Sine and Cosine Curves')       % title the plot
```

You can add a label or any other text string to any specific location on your plot with the **text** command. The format is **text(x,y,'string')**, where **(x,y)** represents the coordinates of the center left edge of the text string in units taken from the plot axes. To add a label identifying the sine curve at the location **(2.5,0.7)**,

```
» text(2.5,0.7,'sin(x)')
```

If you want to add a label, but don't want to stop to figure out the coordinates to use, you can place a text string with the mouse. The **gtext** command switches to the current *Figure* window, puts up a cross-hair that follows the mouse, and waits for a mouse click or keypress. When either one occurs, the text is placed with the lower left corner of the first character at that location. Try labeling the second curve in the plot.

```
» gtext('cos(x)')
```

(See page 200 example plot.)

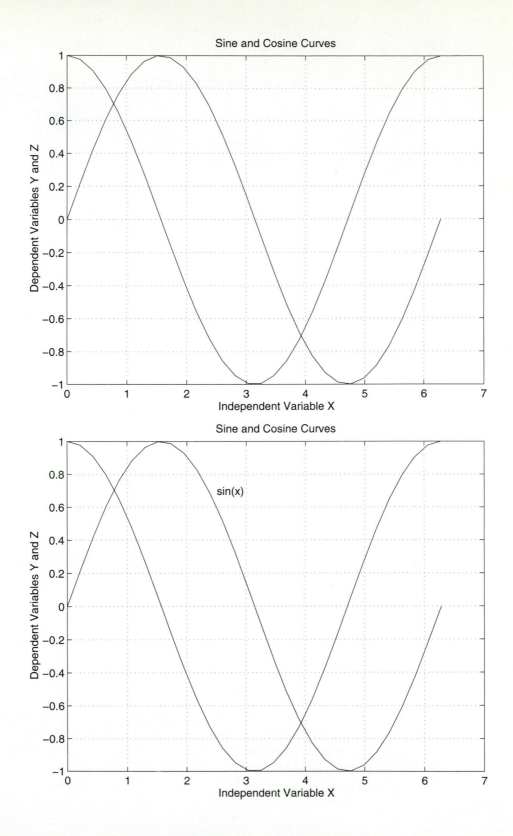

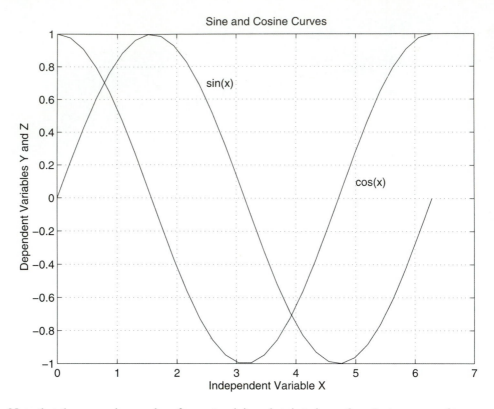

Note that the general procedure for customizing plots is to issue the **plot** command to generate the default plot and then customize it with ensuing commands.

17.4 ADDING A LEGEND

In addition to adding a title, labels, and text to a plot, MATLAB offers the function **legend**, which identifies each data set with text.

```
» x=linspace(0,2*pi,30);

» y=sin(x); z=cos(x);

» plot(x,y,x,z)

» legend('sin(x)','cos(x)')
```

The function **legend** uses MATLAB's Handle Graphics features, which are discussed in later chapters, to place the legend in an open area of the plot. If you wish to move the legend, simply click and hold down the mouse button near the lower left corner of the legend and drag the legend to the desired place. » **legend off** deletes the legend.

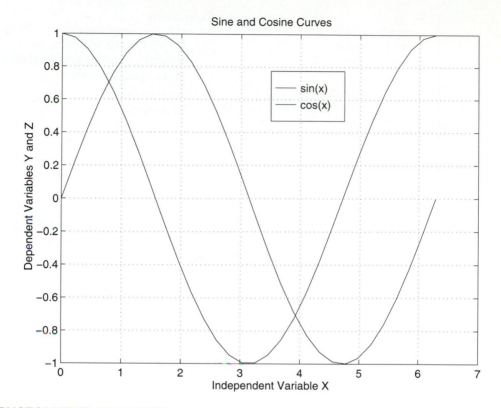

17.5 CUSTOMIZING PLOT AXES

If you are not satisfied with the scaling and appearance of both the horizontal and vertical axes of your plot, MATLAB gives you control over them with the **axis** command. Because this command has so many features, only the most useful will be described here. For a more complete description, see the *MATLAB Reference Guide* or use on-line help. The primary features of the **axis** command are given in the table below.

---

axis **Command**

---

axis([xmin xmax ymin ymax])	Set the maximum and minimum values of the axes using values given in the row vector. If **xmin** or **ymin** is set to **-inf**, the minimum is autoscaled. If **xmax** or **ymax** is set to **inf**, the maximum is autoscaled.

`axis` Command

`axis auto` `axis('auto')`	Return the axis scaling to its automatic defaults: `xmin=min(x)`, `xmax=max(x)`, etc.
`axis(axis)`	Freeze scaling at the current limits, so that if **hold** is turned on, subsequent plots use the same axis limits.
`axis xy` `axis('xy')`	Use the (default) *Cartesian* coordinate form, where the *system origin* (the smallest coordinate pair) is at the lower left corner. The horizontal axis increases left to right, and the vertical axis increases bottom to top.
`axis ij` `axis('ij')`	Use the *matrix* coordinate form, where the *system origin* is at the top left corner. The horizontal axis increases left to right, but the vertical axis increases top to bottom.
`axis square` `axis('square')`	Set the current plot to be a square rather than the default rectangle.
`axis equal` `axis('equal')`	Set the scaling factors for both axes to be equal.
`axis image` `axis('image')`	Set the aspect ratio and the axis limits so the image in the current axes has square pixels.
`axis normal` `axis('normal')`	Turn off **axis equal** and **axis square**.
`axis off` `axis('off')`	Turn off all axis labeling, grid, and tic marks. Leave the title and any labels placed by the text and **gtext** commands.
`axis on` `axis('on')`	Turn on axis labeling, tic marks, and grid.
`v=axis`	Return the current axis limits in the vector **v**.

Multiple commands to **axis** can be given at once and the default axis scaling is **axis('auto','on','xy')**. The **axis** command affects the current plot only. Therefore, it is issued after the **plot** command just as **grid**, **xlabel**, **ylabel**, **title**, **text**, etc. are issued after the plot is on the screen. For example,

```
» x=linspace(0,2*pi,30);

» y=sin(x);

» plot(x,y)

» axis([0 2*pi -1.5 2]) % change axis limits
```

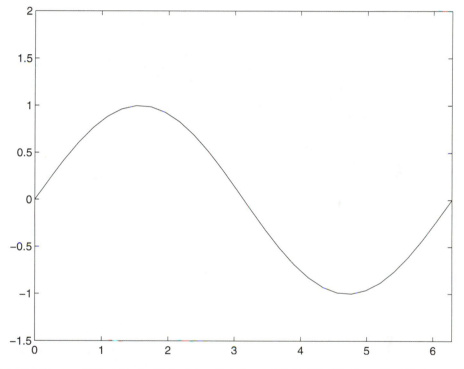

Note that by specifying the maximum x-axis value to be **2*pi**, the plot axis ends at exactly **2*pi**, rather than rounding the axis limit up to 7.

17.6 HOLDING PLOTS

You can add lines to an existing plot using the **hold** command. When you issue the command » **hold on**, MATLAB does not remove the existing curves when new **plot** commands are issued. Instead, it adds new curves to the current axes. However, if the new data do not fit within the current axes limits, the axes are rescaled. Setting **hold off** releases the current *Figure* window for new plots. The **hold** command without arguments toggles the **hold** setting. Going back to our previous example,

```
» x=linspace(0,2*pi,30);

» y=sin(x); z=cos(x);

» plot(x,y)
```

Now hold the plot and add a cosine curve.

```
» hold on

» ishold % this logical function returns True if hold is ON
ans =
      1

» plot(x,z,'m')

» hold off

» ishold % hold is no longer ON
ans =
      0
```

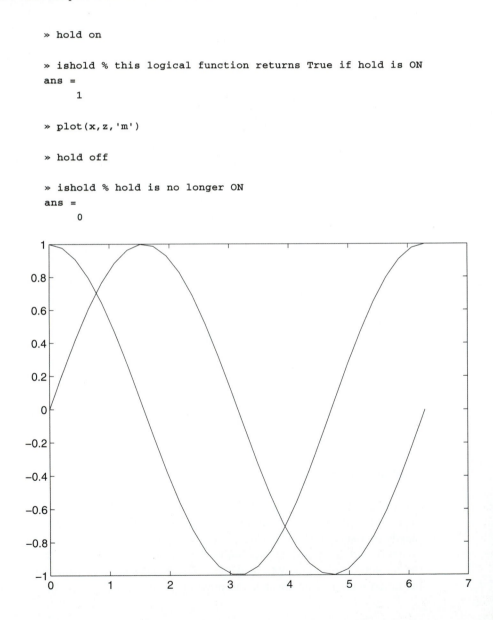

Notice that this example specified the color of the second curve. Since there is only one set of data arrays in each `plot` command, the line color for each `plot` command would otherwise default to yellow, resulting in two yellow lines on the plot. Note also that the function `ishold` provides a means for testing the hold state.

17.7 SUBPLOTS

In many situations, it is more appropriate to plot several data sets on separate axes in one *Figure* window rather than plot multiple data sets on a single axis. The `subplot(m,n,p)` command subdivides the current *Figure* window into an **m**-by-**n** matrix of plotting areas, and chooses the **p**-th area to be active. The subplots are numbered left to right along the top row, then the second row, etc. For example,

```
» x=linspace(0,2*pi,30);

» y=sin(x); z=cos(x);

» a=2*sin(x).*cos(x);

» b=sin(x)./(cos(x)+eps);

» subplot(2,2,1) % pick the upper left of 4 subplots

» plot(x,y),axis([0 2*pi -1 1]),title('sin(x)')

» subplot(2,2,2) % pick the upper right of 4 subplots

» plot(x,z),axis([0 2*pi -1 1]),title('cos(x)')

» subplot(2,2,3) % pick the lower left of 4 subplots

» plot(x,a),axis([0 2*pi -1 1]),title('2sin(x)cos(x)')

» subplot(2,2,4) % pick the lower right of 4 subplots

» plot(x,b),axis([0 2*pi -20 20]),title('sin(x)/cos(x)')
```

(See next page for plot.)

Note that when a particular subplot is active, it is the only subplot or axis that is responsive to `axis`, `hold`, `xlabel`, `ylabel`, `title`, and `grid` commands. The other subplots are not affected. In addition, the active subplot remains active until another `subplot` command is issued. If the new subplot command changes the number of subplots in the *Figure* window, previous subplots are erased to make room for the new orientation. When you want to return to the default of having one axis in a *Figure* window, you must issue the `subplot` command as

```
» subplot(1,1,1) % return to a single plot in the figure window
```

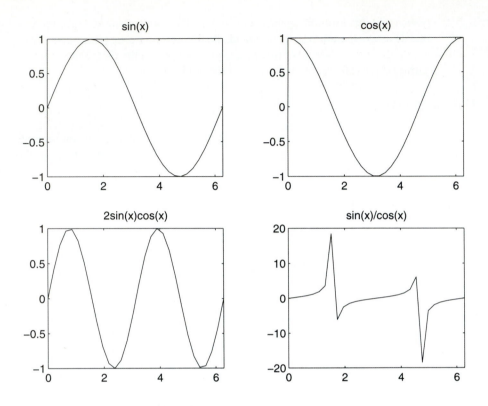

17.8 MULTIPLE FIGURE WINDOWS

As an alternative to placing several axes in one *Figure* window using the **subplot** command, it is possible to create multiple *Figure* windows and then plot different data sets in each one. The MATLAB command **figure** creates *Figure* windows. For example, the above subplot example can be redone using multiple *Figure* windows as

```
» x=linspace(0,2*pi,30);

» y=sin(x); z=cos(x);

» a=2*sin(x).*cos(x);

» b=sin(x)./(cos(x)+eps);

» h1=figure; % create a new figure window and return handle to it

» plot(x,y),axis([0 2*pi -1 1]),title('sin(x)') % first figure window

» h2=figure; % create a new figure window and return handle to it
```

```
» plot(x,z),axis([0 2*pi -1 1]),title('cos(x)')

» h3=figure; % create a new figure window and return handle to it

» plot(x,a),axis([0 2*pi -1 1]),title('2sin(x)cos(x)')

» h4=figure; % create a new figure window and return handle to it

» plot(x,b),axis([0 2*pi -20 20]),title('sin(x)/cos(x)')
```

Every time the command **figure** is issued as shown above, a new *Figure* window is created and a number identifying it, i.e., its ***handle,*** is returned and stored for future use. The figure handle is also displayed in the *Figure* window title bar. When a *Figure* window is created, it is placed in the default figure position on the screen. As a result, when more than one *Figure* window is created, each new window covers all preceeding windows. To see the windows simultaneously, simply drag them around using the mouse on the *Figure* window title bar.

To reuse a *Figure* window for a new plot, it must be made the active or *current* figure. Clicking on the figure of choice with the mouse makes it the current figure. From within MATLAB, » **figure(h)** where **h** is the figure handle makes the corresponding figure active or current. Similar to the **subplot** command, only the current figure is responsive to **axis**, **hold**, **xlabel**, **ylabel**, **title**, and **grid** commands.

Figure windows can be deleted by closing them with a mouse in the way you are familiar with closing windows in your operating environment, if such a feature exists. Alternatively, the command **close** can be issued.

```
» close
```

closes the current *Figure* window.

```
» close(h)
```

closes the *Figure* window having handle **h**.

```
» close all
```

closes all *Figure* windows.

If you simply want to erase the contents of a *Figure* window without closing it, use the command **clf**.

```
» clf
```

clears the current *Figure* window.

```
» clf reset
```

clears the current *Figure* window and resets all properties, such as **hold**, to their default state.

Since each *Figure* window is independent of all others, it is possible to place subplots in some *Figure* windows and not in others, and to vary the number of subplots in each *Figure* window. The **subplot** command always creates subplots in the current *Figure* window.

17.9 SCREEN RENDERING

Because screen rendering is relatively time consuming, MATLAB does not always update the screen after each graphics command. For example, if the following commands are entered at the MATLAB prompt, MATLAB does update the screen after each graphics command.

```
» x=linspace(0,2*pi);

» y=sin(x);

» plot(x,y)

» axis([0 2*pi -1.2 1.2])

» grid
```

However, if the same graphics commands are entered on a single line, i.e.,

```
» plot(x,y), axis([0 2*pi -1.2 1.2]), grid
```

MATLAB renders the figure only once—when the MATLAB prompt reappears. A similar procedure occurs when graphics commands appear as part of a script or function M-file. In this case, even if the commands appear on separate lines in the file, the screen is rendered only once—when all commands are completed and the MATLAB prompt reappears.

In general, five events cause MATLAB to render the screen.

1. a return to the MATLAB prompt,
2. encountering a command that temporarily stops execution, such as the commands **pause**, **keyboard**, **input**, and **waitforbuttonpress**,
3. execution of a **getframe** command,
4. execution of a **drawnow** command, and
5. resizing of a *Figure* window.

Of these, the **drawnow** command specifically allows one to force MATLAB to render the screen at arbitrary times.

17.10 THE ZOOM COMMAND

MATLAB provides an interactive tool to expand sections of a 2-D plot to see more detail, or to **zoom in** on a region of interest. The command » `zoom on` turns on zoom mode for the current figure. Clicking the mouse button on a Macintosh, or left mouse button on most other systems, while within the *Figure* window expands the plot by a factor of 2 centering around the point under the mouse pointer. Each time you click, the plot expands. Click the right mouse button on an MS Windows computer or shift-click on the Macintosh to zoom out by a factor of 2. You can also click-and-drag to zoom into the specific area enclosed by your click and zoom rectangle (intuitively called a *rubberband box*). » `zoom out` returns the plot to its initial state. » `zoom off` turns off zoom mode. Issuing `zoom` with no arguments toggles the on/off zoom state of the current *Figure* window.

The on-line help text for the zoom command is

```
» help zoom

ZOOM Zoom in and out on a 2-D plot.
    ZOOM ON turns zoom on for the current figure. Click
    the left mouse button to zoom in on the point under the
    mouse. Click the right mouse button to zoom out
    (shift-click on the Macintosh). Each time you click,
    the axes limits will be changed by a factor of 2 (in or out).
    You can also click and drag to zoom into an area.

    ZOOM OFF turns zoom off. ZOOM with no arguments
    toggles the zoom status. ZOOM OUT returns the plot
    to its initial (full) zoom.
```

Not shown in the above help text are two other forms: » `zoom xon` enables zooms on the x-axis only, and » `zoom yon` enables zooms on the y-axis only.

For those interested in knowing how the **zoom** command works, print a copy of its M-file. The **zoom** command makes use of MATLAB's Handle Graphics features, which are discussed in later chapters in this text. Based on mouse input, **zoom** simply modifies the plot axes through use of the **axis** command.

Since the **legend**, **zoom**, and **ginput** commands respond to mouse clicks in a *Figure* window, they interfere with each other. **As a result, before using zoom, ginput and legend must be turned off.**

17.11 THE GINPUT FUNCTION

In some situations it is convenient to select coordinate points from a plot in a *Figure* window. In MATLAB this feature is embodied in the **ginput** command. The command » `[x,y]=ginput(n)` gets n points from the current plot or subplot based on mouse click positions within the plot or subplot. If you press the **Return** key before all **n** points are selected, `ginput` terminates with fewer points. The points returned in the vectors **x** and **y** are the respective x and y data coordinate points selected. The returned data are not points from the data used to create the plot, but rather the explicit x and y coordinate values where the mouse was clicked. If points are selected outside the plot or subplot axes limits, e.g., outside the plot box, the points returned are extrapolated values based on the axis scaling.

This command can be somewhat confusing when used in a *Figure* window containing subplots. The data are returned with respect to the current or active subplot. Thus, if `ginput` is issued after a `subplot(2,2,3)` command, the data returned are with respect to the axes of the data plotted in `subplot(2,2,3)`. If points are selected from other subplots, the data are still with respect to the axes of the data in `subplot(2,2,3)`. When an unspecified number of data points are desired, the command `[x,y]=ginput` without an input argument can be used. Here data points are gathered until the **Return** key is pressed.

Since the `legend`, `zoom`, and `ginput` commands respond to mouse clicks in a *Figure* window, they interfere with each other. **As a result, before using `ginput`, `zoom` and `legend` must be turned off.**

The `ginput` command has several other features that are described in its on-line help.

```
» help ginput

GINPUT Graphical input from a mouse or cursor.
    [X,Y] = GINPUT(N) gets N points from the current axes and returns
    the X- and Y-coordinates in length N vectors X and Y. The cursor
    can be positioned using a mouse (or by using the Arrow Keys on some
    systems). Data points are entered by pressing a mouse button
    or any key on the keyboard. A carriage return terminates the
    input before N points are entered.

    [X,Y] = GINPUT gathers an unlimited number of points until the
    return key is pressed.

    [X,Y,BUTTON] = GINPUT(N) returns a third result, BUTTON, that
    contains a vector of integers specifying which mouse button was
    used (1,2,3 from left) or ASCII numbers if a key on the keyboard
    was used.
```

The `gtext` function described earlier in this chapter utilizes the function `ginput` along with the function `text` for placing text with the mouse.

17.12 OTHER BASIC 2-D PLOTS

Up to this point the basic plotting function `plot` has been illustrated. In many situations, plotting lines or points on linearly-scaled axes does not convey the desired information. As a result, MATLAB offers several other basic 2-D plotting functions, as well as numerous specialized plotting functions that are embodied in function M-files. These specialized plotting functions are discussed in the next section.

In addition to the basic plotting command `plot`, MATLAB provides the functions `semilogx` for plotting with a logarithmically-scaled x-axis, `semilogy` for plotting with a logarithmically-scaled y-axis, and `loglog` for plotting with both axes logarithmically spaced. All the features discussed previously with respect to the function `plot` apply to these functions as well.

A fundamentally different plotting function is the function `fill`, which draws filled 2-D polygons.

```
» x=[0 1 1 0 0];y=[0 0 1 1 0]; % vertices of a square

» fill(x,y,'y') % plot and fill with yellow

» axis([-1 2 -1 2]) % change axis
```

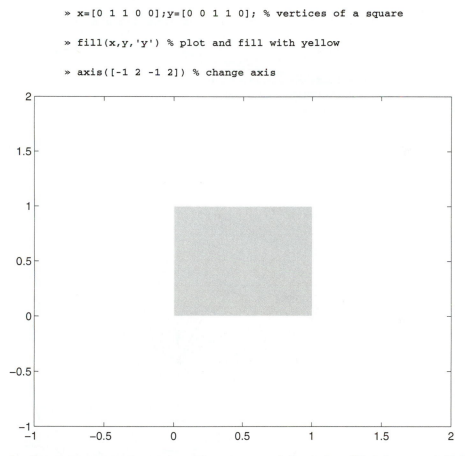

In the above statements, the vertices of a square are defined, then filled. In general, if the polygon is not closed, i.e., its first and last points are not equal, `fill` closes it. As with the plot function, `fill` can have any number of pairs of vertices and associated colors. Moreover, if **x** and **y** are matrices of the same dimension, each column of **x** and **y** is assumed to describe separate polygons.

17.13 SPECIALIZED 2-D PLOT FUNCTIONS

In addition to the basic plotting functions already discussed, MATLAB offers numerous specialized plotting functions that exist as M-file functions. Examples of these plotting functions and their on-line help text include

```
» theta=linspace(0,2*pi);

» rho=sin(2*theta).*cos(2*theta);

» polar(theta,rho,'g') % a polar plot of rho versus angle

» title('Polar Plot of sin(2*theta)cos(2*theta)')

» help polar

 POLAR Polar coordinate plot.
     POLAR(THETA, RHO) makes a plot using polar coordinates of
     the angle THETA, in radians, versus the radius RHO.
     POLAR(THETA,RHO,S) uses the linestyle specified in string S.
     See PLOT for a description of legal linestyles.
```

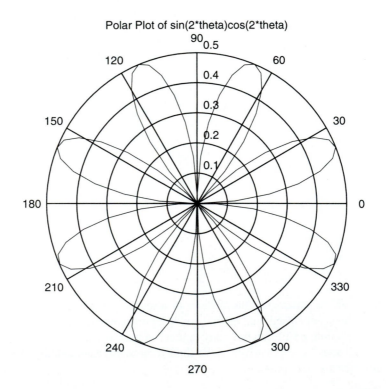

```
» x=linspace(-2.5,2.5,20);

» y=exp(-x.*x);

» bar(x,y) % a Bar chart

» title('Bar Chart of a Bell Curve')

» help bar

  BAR      Bar graph.
       BAR(Y) draws a bar graph of the elements of vector Y. BAR(X,Y)
       draws a bar graph of the elements of vector Y at the locations
       specified in vector X. The X-values must be in ascending order.
       If the X-values are not evenly spaced, the interval chosen is
       not symmetric about each data point. Instead, the bars are drawn
       midway between adjacent X-values. The endpoints simply adopt the
       internal intervals for the external ones needed.

       If X and Y are matrices of the same size, one bar graph per
       column is drawn.

       [XX,YY] = BAR(X,Y) does not draw a graph, but returns vectors
       X and Y such that PLOT(XX,YY) is the bar chart.
       BAR(X,'linetype') or BAR(X,Y,'linetype') uses the plot linetype
       specified. See PLOT for details.
```

Note that the plot on the next page shows unfilled bars. *The Mathworks Inc.* ftp site (see Chapter 23) offers an improved **bar** function that draws filled bars. Perhaps this new version will be standard in the next version of MATLAB.

```
» x=linspace(-2.5,2.5,20);

» y=exp(-x.*x);

» stairs(x,y) % a Stairs or Zero Order Hold Plot

» title('Stairs Plot of a Bell Curve')

» help stairs

  STAIRS Stairstep graph (bar graph without internal lines).
       Stairstep plots are useful for drawing time history plots of
       digital sampled-data systems.
       STAIRS(Y) draws a stairstep graph of the elements of vector Y.
       STAIRS(X,Y) draws a stairstep graph of the elements in vector Y at
       the locations specified in X. The X-values must be in
```

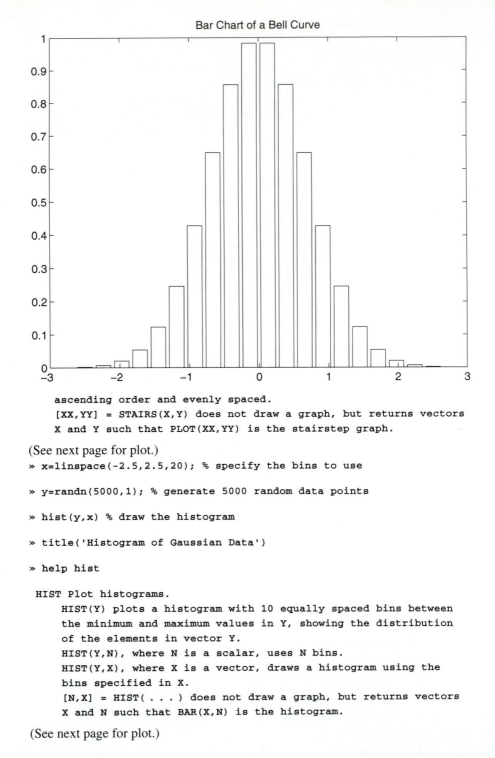

ascending order and evenly spaced.
[XX,YY] = STAIRS(X,Y) does not draw a graph, but returns vectors
X and Y such that PLOT(XX,YY) is the stairstep graph.

(See next page for plot.)

» x=linspace(-2.5,2.5,20); % specify the bins to use

» y=randn(5000,1); % generate 5000 random data points

» hist(y,x) % draw the histogram

» title('Histogram of Gaussian Data')

» help hist

  HIST Plot histograms.
      HIST(Y) plots a histogram with 10 equally spaced bins between
      the minimum and maximum values in Y, showing the distribution
      of the elements in vector Y.
      HIST(Y,N), where N is a scalar, uses N bins.
      HIST(Y,X), where X is a vector, draws a histogram using the
      bins specified in X.
      [N,X] = HIST(...) does not draw a graph, but returns vectors
      X and N such that BAR(X,N) is the histogram.

(See next page for plot.)

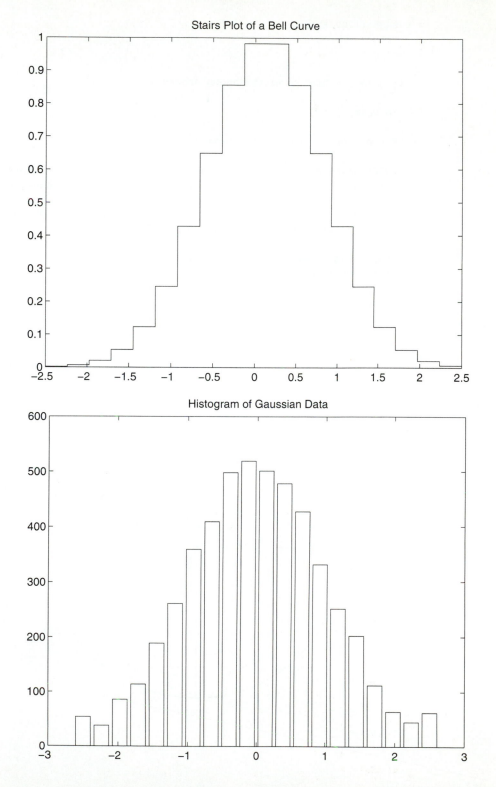

```
» t=randn(1000,1)*pi;

» rose(t)

» title('Angle Histogram of Random Angles')

» help rose
```

 ROSE Plot rose or angle histogram.

     ROSE(THETA) plots the angle histogram for the angles in THETA.
     The angles in the vector THETA must be specified in radians.

     ROSE(THETA,N) where N is a scalar, uses N equally spaced bins
     from 0 to 2*PI. The default value for N is 20.

     ROSE(THETA,X) where X is a vector, draws the histogram using the
     bins specified in X.

     [T,R] = ROSE(. . .) returns the vectors T and R such that
     POLAR(T,R) is the histogram. No plot is drawn.

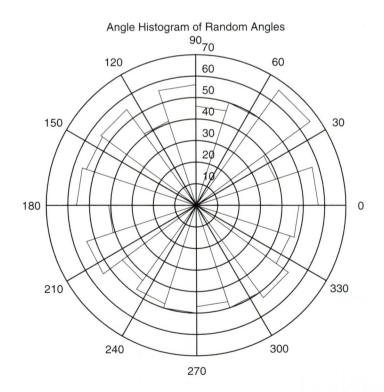

Angle Histogram of Random Angles

```
» y=randn(50,1); % create some random data

» stem(y,':') % draw a stem plot using dotted linestyle

    » title('Stem Plot of Random Data')

    » help stem

      STEM Plot discrete sequence data.
            STEM(Y) plots the data sequence Y as stems from the x-axis
            terminated with circles for the data value.
            STEM(X,Y) plots the data sequence Y at the values specfied
            in X.
            There is an optional final argument to specify a line-type
            for the stems of the data sequence. E.g. STEM(X,Y,'-.') or
            STEM(Y,':').
```

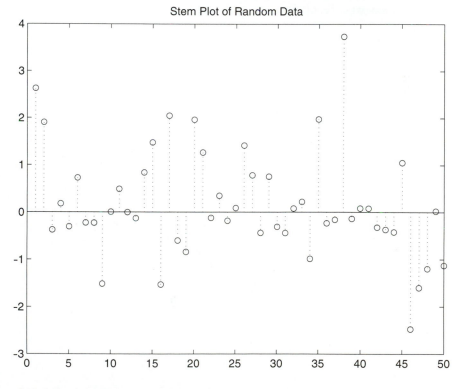

```
» x=0:0.1:2; % create a vector

» y=erf(x); % y is the error function of x

» e=rand(size(x))/10; % e contains random error values
```

» errorbar(x,y,e) % create the plot

» title('Errorbar Plot')

» help errorbar

ERRORBAR Plot graph with error bars.
    ERRORBAR(X,Y,L,U,SYMBOL) plots the graph of vector X vs. vector Y with
    error bars specified by the vectors L and U. The vectors X,Y, L and U
    must be the same length. If X,Y, L and U are matrices then each column
    produces a separate line. The error bars are each drawn a distance
    of U(i) above and L(i) below the points in (X,Y) so that each bar
    is L(i) + U(i) long. SYMBOL is a string that controls the linetype,
    plotting symbols and color for the X-Y plot.
    ERRORBAR(X,Y,L) plots X versus Y with an error bar that is symmetric
    about Y and is 2*L(i) long.

    ERRORBAR(Y,L) plots Y with error bars [Y-L Y+L].

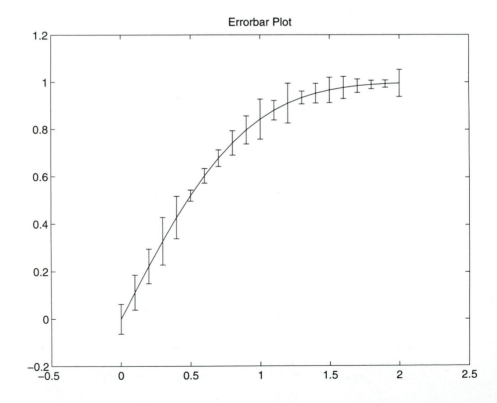

```
» z=eig(randn(20,20));
```

```
» compass(z)
```

```
» title('Compass Plot of Eigenvalues of a Random Matrix')
```

```
» help compass
```

 COMPASS Compass plot.
    COMPASS(Z) draws a graph that displays the angle and magnitude
    of the complex elements of Z as arrows emanating from the origin.

    COMPASS(X,Y) is equivalent to COMPASS(X+i*Y). It displays the
    compass plot for the angles and magnitudes of the elements of
    matrices X and Y.

    COMPASS(Z,'S') and COMPASS(X,Y,'S') use line style 'S' where
    'S' is any legal linetype as described under the PLOT command.

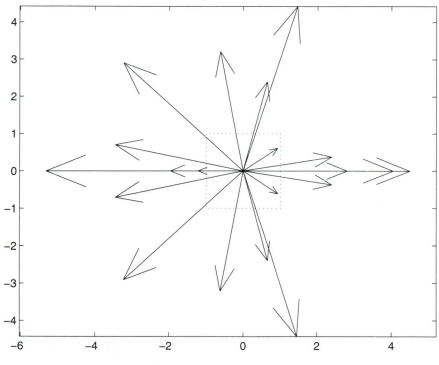

Compass Plot of Eigenvalues of a Random Matrix

```
    » z=eig(randn(20,20));
```

```
    » feather(z)
```

```
» title('Feather Plot of Eigenvalues of a Random Matrix')

» help feather

 FEATHER Feather plot.
      FEATHER(Z) draws a graph that displays the angle and magnitude
      of the complex elements of Z as arrows emanating from equally
      spaced points along a horizontal axis.

      FEATHER(X,Y) is equivalent to FEATHER(X+i*Y). It displays the
      feather plot for the angles and magnitudes of the elements of
      matrices X and Y.

      FEATHER(Z,'S') and FEATHER(X,Y,'S') use line style 'S' where
      'S' is any legal linetype as described under the PLOT command.
```

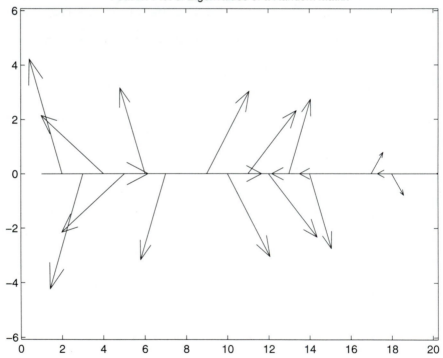

Finally, it is possible to display images. In MATLAB images are stored in matrices, with each matrix element identifying the color of its corresponding pixel. The MATLAB function **image** displays images. For example,

```
» load clown
» image(X), colormap(map), axis off
```

displays the clown image shown in color plate number 12 in the *MATLAB User's Guide*. For further information see the *MATLAB User's Guide*.

17.14 M-FILE EXAMPLES

Although the graphics features in MATLAB are extensive and flexible, MATLAB offers essentially unlimited access to its underlying graphics properties. The functions that provide this access are called ***Handle Graphics*** and are thoroughly discussed in Chapters 20 and 21. By using these functions, it is possible to modify any graphical image in just about any way imaginable. It is also possible to add graphics functions that don't exist as standard MATLAB functions.

In this section a number of graphics utility functions from the *Mastering MATLAB Toolbox* will be illustrated. Most of these utilize the Handle Graphics capabilities of MATLAB and will be examined in detail in Chapters 20 and 21. Here only their syntax and uses will be shown. Before considering each function individually, they are summarized in the table below.

Mastering MATLAB Toolbox Plotting Utilities	
`mmaxes prop value ...`	Modify properties of plot axes
`mmcxy (or) xy=mmcxy`	Show x-y coordinates of mouse over a plot
`mmdraw prop value ...`	Draw straight lines on a plot
`mmfill(x,y,z,c,lb,ub)`	Fill area between two curves
`mmgetxy(N)`	Get x-y coordinates using the mouse
`mmline prop value ...`	Modify properties of plotted lines
`mmtile`	Tile multiple *Figure* windows
`mmtext('optional text')`	Place and/or drag text on a plot
`mmzoom`	Zoom axis in using a rubberband box
`mmzap object`	Delete text, lines, or axes using the mouse
`mmfont prop value`	Modify font properties of text

The help text for the function `mmfill` is as follows:

```
» help mmfill

  MMFILL Fill plot of area between two curves.
   MMFILL(X,Y,Z,C,LB,UB) plots y=f(x) and z=g(x) and fills the
   area between the two curves from LB<= X <=UB with colorspec C.
```

```
X,Y and Z are data vectors of the same length.
Missing arguments take on default values. Examples:
MMFILL(X,Y) fills area under y=f(x) with red.
MMFILL(X,Y,C) fills area under y=f(x) with colorspec C.
MMFILL(X,Y,Z) fills area between y=f(x) and z=g(x) with red.
MMFILL(X,Y,Z,C) fills area between y=f(x) and z=g(x) with C.
MMFILL(X,Y,LB,UB) fills area under y=f(x) in red between bounds.
MMFILL(X,Y,C,LB,UB) fills area under y=f(x) with C between bounds.
MMFILL(X,Y,Z,LB,UB) fills area between curves with red between bounds.

A=MMFILL( ... ) returns the approximate area filled by calling TRAPZ.
```

mmfill does not utilize Handle Graphics functions. It simply automates the process of creating the polygon indices required for the specialized plotting function **fill**. Consider the following examples of its use:

```
» x=linspace(0,1);

» y=sin(2*pi*x);

» z=cos(3*pi*x);

» mmfill(x,y,.3,.5) % fill area under y for .3 <= x <= .5
```

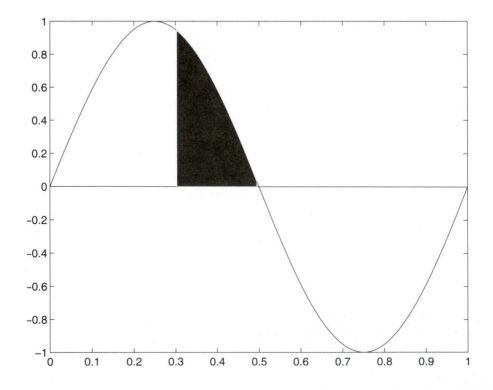

```
» mmfill(x,y,z,'y',-inf,.3) % fill area with yellow up to x=.3
```

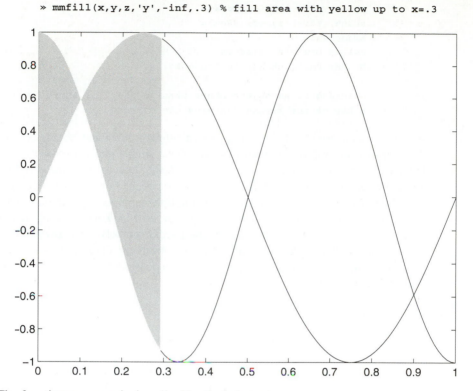

The function **mmaxes** is described by the help text.

```
» help mmaxes

 MMAXES Set Axes Properties Using Mouse.
  MMAXES waits for a mouse click on an axes then
  applies the desired properties to the selected axes.
  Properties are given in pairs, e.g. MMAXES name value ...
  Properties:
  NAME VALUE {default}
  box [{on} off] for axes bounding box
  color [y m c r g b {w} k]
  width [points] for axes linewidth {0.5}
  visible    [{on} off]    for axes and grid visibility
  name       [any valid font name]  font for tick labels
  size       [size in points {12}]  font size for tick labels
  tdir {in} out] for tick direction
  xtick [off {on}] to hide X-axis labels, nonreversible
  xdir [{norm} rev] for X-axis direction (norm)
  xgrid [on {off}] for X-axis grid
  xscale [{lin} log] for X-axis scaling
  zap (n.a.) to delete axes and plot contents

  xtick, xdir, xgrid, xscale have y and z axis counterparts:
```

```
ytick, ydir, ygrid, yscale
ztick, zdir, zgrid, zscale
Examples:
MMAXES box off ygrid on
MMAXES tdir out zscale log color y

Clicking on an object other than an axes, or striking
a key on the keyboard aborts the command.
```

As can be seen from the help text, **mmaxes** offers options and features not found using the standard function **axis**. **mmaxes** provides a simple and convenient interface to the Handle Graphics function **axes** (not **axis**!). Rather than apply to the current set of axes, **mmaxes** applies the desired properties to the axis clicked on using the mouse. Therefore, if multiple *Figure* windows are open or if multiple subplots exist, the user can choose to which set of axes to apply the desired properties. In MATLAB clicking on axis with the mouse means placing the mouse pointer near (within 5 pixels or so) any part of an axis (including the numbers at the tick marks) but away from any lines plotted within the axis.

When lines already exist on a plot, it is also possible to change their properties using the function **mmline**.

```
» help mmline

  MMLINE Set Line Properties Using Mouse.
   MMLINE waits for a mouse click on a line then
   applies the desired properties to the selected line.
   Properties are given in pairs, e.g., MMLINE name value ...
   Properties:
   NAME VALUE {default}
   color [y m c r g b w k] or an rgb in quotes: '[r g b]'
   style [- — : -.]
   mark [o + . * x)]
   width points for linewidth {0.5}
   size points for marker size (6)
   zap (n.a.) delete selected line
   Examples:
   MMLINE color r width 2 sets color to red and width to 2 points
   MMLINE mark + size 8 sets marker type to + and size to 8 points
   MMLINE color '[1.50]' sets color to orange

   Clicking on an object other than a line, or striking
   a key on the keyboard aborts the command.
```

This function operates similar to **mmaxes** described above, except here it is line properties rather than axes properties that are modified. Similar to **mmaxes**, **mmline** provides a simple and convenient interface to the Hande Graphics function **line**.

The function **mmdraw** takes the capabilities of **mmline** above and adds line-drawing capability.

```
» help mmdraw

 MMDRAW Draw a Line and Set Properties Using Mouse.
  MMDRAW draws a line in the current axes using the mouse,
  Click at the starting point and drag to the end point.
  In addition, properties can be given to the line.
  Properties are given in pairs, e.g., MMDRAW name value ...
  Properties:
  NAME   VALUE          {default}
  color [y m c r g b {w} k] or an rgb in quotes: '[r g b]'
  style [- — {:} -.]
  mark  [o + . * x)]
  width points for linewidth {0.5}
  size  points for marker size {6}
  Examples:
  MMDRAW color r width 2 sets color to red and width to 2 points
  MMDRAW mark + size 8 sets marker type to + and size to 8 points
  MMDRAW color '[1.50]' set color to orange
```

This function shows how the Handle Graphics features in MATLAB can be used to implement a line-drawing function common to computer drawing or drafting programs. The properties available in this function correspond to those available in the Handle Graphics function **line**, which is called internally by high-level functions such as **plot**.

The function **mmcxy** is described by the help text.

```
» help mmcxy

 MMCXY Show x-y Coordinates Using Mouse.
  MMCXY places the x-y coordinates of the mouse in the
  lower left hand corner of the current 2-D figure window.
  When the mouse is clicked, the coordinates are erased.
  XY=MMCXY returns XY=[x,y] coordinates where mouse was clicked.
  XY=MMCXY returns XY=[] if a key press was used.
```

This function again uses Handle Graphics features in MATLAB. **mmcxy** allows one to *see* the x-y coordinates where the mouse pointer is located in a plot and optionally return the coordinates where the mouse is clicked.

The function **mmgetxy** uses the function **mmcxy** to mimic some of the features of the standard MATLAB function **ginput**. The M-file **mmgetxy.m** is as follows:

```
function xy=mmgetxy(n)
%MMGETXY Graphical Input Using Mouse.
% XY=MMGETXY(N) gets N points from the current axes at
% points selected with a mouse button press.
% XY=[x,y] matrix having 2 columns and N rows.
```

```
% Striking ANY key on the keyboard aborts the process.
% XY=MMGETXY gathers any number of points until a
% key on the keyboard is pressed.

% Copyright (c) 1996 by Prentice-Hall, Inc.

if nargin==0,n=1000;end
xy=[];
for i=1:n
    tmp=mmcxy;
    if isempty(tmp)
        return
    else
        xy=[xy;tmp];
    end
end
```

Based on the help text at the start of the M-file, use of **mmgetxy** is straightforward and very similar to **ginput**. Here, **mmgetxy** simply calls the preceding function **mmcxy**. The advantage of using **mmgetxy** over **ginput** is that the numerical coordinates of the selected point(s) are displayed on the graphics screen during the selection process.

As an alternative to the standard MATLAB function **gtext**, which places text at a point selected by the mouse, **mmtext** places text in the current figure and then allows one to drag it to the desired place with the mouse. The help text for **mmtext** is

```
» help mmtext

MMTEXT Place and drag text with mouse.
 MMTEXT waits for a mouse click on a text object
 in the current figure then allows it to be dragged
 while the mouse button remains down.
 MMTEXT('whatever') places the string 'whatever' on
 the current axes and allows it to be dragged.

 MMTEXT becomes inactive after the move is completed or
 no text object is selected.
```

Additionally, with no input arguments, **mmtext** allows one to drag text, including standard axis labels and plot titles, that already exists in the current figure.

When multiple *Figure* windows are created, by default they are placed on top of each other in the upper middle region of the computer screen. To uncover buried *Figure* windows, the mouse can be used to drag them around the screen. Alternatively, Handle Graphics can be used to place individual *Figure* windows in different locations so that they can all be seen simultaneously. One *Mastering MATLAB Toolbox* solution is the function **mmtile**.

```
» help mmtile

 MMTILE Tile Figure Windows.
  MMTILE with no arguments, tiles the current figure windows
  and brings them to the foreground.
  Figure size is adjusted so that 4 figure windows fit on the screen.
  Figures are arranged in a clockwise fashion starting in the
  upper left corner of the display.
  MMTILE(N) makes tile N the current figure if it exists.
  Otherwise, the next tile is created for subsequent plotting.

  Tiled figure windows are titled TILE #1, TILE #2, TILE #3, TILE #4.
```

As described above, **mmtile** tiles up to four *Figure* windows within the computer screen. On most screens it shrinks the default *Figure* window size so that four tiles fit on the screen. Because of slight differences between computer platforms, the tiled windows might be slightly overlapped or slightly separated. If this is a problem, simply open the M-file and adjust the first two parameters, **HT** and **WD**.

The **zoom** function described earlier in this chapter is extremely powerful. It is also a very complex and long M-file function. As a result, the function **mmzoom** was written as a simple application of the Handle Graphics function **rbbox**, which draws the dynamic rubberband box that appears when the **zoom** function is called upon to zoom into a specific area of the display. The help text for **mmzoom** is

```
» help mmzoom

  MMZOOM Simple 2-D Zoom-In Function Using RBBOX.
   MMZOOM zooms in on a plot based on the size of a
   rubberband box drawn by the user with the mouse.
   MMZOOM x zooms the x-axis only.
   MMZOOM y zooms the y-axis only.
   MMZOOM reset or
   MMZOOM out restores original axis limits.
   Striking a key on the keyboard aborts the command.
   MMZOOM becomes inactive after zoom is complete or aborted.
```

From the help text it is clear that **mmzoom** is similar to **zoom**, but has fewer features. **mmzoom** only allows one to zoom using the rubberband box approach. Furthermore, **mmzoom** becomes inactive, i.e., shuts itself off, as soon as one zoom is complete.

The final *Mastering MATLAB Toolbox* utility described here provides the ability to delete various graphics objects. This utility, **mmzap**, is described by the help text.

```
» help mmzap

 MMZAP Delete graphics object using mouse.
  MMZAP waits for a mouse click on an object in
  a figure window and deletes the object.
```

```
MMZAP or MMZAP text erases text objects.
MMZAP axes erases axes objects.
MMZAP line erases line objects.
MMZAP surf erases surface objects.
MMZAP patch erases patch objects.

Clicking on an object other than the selected type or striking
a key on the keyboard aborts the command.
```

mmzap works in a similar way to **mmaxes** and **mmline** in that the mouse is used to click on the desired object to be deleted.

The functions described in this section form a core group of useful graphics utilities. In later chapters on Handle Graphics, the contents of some of these functions are described in more detail as a means to show how Handle Graphics features are used.

17.15 SUMMARY

The functions and features considered in this chapter are summarized in the following tables.

<table>
<tr><td colspan="2" align="center">**2-D Plotting Functions**</td></tr>
<tr><td>`bar`</td><td>Bar graph</td></tr>
<tr><td>`compass`</td><td>Plot of vectors emanating from the origin</td></tr>
<tr><td>`errorbar`</td><td>Linear plot with errorbars</td></tr>
<tr><td>`feather`</td><td>Plot of vectors emanating from a line</td></tr>
<tr><td>`fill`</td><td>Draw filled 2-D polygons (basic)</td></tr>
<tr><td>`hist`</td><td>Histogram plot</td></tr>
<tr><td>`image`</td><td>Display image</td></tr>
<tr><td>`loglog`</td><td>Plot with both axes scaled logarithmically (basic)</td></tr>
<tr><td>`plot`</td><td>Simple linear plot (basic)</td></tr>
<tr><td>`polar`</td><td>Plot in polar coordinates</td></tr>
<tr><td>`rose`</td><td>Angle histogram plot</td></tr>
<tr><td>`semilogx`</td><td>Plot with x-axis scaled logarithmically (basic)</td></tr>
<tr><td>`semilogy`</td><td>Plot with y-axis scaled logarithmically (basic)</td></tr>
<tr><td>`stairs`</td><td>Staircase or zero-order-hold plot</td></tr>
<tr><td>`stem`</td><td>Discrete sequence plot</td></tr>
</table>

Basic Plot Line Types and Colors

Symbol	Color	Symbol	Linestyle
y	yellow	.	point
m	magenta	o	circle
c	cyan	x	x-mark
r	red	+	plus
g	green	*	star
b	blue	–	solid line
w	white	:	dotted line
k	black	–.	dash-dot line
		– –	dashed line

2-D Plotting Tools

`axis`	Modify axis properties
`clf`	Clear *Figure* windows
`close`	Close *Figure* windows
`figure`	Create or select *Figure* windows
`grid`	Place grid
`gtext`	Place text with mouse
`hold`	Hold current plot
`subplot`	Create subplots within a *Figure* window
`text`	Place text at given location
`title`	Place title
`xlabel`	Place x-axis label
`ylabel`	Place y-axis label
`zoom`	Magnify or shrink plot axes

`axis` **Command**

`axis([xmin xmax ymin ymax])`	Set the maximum and minimum values of the axes using values given in the row vector. If **xmin** or **ymin** is set to **-inf**, the minimum is autoscaled. If **xmax** or **ymax** is set to **inf**, the maximum is autoscaled.
`axis auto` `axis('auto')`	Return the axis scaling to its automatic defaults: **xmin=min(x)**, **xmax=max(x)**, etc.
`axis(axis)`	Freeze scaling at the current limits, so that if **hold** is turned on, subsequent plots use the same axis limits.
`axis xy` `axis('xy')`	Use the (default) *Cartesian* coordinate form, where the *system origin* (the smallest coordinate pair) is at the lower left corner. The horizontal axis increases left to right, and the vertical axis increases bottom to top.
`axis ij` `axis('ij')`	Use the *matrix* coordinate form, where the *system origin* is at the top left corner. The horizontal axis increases left to right, but the vertical axis increases top to bottom.
`axis square` `axis('square')`	Set the current plot to be a square rather than the default rectangle.
`axis equal` `axis('equal')`	Set the scaling factors for both axes to be equal.
`axis image` `axis('image')`	Set the aspect ratio and the axis limits so the image in the current axes has square pixels.
`axis normal` `axis('normal')`	Turn off **axis equal** and **axis square.**
`axis off` `axis('off')`	Turn off all axis labeling, grid, and tic marks. Leave the title and any labels placed by the text and **gtext** commands.

`axis on`	Turn on axis labeling, tic marks,
`axis('on')`	and grid.
`v=axis`	Return the current axis limits in the vector **v**.

Mastering MATLAB Toolbox Plotting Utilities

`mmaxes prop value ...`	Modify properties of plot axes
`mmcxy` (or) `xy=mmcxy`	Show x-y coordinates of mouse over a plot
`mmdraw prop value ...`	Draw straight lines on a plot
`mmfill(x,y,z,c,lb,ub)`	Fill area between two curves
`mmgetxy(N)`	Get x-y coordinates using the mouse
`mmline prop value ...`	Modify properties of plotted lines
`mmtile`	Tile multiple *Figure* windows
`mmtext('optional text')`	Place and/or drag text on a plot
`mmzoom`	Zoom axis in using a rubberband box
`mmzap object`	Delete text, lines, or axes using the mouse
`mmfont prop value`	Modify font properties of text

18

3-D Graphics

MATLAB provides a variety of functions to display 3-D data. Some functions plot lines in three dimensions, while others draw surfaces and wire frames. In addition, color can be used to represent a fourth dimension. When color is used in this manner, it is called *pseudocolor* since color is not an inherent or natural property of the underlying data in the way that color in a photograph is a natural characteristic of the information displayed. To simplify the discussion of 3-D graphics, the use of color is postponed until the next chapter. In this chapter, the fundamental concepts of producing useful 3-D plots is discussed.

18.1 THE `PLOT3` FUNCTION

The **plot3** command extends the features of the 2-D **plot** function into three dimensions. The format is the same as the 2-D **plot**, except the data include information for the third dimension, i.e., the z-direction. The general syntax of **plot3** is **plot3($x_1, y_1, z_1, S_1, x_2, y_2, z_2, S_2, \ldots$)**, where x_n, y_n, and z_n are vectors or matrices, and S_n's are optional character strings specifying color, marker symbol, and/or linestyle.

232

Commonly **plot3** is used to plot a 3-dimensional function of a single variable. An example of a 3-D helix follows:

```
» t=0:pi/50:10*pi;

» plot3(sin(t),cos(t),t)

» title('Helix'), xlabel('sin(t)'), ylabel('cos(t)'), zlabel('t')

» text(0,0,0,'Origin')

» grid

» v=axis
v =
     -1     1    -1     1     0    40
```

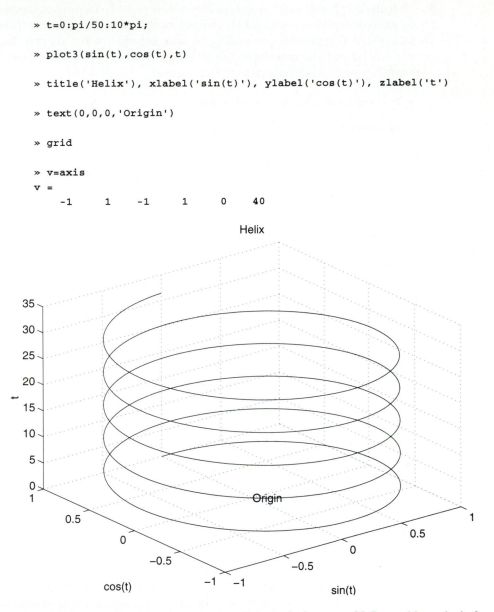

From this simple example it is apparent that all the basic features of 2-D graphics exist in 3-D also. The **axis** command extends to 3-D by returning the z-axis limits (0 and 40)as two additional elements in the axis vector. There is a **zlabel** function for labeling the z-axis. **grid** draws a 3-D grid underneath the plot. The function **text(x,y,z,'string')**

places a character string at the postion identified by the triplet **x,y, z**. In addition, subplots and multiple *Figure* windows apply directly to 3-D graphics functions.

In the last chapter, multiple lines or curves were plotted on top of one another by specifying multiple arguments to the **plot** command or by using the **hold** command. **plot3** and the other 3-D graphics functions offer the same capabilities. For example, the added dimension of **plot3** allows multiple 2-D plots to be stacked next to one another along one dimension, rather than directly on top of one another.

```
» x=linspace(0,3*pi); % x-axis data

» z1=sin(x);      % plot in x-z plane

» z2=sin(2*x);

» z3=sin(3*x);

» y1=zeros(size(x));     % spread out along y-axes

» y3=ones(size(x));      % by giving each curve different y-axis values

» y2=y3/2;

» plot3(x,y1,z1,x,y2,z2,x,y3,z3)

» grid, xlabel('X-axis'), ylabel('Y-axis'), zlabel('Z-axis')

» title('sin(x), sin(2x), sin(3x)')
```

(See top of next page for plot.)
The above can also be stacked in other directions, e.g.,

```
» plot3(x,z1,y1,x,z2,y2,x,z3,y3)

» grid, xlabel('X-axis'), ylabel('Y-axis'), zlabel('Z-axis')

» title('sin(x), sin(2x), sin(3x)')
```

(See bottom of next page for plot.)

18.2 CHANGING VIEWPOINTS

Note from the two plots that you are looking down at the z=0 plane at an angle of 30 degrees and that you are looking up at the x=0 plane at an angle of 37.5 degrees. This is the default viewpoint for all 3-D graphics. The angle of orientation with respect to the z=0 plane is called *elevation,* and the angle with respect to the x=0 plane is called *azimuth.* Thus the default 3-D viewpoint is elevation = 30 degrees and azimuth = −37.5 degrees. The default 2-D viewpoint is elevation = 90 degrees and azimuth = 0 degrees. The concepts of azimuth and elevation are described visually on page 236.

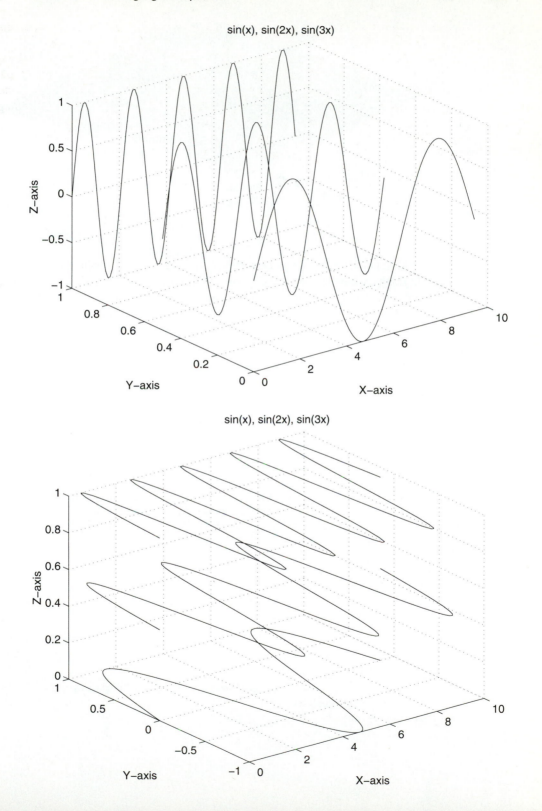

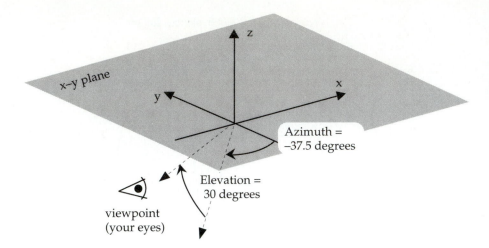

In MATLAB, the function **view** changes the graphical viewpoint for all types of 2-D and 3-D graphs. **view(az,el)** and **view([az,el])** change the viewpoint to the specified azimuth **az** and elevation **el**. Consider the following examples in script M-file form:

```
% viewpoint example using subplots

x=linspace(0,3*pi).';

Z=[sin(x) sin(2*x) sin(3*x)]; % create Y and Z axes as matrices

Y=[zeros(size(x)) ones(size(x))/2 ones(size(x))];

subplot(2,2,1)
plot3(x,Y,Z)  % plot3 works with column-oriented matrices too
grid, xlabel('X-axis'), ylabel('Y-axis'), zlabel('Z-axis')
title('Default Az = -37.5, El = 30')
view(-37.5,30)

subplot(2,2,2)
plot3(x,Y,Z)
grid, xlabel('X-axis'), ylabel('Y-axis'), zlabel('Z-axis')
title('Az Rotated to 52.5')
view(-37.5+90,30)

subplot(2,2,3)
plot3(x,Y,Z)
grid, xlabel('X-axis'), ylabel('Y-axis'), zlabel('Z-axis')
title('El Increased to 60')
view(-37.5,60)
```

```
subplot(2,2,4)
plot3(x,Y,Z)
grid, xlabel('X-axis'), ylabel('Y-axis')
title('Az = 0, El = 90')
view(0,90)
```

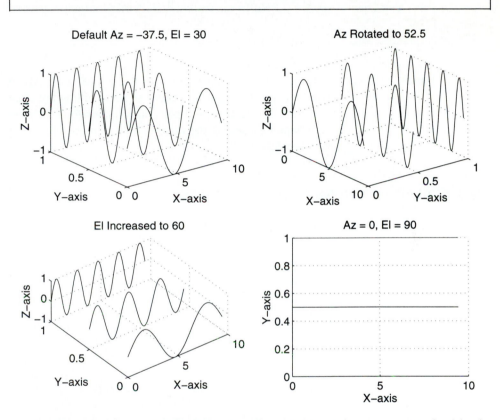

In addition to the above form, **view** also offers additional features summarized in the table below.

view **Function**	
view(az,el) view([az,el])	Set view to azimuth **az** and elevation **el**
view([x,y,z])	Set view looking toward the origin along the vector **[x,y,z]** in Cartesian coordinates, e.g., view([0 0 1])=view(0,90)

`view(2)`	Sets the default 2-D view, `az = 0, el = 90`
`view(3)`	Sets the default 3-D view, `az = -37.5, el = 30`
`[az,el]=view`	Returns the current azimuth `az` and elevation `el`
`view(T)`	Uses the 4-by-4 transformation matrix `T` to set the view
`T=view`	Returns the current 4-by-4 transformation matrix

Finally, as a preview showing the Handle Graphics capabilities of MATLAB, the *Mastering MATLAB Toolbox* contains the function **mmview3d**. Calling this function, i.e., » **mmview3d**, after generating a 2-D or 3-D plot, places azimuth and elevation sliders (scroll bars) on the current plot for setting the viewpoint. See on-line help for more information about using **mmview3d**.

18.3 SCALAR FUNCTIONS OF TWO VARIABLES

As opposed to line plots generated with **plot3**, it is often desirable to visualize a scalar function of two variables, i.e.,

$$z = f(x,y)$$

Here each pair of values for x and y produces a value for z. A plot of z as a function of x and y is a surface in 3-dimensions. To plot this surface in MATLAB, the values for z are stored in a matrix. As described in the section on 2-dimensional interpolation, given that **x** and **y** are the independent variables, **z** is a matrix of the dependent variable, and the association of **x** and **y** to **z** is

$$z(i,:) = f(x,y(i)) \text{ and } z(:,j) = f(x(j),y).$$

That is, the ith row of **z** is associated with i^{th} element of **y**, and the jth column of **z** is associated with j^{th} element of **x**. In other words, **y** varies down the columns of **z**, and **x** varies across the rows of **z**.

When z = f(x,y) can be expressed simply, it is convenient to use array operations to compute all the values of z in a single statement. To do so requires that we create matrices of all x and y values in the proper orientation. (This orientation is sometimes called *plaid* by *The Mathworks Inc.*) MATLAB provides the function **meshgrid** to perform this step.

```
» x=-3:3; % choose x-axis values

» y=1:5;   % y-axis values

» [X,Y]=meshgrid(x,y)
```

```
X =

        -3      -2      -1      0      1      2      3
        -3      -2      -1      0      1      2      3
        -3      -2      -1      0      1      2      3
        -3      -2      -1      0      1      2      3
        -3      -2      -1      0      1      2      3
Y =

         1       1       1      1      1      1      1
         2       2       2      2      2      2      2
         3       3       3      3      3      3      3
         4       4       4      4      4      4      4
         5       5       5      5      5      5      5
```

As you can see, **meshgrid** duplicated **x** for each of the five rows in **y**. Similarly it duplicated **y** as a column for each of the seven columns in **x**. This orientation agrees with the earlier statement where **y** varies down its columns and **x** varies across its rows. Given **X** and **Y**, if $z = f(x,y) = (x + y)^2$, then the matrix of data values defining the 3-dimensional surface is

```
» Z=(X+Y).^2
Z =

         4       1       0      1      4      9     16
         1       0       1      4      9     16     25
         0       1       4      9     16     25     36
         1       4       9     16     25     36     49
         4       9      16     25     36     49     64
```

When a function cannot be expressed simply as shown above, one must use For Loops or While Loops to compute the elements of **Z**. In many cases it may be possible to compute the elements of **Z** row-wise or column-wise. For example, if it is possible to compute **Z** row-wise, the following script file fragment can be helpful:

```
x= ??? % statement defining vector of x-axis values
y= ??? % statement defining vector of y-axis values
nx=length(x); % length of x is no. of rows in Z
ny=length(y); % length of y is no. of columns in Z

Z=zeros(nx,ny); % initialize Z matrix for speed

for r=1:nx
      {preliminary commands}
      Z(r,:)= {a function of y and x(r) defining r-th row of Z}
end
```

On the other hand, if **z** can be computed column-wise, the following script file fragment can be helpful:

```
x= ??? % statement defining vector of x-axis values
y= ??? % statement defining vector of y-axis values

nx=length(x); % length of x is no. of rows in Z
ny=length(y); % length of y is no. of columns in Z

Z=zeros(nx,ny); % initialize Z matrix for speed

for c=1:ny
      {preliminary commands}
      Z(:,c)= {a function of y(c) and x defining c-th column of Z}
end
```

Only when the elements of **z** must be computed element-by-element does the computation usually require a nested For Loop such as the following script file fragment.

```
x= ??? % statement defining vector of x-axis values
y= ??? % statement defining vector of y-axis values

nx=length(x); % length of x is no. of rows in Z
ny=length(y); % length of y is no. of columns in Z

Z=zeros(nx,ny); % initialize Z matrix for speed

for r=1:nx
    for c=1:ny
        {preliminary commands}
        Z(r,c)= {a function of y(c) and x(r) defining z(r,c)}
    end
end
```

18.4 INTERPOLATION OF ROUGH OR SCATTERED DATA

In some circumstances, the values of a scalar function of two variables, $z=f(x,y)$, either cannot be easily computed, are given nonuniformly spaced (at worst random) values of x and y, or are defined using a different coordinate system, e.g., a nonrectangular grid. When any of these cases appear, the MATLAB function **griddata** is useful for generating uniformly-spaced, interpolated values for plotting. First, consider the previous example and assume that higher resolution is desired, but we do not wish to reevaluate the function to compute the new values.

```
» x=-3:3;  % original x-axis values

» y=1:5;   % original y-axis values

» [X,Y]=meshgrid(x,y);  % create plaid data matrices

» Z=(X+Y).^2;  % original z values

» size(Z)  % original array size
ans =
     5     7

» xi=-3:.5:4;  % interpolated x-axis values

» length(xi)  % get new x-axis length
ans =
     15

» yi=0:.2:5;  % interpolated y-axis values

» length(yi)  % get new y-axis length
ans =
     26

» [Xi,Yi]=meshgrid(xi,yi);  % make new data plaid

» Zi=griddata(X,Y,Z,Xi,Yi);  % interpolated Z data using original data

» size(Zi)  % interpolated size is correct
ans =
     26    15
```

Here **griddata** takes the three original matrices **X**, **Y**, **Z**, as well as plaid matrices of the desired interpolation values, and creates a new dependent variable matrix **Zi**. Note that the interpolation values need not be within those of the original data, e.g., **x** varies between **-3** and **3**, whereas **xi** varies between **-3** and **4**.

As opposed to 2-D interpolation described in Chapter 11, `griddata` also works when the data are not monotonic or are irregularly spaced. For example, consider random data.

```
» x=2*rand(1,20);  % nonmonotonic x-axis

» y=4*rand(1,20)-2;  % nonmonotonic y-axis

» [X,Y]=meshgrid(x,y);  % make data plaid

» Z=(X+Y).^2;  % compute function

» xi=linspace(0,2,50);  % interpolated monotonic x-axis values

» yi=linspace(-2,2,30);  % interpolated monotonic y-axis values

» [Xi,Yi]=meshgrid(xi,yi);  % make data plaid

» Zi=griddata(X,Y,Z,Xi,Yi);  % interpolate on monotonic plaid data
```

Here, random data are interpolated to give monotonic data that are more useful for plotting. This is particularly true for the contour plots discussed later in this chapter, which require data defined on uniformly-spaced grids.

In both of the examples shown above, it is easier to simply recompute the function of interest with the desired interpolated values. As a general rule, when it is easy to recompute the function of interest, do so and avoid using `griddata`. On the other hand, if the function of interest is unavailable or requires a great deal of computational effort, `griddata` provides a tool for generating estimates of the underlying function at uniformly-spaced interpolated data points.

18.5 MESH PLOTS

MATLAB defines a *mesh* surface by the **z**-coordinates of points above a rectangular grid in the **x-y** plane. MATLAB forms the mesh plot by joining adjacent points with straight lines. The result looks like a fishing net with knots at the data points. For example, using the MATLAB **peaks** function, which describes a simple surface,

```
» [X,Y,Z]=peaks(30);

» mesh(X,Y,Z)

» grid, xlabel('X-axis'), ylabel('Y-axis'), zlabel('Z-axis')

» title('MESH of PEAKS')
```

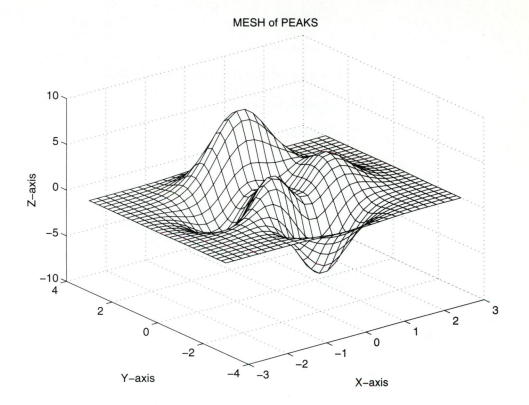

Note on your monitor how the line colors are related to the height of the mesh. In general, **mesh** will accept optional arguments to control color use in the plot. This ability to change how MATLAB uses color will be discussed in the next chapter. In any case, this use of color is called pseudocolor since color is used to add a fourth effective dimension to the graph.

In addition to the above input arguments, **mesh** and most 3-D plot functions can also be called with a variety of input arguments. The syntax used here is the most specific in that information for all three axes is given. The most common variation is to use the vectors that were passed to **meshgrid** for the x- and y- axes, e.g., **mesh(x,y,Z)**. For information regarding other syntax forms, see the *MATLAB Reference Guide* or on-line help.

As shown in the above figure, the areas between the mesh lines are opaque rather than transparent. The MATLAB command **hidden** controls this aspect of **mesh** plots. For example, using the MATLAB **sphere** function to generate two spheres gives

```
» [X,Y,Z]=sphere(12);

» subplot(1,2,1)

» mesh(X,Y,Z), title('Opaque')
```

```
» hidden on

» axis off

» subplot(1,2,2), title('Transparent')

» mesh(X,Y,Z)

» hidden off

» axis off
```

Opaque   Transparent

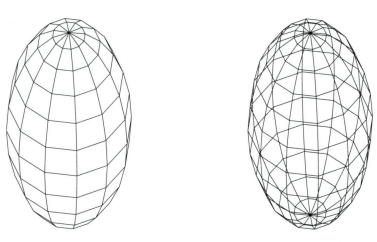

The sphere on the left is opaque (lines are hidden), whereas the one on the right is transparent (lines are not hidden).

The MATLAB **mesh** function has two siblings: **meshc**, which is a mesh plot and underlying contour plot, and **meshz**, which is a mesh plot including a zero plane.

```
» meshc(X,Y,Z) % mesh plot with underlying contour plot

» title('MESHC of PEAKS')

» meshz(X,Y,Z) % mesh plot with zero plane

» title('MESHZ of PEAKS')

» hidden off % make mesh transparent so minimums can be seen
```

MESHC of PEAKS

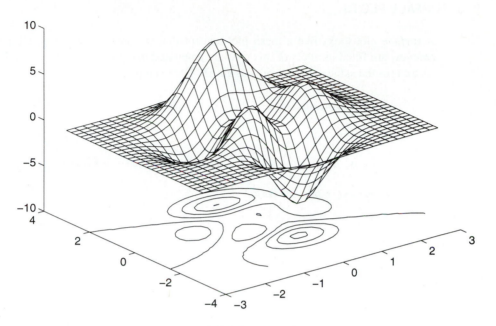

MESHZ of PEAKS

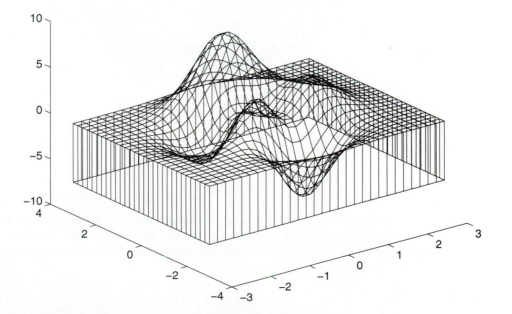

For further information regarding the above functions see the *MATLAB Reference Guide* or use on-line help.

18.6 SURFACE PLOTS

A *surface* plot looks like a mesh plot, except that the spaces between the lines (called *patches*) are filled in. Plots of this type are generated using the **surf** function. Naturally, **surf** uses the same calling syntax as **mesh**. For example,

```
» [X,Y,Z]=peaks(30);

» surf(X,Y,Z)

» grid, xlabel('X-axis'), ylabel('Y-axis'), zlabel('Z-axis')

» title('SURF of PEAKS')
```

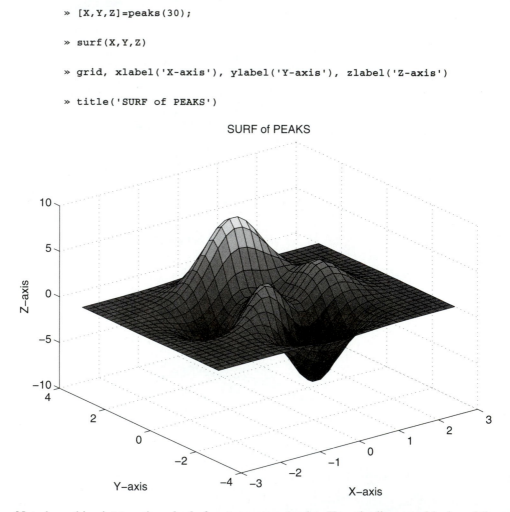

Note how this plot type is a *dual* of sorts to a **mesh** plot. Here the lines are black and the patches have color, whereas in **mesh**, the patches are black and the lines have color. As with **mesh**, color varies along the z-axis with each patch or line having constant color.

In a surface plot one does not think about hidden line removal as in a **mesh** plot, but rather one thinks about different ways to shade the surface. In the previous **surf** plot, the shading is *faceted* like a stained-glass window or object, where the black lines are the joints between the constant-color stained-glass patches. In addition to faceted, MATLAB provides *flat* shading and *interpolated* shading. These are applied by using the function **shading**.

```
» [X,Y,Z]=peaks(30);

» surf(X,Y,Z)   % same plot as above

» grid, xlabel('X-axis'), ylabel('Y-axis'), zlabel('Z-axis')

» title('SURF of PEAKS')

» shading flat
```

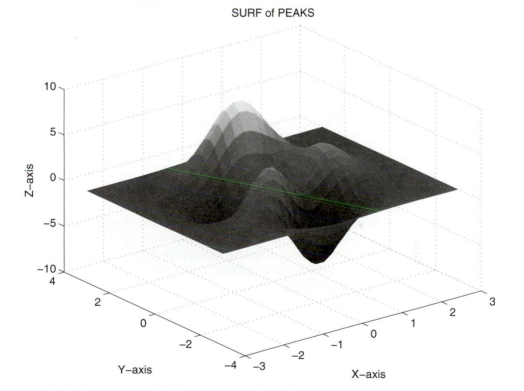

In the flat shading shown above, the black lines are removed and each patch retains its single color.

```
» shading interp
```

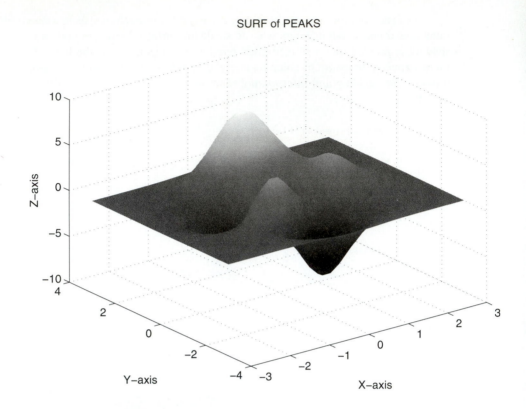

In the interpolated shading shown above, the lines are also removed but each patch is given interpolated shading. That is, the color of each patch is interpolated over its area based on the color values assigned to its vertices. Needless to say, interpolated shading requires much more computation than faceted and flat shading. **On some computer systems, interpolated shading creates extremely long printing delays or at worst printing errors. These problems are not due to the size of the PostScript data file, but rather due to the enormous amount of computation required in the printer to generate shading that continually changes over the surface of the plot. Often the easiest solution to this problem is to use flat shading for printouts.**

While shading has a significant visual impact on **surf** plots, it also applies to **mesh** plots, although in this case the visual impact is relatively minor since only the lines have color.

Since surface plots cannot be made transparent, in some situations it may be convenient to remove part of a surface so that underlying parts of the surface can be seen. In MATLAB this is accomplished by setting data values to the special value **NaN** where holes are desired. **Since NaNs have no value, all MATLAB plotting functions simply ignore NaN data points, leaving a hole in the plot where one appears.** For example,

```
» [X,Y,Z]=peaks(30);

» x=X(1,:);   % vector of x axis

» y=Y(:,1);   % vector of y axis

» i=find(y>.8 & y<1.2);   % find y-axis indices of hole

» j=find(x>-.6 & x<.5);   % find x-axis indices of hole

» Z(i,j)=nan*Z(i,j);   % set values at hole indices to NaNs

» surf(X,Y,Z)
```

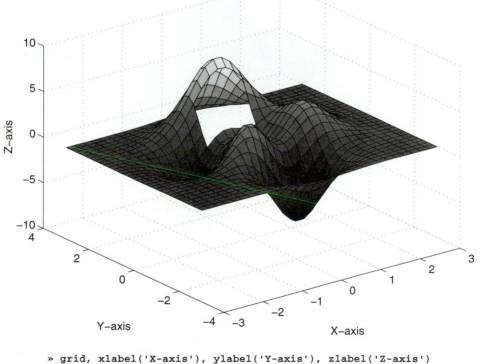

SURF of PEAKS with a Hole

```
» grid, xlabel('X-axis'), ylabel('Y-axis'), zlabel('Z-axis')

» title('SURF of PEAKS with a Hole')
```

The MATLAB **surf** function also has two siblings: **surfc**, which is a surface plot with underlying contour plot, and **surfl**, which is a surface plot with lighting. For example,

```
» [X,Y,Z]=peaks(30);

» surfc(X,Y,Z)  % surf plot with contour plot

» grid, xlabel('X-axis'), ylabel('Y-axis'), zlabel('Z-axis')

» title('SURFC of PEAKS')
```

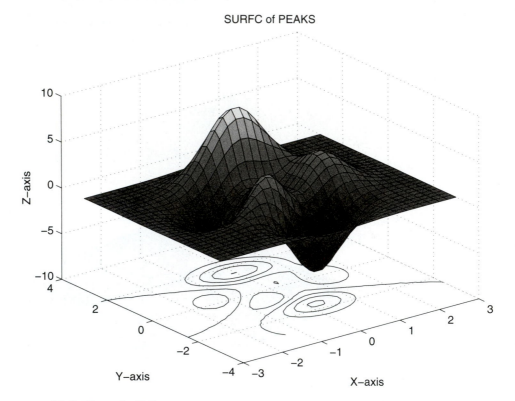

```
» [X,Y,Z]=peaks(30);

» surfl(X,Y,Z)  % surf plot with lighting

» shading interp  % surfl plots look best with interp shading

» colormap pink % they also look better with shades of a single color

» grid, xlabel('X-axis'), ylabel('Y-axis'), zlabel('Z-axis')

» title('SURFL of PEAKS')
```

The function **surfl** makes a number of assumptions regarding the light applied to the surface. For information regarding how to set light properties, see the *MATLAB Reference*

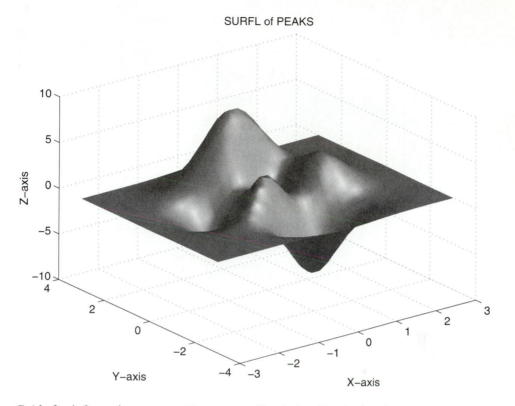

Guide for information on **surfl** or use on-line help. Also in the above executed commands, **colormap** is a MATLAB function for applying a different set of colors to a figure. This function is discussed in the next chapter.

18.7 CONTOUR PLOTS

MATLAB provides another basic 3-D plot type, namely 3-D contour plots. These plots are generated using the **contour3** function.

```
» [X,Y,Z]=peaks(30);

» contour3(X,Y,Z,16) % draw sixteen contour lines

» grid, xlabel('X-axis'), ylabel('Y-axis'), zlabel('Z-axis')

» title('CONTOUR3 of PEAKS')
```

Note that the color of each line follows the color order used in the 2-D function **plot**. This ordering exhibits a lot of contrast, but often obscures important features of the underlying

CONTOUR3 of PEAKS

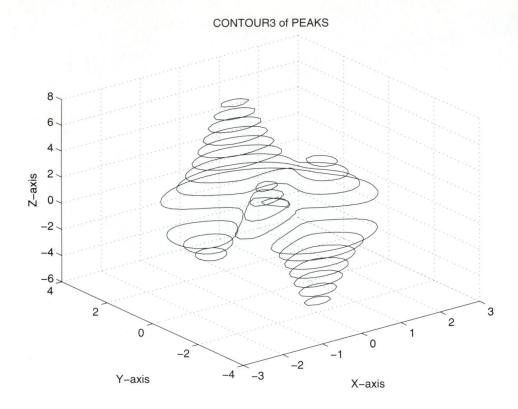

data. It would be much better if the individual lines followed a coloring scheme in the same way that **mesh** and **surf** plots do. Perhaps in the next version of MATLAB this will become the default. However, using the Handle Graphics capabilities of MATLAB that are discussed in later chapters, it is possible to fix this problem.

```
» [X,Y,Z]=peaks(30);

» N=16;    % number of contour lines and their colors

» clf      % clear the current figure

» view(3)  % set view to 3-D

» hold on  % hold blank screen

» set(gca,'ColorOrder',hsv(N)) % use colors from default hsv colormap

» contour3(X,Y,Z,N)  % draw N contour lines

» grid, xlabel('X-axis'), ylabel('Y-axis'), zlabel('Z-axis')
```

```
» title('CONTOUR3 of PEAKS')

» hold off
```

CONTOUR3 of PEAKS

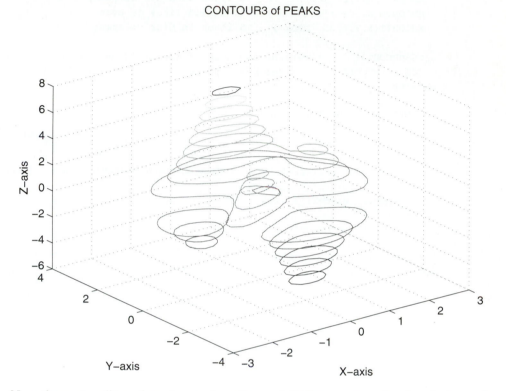

Now, the contour line colors change along the z-axis in the same way that they do in **mesh** and **surf** plots. For convenience, the above strategy is embodied in the *Mastering MATLAB Toolbox* function **mmcont3**. **mmcont3** accepts the same calling syntax variations that **contour3** does and allows one to select a colormap for use. For example, » **mmcont3(X,Y,Z,N,'hsv')** duplicates the above figure. The on-line help text for **mmcont3** is

```
» help mmcont3

MMCONT3 3-D contour plot using a colormap.
  MMCONT3(X,Y,Z,N,C) plots N contours of Z in 3-D using the color
  specified in C. C can be a linestyle and color as used in plot,
  e.g., 'r-', or C can be the string name of a colormap. X and Y
  define the axis limits.
  If not given default argument values are: N = 10, C = 'hot',
  X and Y = row and column indices of Z. Examples:
  MMCONT3(Z)                  10 lines with hot colormap
```

```
MMCONT3(Z,20)                    20 lines with hot colormap
MMCONT3(Z,'copper')              10 lines with copper colormap
MMCONT3(Z,20,'gray')             20 lines with gray colormap
MMCONT3(X,Y,Z,'jet')            10 lines with jet colormap
MMCONT3(Z,'c--')                10 dashed lines in cyan
MMCONT3(X,Y,Z,25,'pink')        25 lines in pink colormap
```

```
CS=MMCONT3( ... ) returns the contour matrix CS as described in CONTOURC.
 [CS,H]=MMCONT3( ... ) returns a column vector H of handles to line
objects.
```

Contour lines can also be given one color.

```
» [X,Y,Z]=peaks(30);

» contour3(X,Y,Z,16,'y')   % draw sixteen contour lines in yellow

» grid, xlabel('X-axis'), ylabel('Y-axis'), zlabel('Z-axis')

» title('CONTOUR3 of PEAKS in Yellow')
```

For further information regarding the use of color, see the next chapter. For further information regarding the functions used above, see the *MATLAB Reference Guide* or use on-line help.

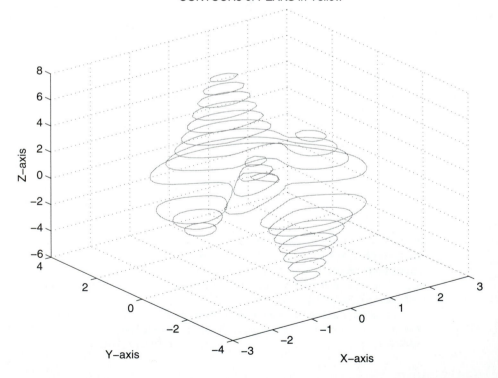

CONTOUR3 of PEAKS in Yellow

18.8 2-D PLOTS OF 3-D DATA

Sometimes it is desirable to have a 2-D representation of 3-D data. In MATLAB this can be accomplished by setting the viewpoint with **view** so that one dimension does not appear. Alternatively, MATLAB offers two functions that give views of **contour3** and **surf** plots looking straight down onto the x-y plane. For example, the 2-D equivalent of **contour3** is simply the function **contour**.

```
» [X,Y,Z]=peaks(30);

» contour(X,Y,Z,16)   % draw sixteen contour lines

» xlabel('X-axis'), ylabel('Y-axis')

» title('CONTOUR of PEAKS')
```

Note how this is equivalent to using **contour3** and changing the viewpoint to look down at the x-y plane. As with **contour3**, the contour lines here cycle through the six basic colors used by the **plot** command. As before, this type of plot is often more confusing than useful because color offers no visual cues. This default behavior may be changed in the next version of MATLAB, but can be changed using Handle Graphics.

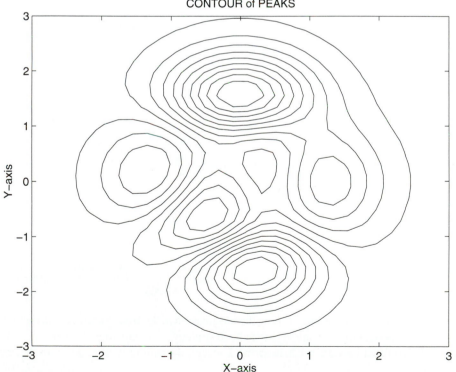

```
» [X,Y,Z]=peaks(30);

» N=16;     % number of contour lines and their colors

» clf       % clear the current figure

» hold on  % hold blank screen

» set(gca,'ColorOrder',hsv(N)) % use colors from default hsv colormap

» contour(X,Y,Z,N)   % draw N contour lines

» xlabel('X-axis'), ylabel('Y-axis'), zlabel('Z-axis')

» title('CONTOUR of PEAKS')

» hold off
```

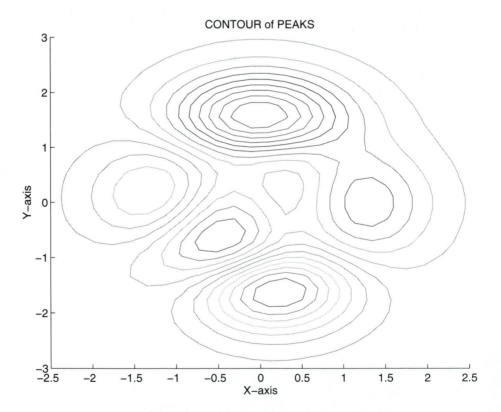

Now the contour lines follow the progression in the **hsv** colormap and color serves a useful purpose! For convenience, the above strategy is embodied in the *Mastering MATLAB Toolbox* function **mmcont2**. **mmcont2** accepts the calling syntax variations that **contour** does

and allows one to select a colormap for use. For example, » `mmcont2(X,Y,Z,N,'hsv')` duplicates the above figure. The on-line help text for `mmcont2` is

```
» help mmcont2

MMCONT2 2-D contour plot using a colormap.
 MMCONT2(X,Y,Z,N,C) plots N contours of Z in 2-D using the color
 specified in C. C can be a linestyle and color as used in plot,
 e.g., 'r-', or C can be the string name of a colormap. X and Y
 define the axis limits.
 If not given default argument values are: N = 10, C = 'hot',
 X and Y = row and column indices of Z. Examples:
 MMCONT2(Z)                 10 lines with hot colormap
 MMCONT2(Z,20)              20 lines with hot colormap
 MMCONT2(Z,'copper')       10 lines with copper colormap
 MMCONT2(Z,20,'gray')      20 lines with gray colormap
 MMCONT2(X,Y,Z,'jet')      10 lines with jet colormap
 MMCONT2(Z,'c--')           10 dashed lines in cyan
 MMCONT2(X,Y,Z,25,'pink')  25 lines in pink colormap

 CS=MMCONT2( . . . ) returns the contour matrix CS as described in CONTOURC.
 [CS,H]=MMCONT2( . . . ) returns a column vector H of handles to line
objects.
```

In both the 3-D and 2-D contour plots discussed above, it was assumed that the data were defined on a rectangular grid or domain. In cases where this is not true, the data must be converted to a rectangular grid so that contour plots look right. As described earlier in Section 18.4 of this chapter, the MATLAB function **griddata** provides the required conversion. See that section for more information.

The 2-D equivalent of the **surf** function is the function **pcolor**, which stands for pseudocolor.

```
» [X,Y,Z]=peaks(30);

» pcolor(X,Y,Z)  % surf plot view from above

» xlabel('X-axis'), ylabel('Y-axis')

» title('PCOLOR of PEAKS')
```

(See next page for plot.)
Since this is a **surf** plot, the **shading** function can be used. In addition, it is sometimes useful to place a single color **contour** plot on top of the **pcolor** plot.

```
» [X,Y,Z]=peaks(30);

» pcolor(X,Y,Z)
```

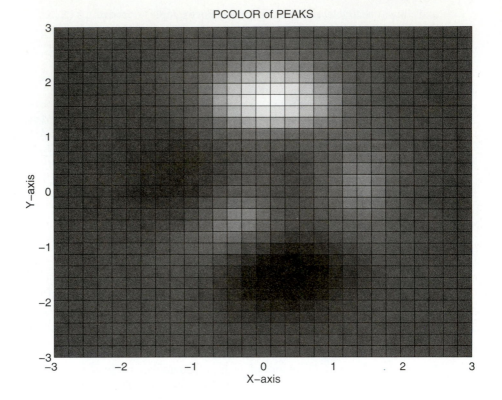

```
» shading interp

» hold on

» contour(X,Y,Z,19,'k')   % add 19 contour lines in black

» xlabel('X-axis'), ylabel('Y-axis')

» title('PCOLOR and CONTOUR of PEAKS')

» hold off
```

18.9 OTHER FUNCTIONS

In addition to the 3-D functions discussed above, MATLAB provides the functions **waterfall**, **quiver**, **fill3**, and **clabel**.

The function **waterfall** is identical to **mesh** except that the mesh lines appear only in the x-direction. For example,

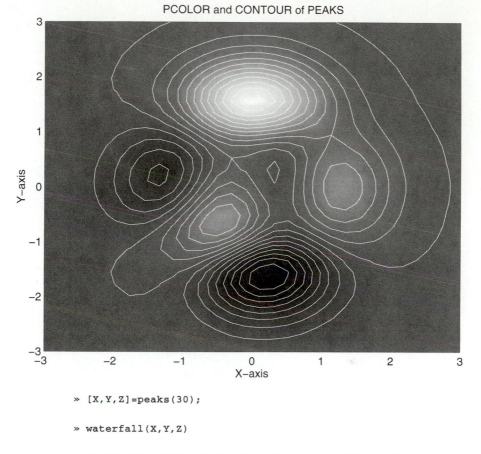

PCOLOR and CONTOUR of PEAKS

```
» [X,Y,Z]=peaks(30);

» waterfall(X,Y,Z)

» xlabel('X-axis'),ylabel('Y-axis'),zlabel('Z-axis')
```

(See next page for plot.)

The function **quiver** draws directional or velocity arrows on a contour plot. For example,

```
» [X,Y,Z]=peaks(16);

» [DX,DY]=gradient(Z,.5,.5);

» contour(X,Y,Z,10)

» hold on

» quiver(X,Y,DX,DY)

» hold off
```

(See next page for plot.)

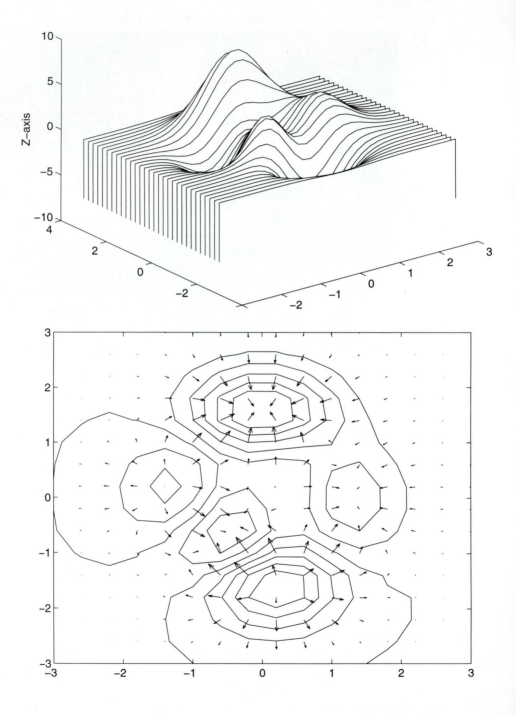

The function **fill3**, being the 3-D equivalent of **fill**, draws filled polygons in 3-D space. **fill3(x,y,z,c)** uses the arrays **x**, **y**, and **z** as the vertices of the polygon and **c** specifies the fill color. For example, the following draws 5 yellow triangles with random vertices:

```
» fill3(rand(3,5),rand(3,5),rand(3,5),'y')
```

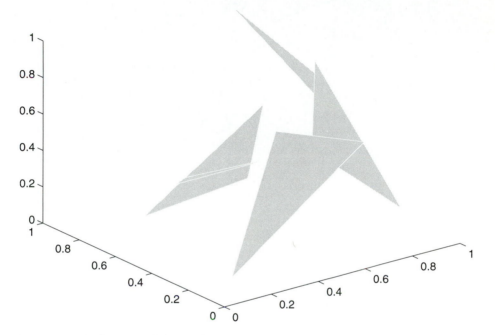

The **clabel** function adds height labels to contour plots. To do so, **clabel** needs the contour-matrix output of **contour**.

```
» [X,Y,Z]=peaks(30);

» cs=contour(X,Y,Z,8);  % request output from contour

» clabel(cs)  % add labels identifying heights

» xlabel('X-axis'), ylabel('Y-axis')

» title('CONTOUR of PEAKS with Labels')
```

(See next page for plot.)

18.10 MOVIES

MATLAB offers the ability to save a sequence of plots of any type, 2-D or 3-D, and then play the sequence back as a movie. In a sense, the motion provided by a movie adds yet

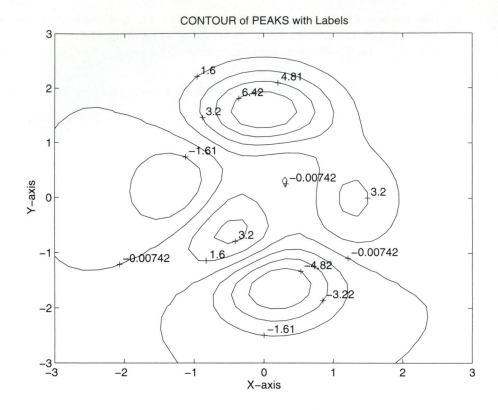

another dimension to a plot. While the sequence of plots does not have to be related in any way, they commonly are. One obvious movie type takes a 3-D plot and slowly rotates it so it can be viewed from a variety of angles. Another would be a sequence of plots showing the solution of some problem as a parameter value changes.

In MATLAB the functions **moviein**, **getframe**, and **movie** provide the tools required to capture and play movies. **moviein** creates a frame matrix to store the movie frames in, **getframe** takes a snapshot of the current figure, and **movie** plays back the sequence of frames. With this understanding, the approach to capturing and playing back movies is (1) create the frame matrix, (2) for each movie frame create the plot and capture it in the frame matrix, and (3) play the movie back from the frame matrix.

Consider the following script M-file example where the **peaks** function is plotted and rotated about the z-axis:

```
% movie making example: rotate a 3-D surface plot

[X,Y,Z]=peaks(30);      % create data
surfl(X,Y,Z)            % plot surface with lighting
```

```
axis([-3 3 -3 3 -10 10])% fix axes so that scaling does not change
axis off              % erase axes because they jump around
shading interp        % make it pretty with interpolated shading
colormap(hot)         % choose a good colormap for lighting

m=moviein(15);  % choose 15 movie frames for frame matrix m

for i=1:15                    % rotate and capture each frame
    view(-37.5+24*(i-1),30)   % change viewpoint for this frame
    m(:,i)=getframe;          % add figure to frame matrix
end

movie(m)  % play the movie!
```

Note that each movie frame occupies a different column in the frame matrix. The size of the frame matrix increases with the number of frames and with the size of the *Figure* window. The frame matrix size is not a function of the complexity of the plotted figure since **getframe** simply captures a bitmap image. By default the function **movie** plays a movie once. By including other input arguments, it can play the movie forward, backward, a specified number of times, and at a specified frame rate. For information regarding these features, see the *MATLAB Reference Guide* or on-line help.

Because the above movie-making strategy is generally useful, it is embodied in the *Mastering MATLAB Toolbox* function **mmspin3d**.

```
function M=mmspin3d(n)
%MMSPIN3D Make Movie by 3D Azimuth Rotation of Current Figure.
% MMSPIN3D(N) captures and plays N frames of the current figure
% through one rotation about the Z-axis at the current elevation.
% M=MMSPIN3D(N) returns the movie in M for later playing with
movie.
% If not given, N=18 is used.
% MMSPIN3D fixes the axis limits and issues axis off.

% Copyright (c) 1996 by Prentice-Hall, Inc.

if nargin<1, n=18; end
n=max(abs(round(n)),2);

axis(axis);
axis off
```

```
incaz=round(360/n);
[az,el]=view;

m=moviein(n);
for i=1:n
    view(az+incaz*(i-1),el)
    m(:,i)=getframe;
end
if nargout,    M=m;
else,          movie(m);
end
```

Using **mmspin3d**, the above script file simplifies to

```
% movie-making example: rotate a 3-D surface plot

[X,Y,Z]=peaks(30);      % create data
surfl(X,Y,Z)            % plot surface with lighting
shading interp          % make it pretty with interpolated shading
colormap(hot)           % choose a good colormap for lighting
mmspin3d(15)
```

18.11 SUMMARY

The functions and features considered in this chapter are summarized in the following tables.

3-D Plotting Functions

contour	2-D contour plot, i.e., contour3 viewed from above
contour3	Contour plot
fill3	Filled polygons
mesh	Mesh plot
meshc	Mesh plot with underlying contour plot
meshz	Mesh plot with zero plane
pcolor	2-D pseudocolor plot, i.e., surf plot viewed from above

`plot3`	Line plot
`quiver`	2-D arrow velocity plot
`surf`	Surface plot
`surfc`	Surface plot with underlying contour plot
`surfl`	Surface plot with lighting
`waterfall`	Mesh plot lacking cross lines

3-D Plotting Tools

`axis`	Modify axis properties
`clf`	Clear *Figure* window
`clabel`	Place contour labels
`close`	Close *Figure* windows
`figure`	Create or select *Figure* windows
`getframe`	Capture movie frame
`grid`	Place grid
`griddata`	Interpolate (potentially scattered) data for plots
`hidden`	Hidden line removal in `mesh` plots
`hold`	Hold current plot
`meshgrid`	Data generation for 3-D plots
`movie`	Play movie
`moviein`	Create frame matrix for movie storage
`shading`	`faceted`, `flat`, `interp` shading in `surf` and `pcolor` plots
`subplot`	Create subplots within a *Figure* window
`text`	Place text at given location
`title`	Place title
`view`	Change viewpoint of plot
`xlabel`	Place x-axis label
`ylabel`	Place y-axis label
`zlabel`	Place z-axis label

`view` **Function**

`view(az,el)` `view ([az,el])`	Set view to azimuth **az** and elevation **el**
`view([x,y,z])`	Set view looking toward the origin along the vector [x,y,z] in Cartesian coordinates, e.g., `view([0 0 1])=view(0,90)`
`view(2)`	Sets the default 2-D **view**, **az** $= 0$, **el** $= 90$
`view(3)`	Sets the default 3-D view, **az** $= -37.5$, **el** $= 30$
`[az,el]=view`	Returns the current azimuth **az** and elevation **el**
`view(T)`	Uses the 4-by-4 transformation matrix **T** to set the view
`T=view`	Returns the current 4-by-4 transformation matrix

Mastering MATLAB High-Level Graphics

`mmcont2(X,Y,Z,N,C)`	2-D contour plot with colormap
`mmcont3(X,Y,Z,N,C)`	3-D contour plot with colormap
`mmspin3d(N)`	Make movie by 3-D azimuth spin of current figure
`mmview3d`	Viewpoint adjustment using sliders

Using Color

MATLAB provides a number of tools for displaying information visually in 2 and 3 dimensions. For example, the plot of a sine curve presents more information at a glance than a set of data points could. The technique of using plots and graphs to present data sets is known as *data visualization.* In addition to being a powerful computational engine, MATLAB excels in presenting data visually in interesting and informative ways.

Often, however, a simple 2-D or 3-D plot cannot display all of the information you would like to present at one time. Color can provide an additional dimension to a plot. Many of the plotting functions discussed in previous chapters accept a *color* argument that you can use to add that additional dimension.

This discussion begins with an investigation of color maps: how to use them, display them, alter them, and create your own. Next, techniques to simulate more than one color map in a *Figure* window or to use only a portion of a color map are illustrated. Finally, lighting models are discussed and examples are presented.

19.1 UNDERSTANDING COLOR MAPS

MATLAB uses a data structure called a **color map** to represent color values. A color map
is defined as a matrix having three columns and some number of rows. Each row in the matrix represents an individual color using numbers in the range 0 to 1. The numbers in any
row specify RGB values, that is the intensity of Red, Green, and Blue, making up a specific
color. Some representative RGB examples are given in the following table:

Simple Colors			
Red	**Green**	**Blue**	**Color**
0	0	0	black
1	1	1	white
1	0	0	red
0	1	0	green
0	0	1	blue
1	1	0	yellow
1	0	1	magenta
0	1	1	cyan
2/3	0	1	violet
1	1/2	0	orange
.5	0	0	dark red
.5	.5	.5	medium gray

There are ten MATLAB functions that generate predefined color maps.

Standard Color Maps	
`hsv`	Hue-saturation-value (begins and ends with red)
`hot`	Black to red to yellow to white
`cool`	Shades of cyan and magenta
`pink`	Pastel shades of pink
`gray`	Linear gray-scale
`bone`	Gray-scale with a tinge of blue
`jet`	A variant of `hsv` (begins with blue and ends with red)

`copper`	Linear copper-tone
`prism`	Prism, alternating red, orange, yellow, green, and blue-violet
`flag`	Alternating red, white, blue, and black

By default, each of the above color maps generates a 64-by-3 matrix specifying the RGB descriptions of 64 colors. Each of these functions accepts an argument specifying the number of rows to generate. For example, `hot(m)` will generate an **m**-by-3 matrix containing the RGB values of colors which range from black through shades of red, orange, and yellow to white.

Most computers can display up to 256 colors at one time in an 8-bit hardware color-lookup table, although some have display cards that can handle many more colors simultaneously. This means that normally up to three or four 64-by-3 color maps can be in use at one time in different figures. If more color map entries are used, the computer must usually swap out entries in its hardware-lookup table. For example, this occurs if you notice the screen background pattern change when plotting MATLAB figures. As a result, it is usually prudent to keep the total number of color map entries used at any one time below 256 unless you have a display card that can display more colors at once.

19.2 USING COLOR MAPS

The statement `colormap(M)` installs the matrix **M** as the color map to be used in the current *Figure* window. For example, `colormap(cool)` installs a 64-entry version of the `cool` color map. `colormap default` installs the default color map (**hsv**).

The **plot**, **plot3**, **contour**, and **contour3** functions do not use color maps; they use the colors listed in the **plot** color and linestyle table. Most other plotting functions, such as **mesh**, **surf**, **fill**, **pcolor**, and their variations, use the current color map.

Plotting functions that accept a *color* argument usually accept the argument in one of three forms: (1) a character string representing one of the colors in the **plot** color and linestyle table, e.g., `'r'` for red, (2) a 3-entry row vector representing a single RGB value, e.g., `[.25  .50  .75]`, or (3) a matrix. If the color argument is a matrix, the elements are scaled and used as indices into the current matrix color map. This last form will be discussed more later.

19.3 DISPLAYING COLOR MAPS

You can display a color map in a number of ways. One way is to view the elements in a color-map matrix.

```
» hot(8)
ans =
      0.3333        0        0
```

```
0.6667        0        0
1.0000        0        0
1.0000   0.3333        0
1.0000   0.6667        0
1.0000   1.0000        0
1.0000   1.0000   0.5000
1.0000   1.0000   1.0000
```

The above shows the first row to be 1/3 red and the last row to be white. In addition, the **pcolor** function can be used to display a color map. For example,

```
» n=16;

» colormap(jet(n))

» pcolor([1:n+1;1:n+1]')

» title('Using Pcolor to Display a Color Map')
```

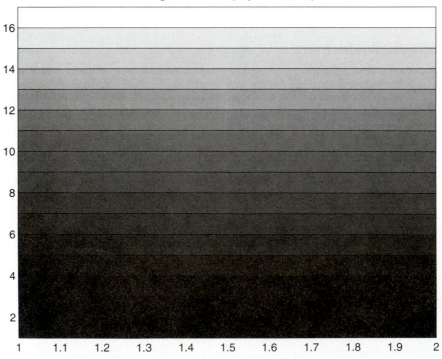

Because of the utility of the above procedure, it has been encapsulated in the *Mastering MATLAB Toolbox* function **mmshow**.

```
» help mmshow

MMSHOW PCOLOR Colormap Display.
MMSHOW uses pcolor to display the current colormap.
MMSHOW(MAP) displays the colormap MAP.
MMSHOW(MAP(N)) displays the colormap MAP having N elements.

Examples: MMSHOW(hot)
          MMSHOW(pink(30))
```

mmshow takes the same input argument as **colormap**, but in this case produces its own **pcolor** display rather than applying the color map to the current figure.

As an alternative to the above, the columns of a color map can be plotted in red, green, and blue, respectively, using the MATLAB function **rgbplot**. For example,

» rgbplot(hot)

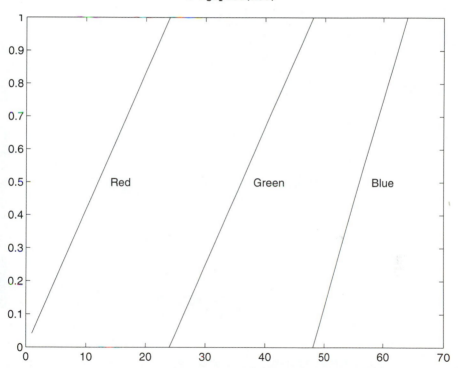

This shows that the red component increases first, then the green, then the blue. **rgbplot(gray)** shows that all three columns increase linearly and equally (all three lines overlap).

Finally, the **colorbar** function adds a vertical or horizontal color scale to the current *Figure* window showing the color map for the current axis. » **colorbar('horiz')** places a color bar horizontally beneath your current plot. » **colorbar('vert')** places a vertical color bar to the right of your plot. `colorbar` without arguments either adds a vertical color bar if no color bars exist, or updates any existing color bar. The following example demonstrates the use of `colorbar`:

```
» [x,y,z]=peaks;

» mesh(x,y,z);

» colormap(hsv)

» axis([-3 3 -3 3 -6 8])

» colorbar
```

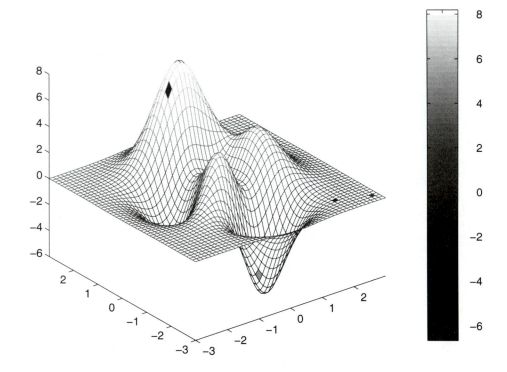

19.4 CREATING AND ALTERING COLOR MAPS

The fact that color maps are matrices means that you can manipulate them exactly like other arrays. The function **brighten** takes advantage of this fact to adjust a given color map to increase or decrease the intensity of the dark colors. » **brighten(n)** brightens (0 < n ≤ 1) or darkens (-1 ≤ n < 0) the current color map. » **brighten(n)** followed by **brighten(-n)** restores the original color map. The command

» **newmap=brighten(n)** creates a brighter or darker version of the current color map without changing the current map. The command » **newmap=brighten(cmap,n)** creates an adjusted version of the specified color map without affecting either the current color map or the specified color map **cmap**.

You can create your own color map by generating an **m**-by-3 matrix **mymap** and installing it with **colormap(mymap)**. Each value in a color-map matrix must be between 0 and 1. If you try to use a matrix with more or less than 3 columns or one containing any values less than 0 or greater than 1, **colormap** will report an error and abort.

You can also combine color maps arithmetically, although the results are sometimes unpredictable. For example, the map called **pink** is simply

» pinkmap=sqrt(2/3*gray+1/3*hot);

Again, the result is a valid color map only if all elements are between 0 and 1 inclusive. The *Mastering MATLAB Toolbox* includes a color map called **rainbow** that spreads the visual spectrum across the entire color map. It is used the same way you use any other color map. The on-line help text for **rainbow** is

```
» help rainbow

  RAINBOW Colormap variant to HSV.
   RAINBOW(M) Rainbow Colormap with M entries.

   Red - Orange - Yellow - Green - Blue - Violet

   RAINBOW by itself is the same length as the current colormap.

  Apply using: colormap(rainbow)
```

The *Mastering MATLAB Toolbox* also includes a function called **mmap** that creates a single-color color map (like **pink**, **gray**, or **copper**) based on a color that you supply. The on-line help text for **mmap** is

```
» help mmap

  MMAP Single Color Colormap.
   MMAP(C,M) makes a colormap of length M starting with the
   basic colorspec C. The map changes from dark to light.
   MMAP(C) is the same length as the current colormap.

   Example: mmap('y') is a yellow colormap
            mmap([.49 1 .83]) is an aquamarine colormap
            mmap('c',20) is a cyan colormap having length 20.

  Apply using: colormap(mmap(c,m))
```

A color map defines the color palate that will be used to render a plot. A default color map allows 64 distinct RGB values to be used for your data. MATLAB uses the function **caxis** to determine which data-values map to individual color-map entries.

Normally, a color map is scaled to extend from the minimum to the maximum values of your data; that is, the entire color map is used to render your plot. You may occasionally wish to change the way these colors are used. The **caxis** function, which stands for color axis since color adds another dimension, allows you to use the entire color map for a subset of your data range, or use only a portion of the current color map for your entire data set.

» **[cmin,cmax]=caxis** returns the minimum and maximum data values that are mapped to the first and last entries of the color map. These will normally be set to the minimum and maximum values of your data. For example, **mesh(peaks)** will create a mesh plot of the **peaks** function and set **caxis** to **[-6.5466, 8.0752]**, the minimum and maximum **Z** values. Data points between these values use colors interpolated from the color map.

» **caxis([cmin,cmax])** uses the entire color map for data in the range between **cmin** and **cmax**. Data points greater than **cmax** will be rendered with the color associated with **cmax**, and data points smaller than **cmin** will be rendered with the color associated with **cmin**. If **cmin** is less than **min(data)** and/or **cmax** is greater than **max(data)**, the colors associated with **cmin** and/or **cmax** will never be used. Only the portion of the color map associated with **data** will be used. » **caxis('auto')** restores the default values of **cmin** and **cmax**.

While the following examples are difficult to show distinctively in grayscale in this text, a more involved sequence of examples can be displayed by running the script M-file **mmcaxisd.m**. The default color range is illustrated by

```
» pcolor([1:17;1:17]'), colormap(hsv(8))

» title('Default Color Range')

» caxis('auto')

» colorbar

» caxis
ans =
      1  17
```

As you can see, all 8 colors in the current color map are used for the entire data set, two bars for each color. If the colors are mapped to values from −3 to 23, only 5 colors will be used in the plot, which is generated by the commands

```
» title('Extended Color Range')

» caxis([-3,23]) % extend the color range

» colorbar     % redraw the color scale
```

(See page 276 for plot.)

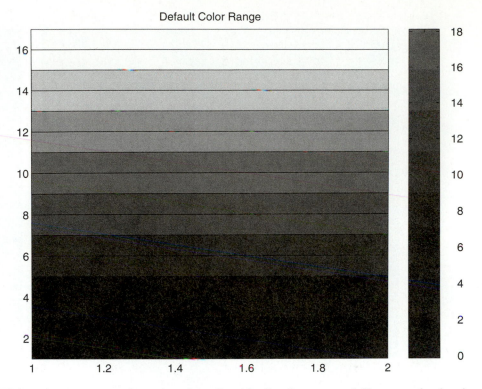

If the colors are mapped to values from 5 to 12, all colors are used. However, the data less than 5 or greater than 12 map to the colors associated with 5 and 12, respectively. This is produced by the commands

```
» title('Restricted Color Range')

» caxis([5,12]) % restrict the color range

» colorbar    % redraw the color scale
```

(See page 276 for plot.)

19.5 USING MORE THAN ONE COLOR MAP IN A FIGURE

Sometimes, it is helpful to use different colors for different portions of a plot. Since a color map is a property of the *Figure* window itself, only one color map can be used in any one *Figure* window. However, you can create your own color map to give the effect you want. For example, the *Mastering MATLAB Toolbox* contains the script M-file **mmcmapd.m** which performs the following:

```
» figure                      % create a figure window

» mymap=[rainbow(32); copper(32)];  % stack two color maps into one
```

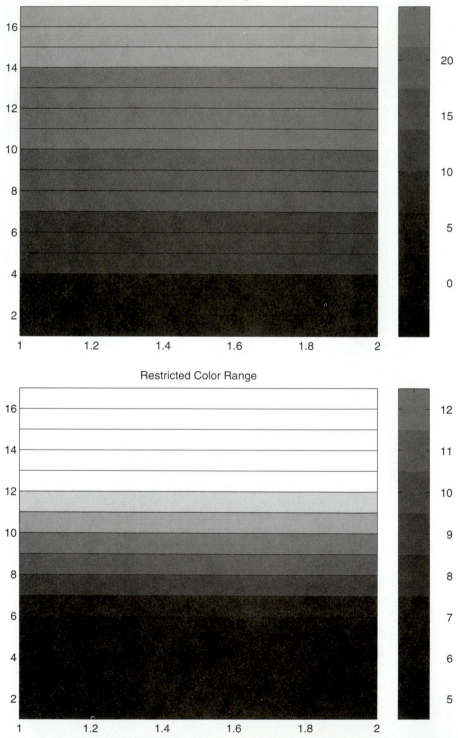

```
» colormap(mymap)                    % install it

» mesh(peaks+8); view(0,0);          % create two sample plots

» hold on; mesh(peaks-8);

» colorbar                           % and add a color scale

» title('Merging two colormaps')

» hold off
```

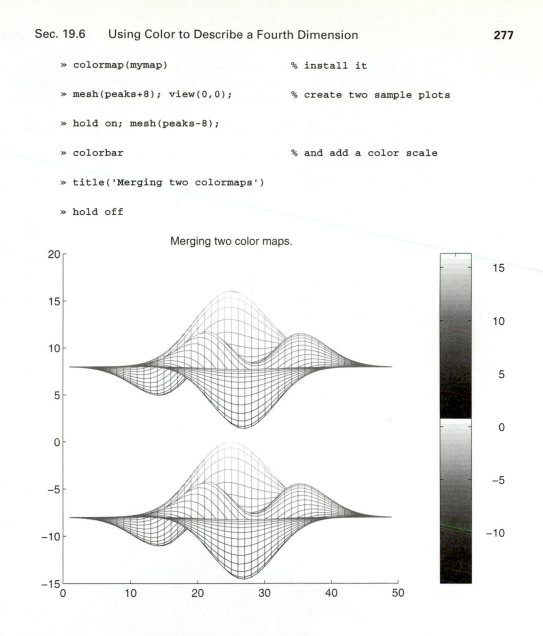

Merging two color maps.

19.6 USING COLOR TO DESCRIBE A FOURTH DIMENSION

Functions such as **mesh** and **surf** vary the color along the z-axis unless a color argument is given, e.g., **surf(X,Y,Z)** is equivalent to **surf(X,Y,Z,Z)**. Applying color to the z-axis produces a colorful plot, but does not provide additional information since the z-axis already exists. To make better use of color, it is suggested that color be used to describe some property of the data that is not reflected by the three axes. To do so requires specifying different data for the color argument to 3-D plotting functions.

If the color argument to a plotting function is a vector or matrix, it is used as an index into the color map. This argument can be any real vector or matrix the same size as the other arguments. Consider the following examples as illustrated on the next several pages.

```
» x=-7.5:.5:7.5; y=x;        % create a data set - the famous sombrero

» [X Y]=meshgrid(x,y);       % create plaid data

» R=sqrt(X.^2 + Y.^2)+eps;

» Z=sin(R)./R;

» surf(X,Y,Z,Z)             % default color order

» surf(X,Y,Z,-Z)            % reverse the default color order

» surf(X,Y,Z,X)             % color varies along the X-axis

» surf(X,Y,Z,X+Y)           % color varies along the XY diagonal

» surf(X,Y,Z,R)             % color varies radially from the center
```

surf(X,Y,Z,Z)

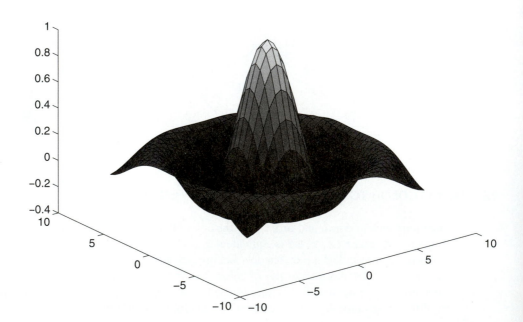

surf(X,Y,Z,−Z)

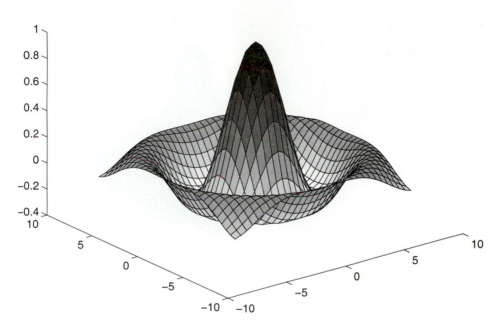

surf(X,Y,Z,X)

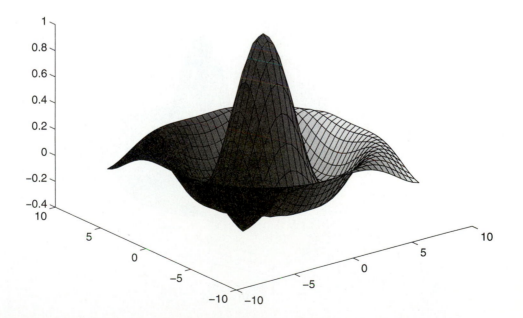

surf(X,Y,Z,X+Y)

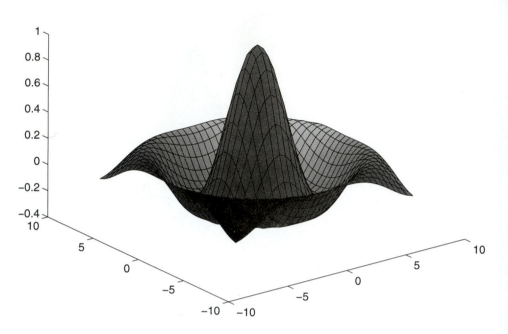

surf(X,Y,Z,R)

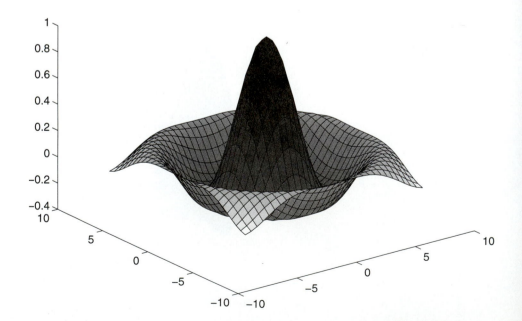

```
» surf(X,Y,Z,abs(del2(Z)))% color varies with absolute value of Laplacian

» [dZdx,dZdy]=gradient(Z);% compute gradient or slope of surface

» surf(X,Y,Z,abs(dZdx)) % color varies with absolute slope in x-direction

» surf(X,Y,Z,abs(dZdy)) % color varies with absolute slope in y-direction

» dZ=sqrt(dZdx.^2 + dZdy.^2);

» surf(X,Y,Z,dZ)        % color varies with magnitude of slope
```

Note how color in these last five examples does indeed provide an additional dimension to the plotted surface. The function **del2** is the discrete Laplacian function that applies color based on the curvature of the surface. **del2** is described by

surf(X,Y,Z,abs(del2(Z)))

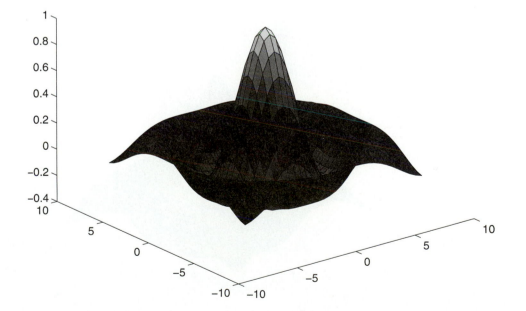

surf(X,Y,Z,abs(dZdx))

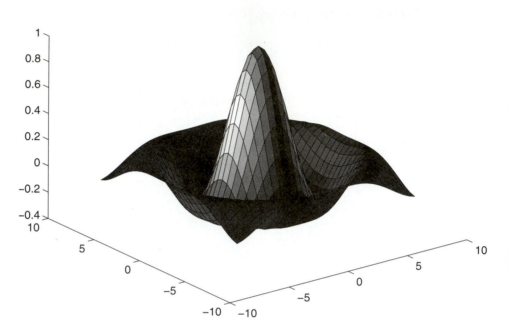

surf(X,Y,Z,abs(dZdy))

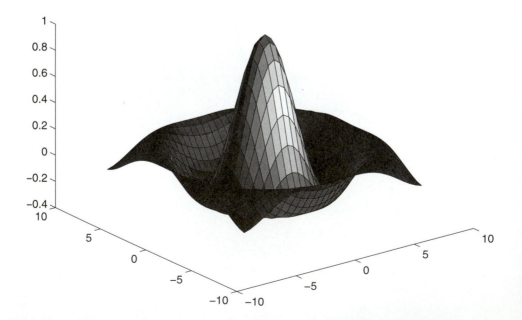

surf(X,Y,Z,dZ)

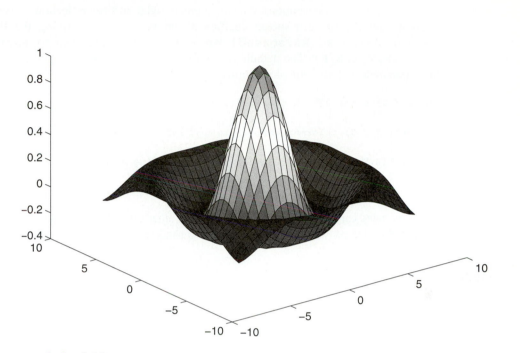

```
» help del2

DEL2 Five-point discrete Laplacian.
    V = del2(U) is a matrix the same size as U with each element
    equal to the difference between an element of U and the average
    of its four neighbors. For the "corners" and "edges", only two
    or three neighbors are used.
    See also GRADIENT, DIFF.
```

As described, the function **gradient** approximates the gradient or slope of the surface. For your convenience, the commands above can be executed by running *Mastering MATLAB Toolbox* script M-file **mm4d**.

19.7 LIGHTING MODELS

The **surfl** function produces a shaded surface plot similar to **surf** based on a combination of diffuse, specular, and ambient lighting models. The surfaces are rendered most effectively with one of the monochromatic color maps (**gray**, **bone**, **copper**, or **pink**) and interpolated shading.

The normal argument list is **surfl(X,Y,Z,S)** where **X, Y,** and **Z** are equivalent to **surf(X,Y,Z)**, and **S** is the direction of the light source in the form **[Sx,Sy,Sz]** or `[az,el]`. If not specified, the default light source is 45 degrees counterclockwise, i.e, to the right, from the current viewpoint.

The relative contributions due to ambient light, diffuse reflection, specular reflection, and the specular-spread coefficient can be set by specifying the five elements of `K=[ka,kd,ks,spread]` where `K` is a fifth argument to `surfl`, i.e., `surfl(X,Y,Z,S,K)`. The default values for `K` are `[.55 .6 .4 10]`. To see how these parameters change the plot lighting, consider these examples:

```
» [X,Y,Z]=peaks(32); % data to plot

» surfl(X,Y,Z),colormap(copper),title('Default Lighting'), shading interp

» surfl(X,Y,Z,[7.5 30],[.55 .6 .4 10]), shading interp

» surfl(X,Y,Z,[-90 30],[.55 .6 2 10]), shading interp
```

Again as stated earlier, interpolated shading can dramatically slow printing since each pixel has a different color value, and the printer must shade each one individually.

19.8 SUMMARY

The functions considered in this chapter are summarized below.

Simple Colors

Red	Green	Blue	Color
0	0	0	black
1	1	1	white
1	0	0	red
0	1	0	green
0	0	1	blue
1	1	0	yellow
1	0	1	magenta
0	1	1	cyan
2/3	0	1	violet
1	1/2	0	orange
.5	0	0	dark red
.5	.5	.5	medium gray

Standard Color Maps

`hsv`	Hue-saturation-value (begins and ends with red)
`hot`	Black to red to yellow to white
`cool`	Shades of cyan and magenta
`pink`	Pastel shades of pink
`gray`	Linear gray-scale
`bone`	Gray-scale with a tinge of blue
`jet`	A variant of `hsv` (begins with blue and ends with red)
`copper`	Linear copper-tone
`prism`	Prism: alternating red, orange, yellow, green, and blue-violet
`flag`	Alternating red, white, blue, and black

Color as a Fourth Dimension in surf, mesh, and pcolor Plots

`surf(X,Y,Z,fun(X,Y,Z))`	Color is applied according to function `fun(X,Y,Z)`
`surf(X,Y,Z)=surf(X,Y,Z,Z )`	Default action, color is applied to z-axis
`surf(X,Y,Z,X)`	Color is applied to x-axis
`surf(X,Y,Z,Y)`	Color is applied to y-axis
`surf(X,Y,Z,X.^2+Y.^2)`	Color is applied according to the radial distance from the `x=0`, `y=0` origin in the `z=0` plane
`surf(X,Y,Z,del2(Z))`	Color is applied according to Laplacian of surface
`[dZdx,dZdy]=gradient(Z);` `surf(X,Y,Z,dZdx)`	Color is applied according to surface slope in x-direction
`dZ=sqrt(dZdx.^2 + dZdy.^2);` `surf(X,Y,Z,dZ)`	Color is applied according to magnitude of surface slope

Color and Lighting Functions

`colormap(map)`	Install a color map in the current *Figure* window
`colorbar`	Display a vertical or horizontal color scale on the current figure
`rgbplot(map)`	Line plot of the red, green, and blue components of a color map
`brighten(a)`	Brighten current color map if `0<a<1`, darken if `-1<a<0`
`m=brighte n(map,a)`	Return brightened color map **m**
`[cmin,cmax]=caxis`	Return color axis limits
`caxis([cmin,cmax])`	Set color axis limits

Handle Graphics

What is Handle Graphics? Handle Graphics is the name of the collection of low-level graphics routines that actually do the work of generating graphics. These details are usually hidden inside graphics M-files, but are available if you want to use them.

The *MATLAB User's Guide* may give the impression that Handle Graphics is very complex and of use to power-users only. However, this is not the case. Handle Graphics can be used by anyone to change the way MATLAB renders graphics, whether you want to make a small change in a single plot or to make global changes that affect all graphical output.

Handle Graphics lets you customize aspects of plots that cannot be addressed using the high-level commands and functions described in previous chapters. For example, suppose you wanted to plot an orange line rather than one of the colors available to the `plot` command. How would you do it? Handle Graphics provides a way.

This chapter is not an exhaustive discussion of Handle Graphics. There's just too much detail involved to do so. Here the goal is to develop a basic understanding of Handle Graphics concepts and present enough practical information so that Handle Graphics is accessible to even casual MATLAB users. With this background, the table of Handle

287

Graphics' object properties and their values, presented near the end of this chapter, is useful and makes sense.

20.1 WHO NEEDS HANDLE GRAPHICS?

To start with, we want to emphasize that this chapter is intended primarily for those readers who want more than the normal graphics features of MATLAB. If you are happy with your plots as they are, skip this discussion for now. Just remember that the information is here if you want to customize your graphics in the future.

Now, for those of you who are sticking around, we want to emphasize that learning to use Handle Graphics is not difficult. If you just want to change the font in the title of a plot or change the background color of a *Figure* window, you can do it without becoming a Handle Graphics expert.

On the other hand, if you like to customize your graphics and like the idea of having control over every possible aspect of your plots, Handle Graphics provides powerful tools to do just that.

The graphics features presented in previous chapters are considered high-level commands and functions. They include `plot`, `mesh`, `axis`, and others. These functions are based on lower-level functions and properties that collectively are called Handle Graphics.

20.2 WHAT ARE HANDLE GRAPHICS OBJECTS?

Handle Graphics is based on the idea that every component of a graph is an ***object,*** that each object has a number or ***handle*** associated with it, and that each object has ***properties*** that can be modified as desired.

One of the most popular buzzwords in computing today is the word *object*. Object-oriented programming languages, data base objects, and operating system and application interfaces all use the concept of objects. An object can be loosely defined as a closely related collection of data structures or functions that form a unique whole. **In MATLAB, a graphics object is a distinct component of a graph that can be manipulated individually.**

Everything created by a graphics command is a graphics object. These include *Figure* windows or simply *figures,* as well as *axes, lines, surfaces, text,* and others. These objects are arranged in a hierarchy with parent objects and child objects. The computer screen itself is the *root* object and the parent of everything else. *Figures* are children of the *root; axes* and user-interface objects (to be discussed in the next chapter) are children of *figures. Line, text, surface, patch,* and *image* objects are children of *axes.* This hierarchy is diagrammed on the next page.

The *root* can contain one or more *figures,* and each *figure* can contain one or more sets of *axes.* All other objects (except *uicontrol* and *uimenu* objects discussed in the next chapter) are children of *axes* and display on those *axes.* All functions that create an object will create the parent object or objects if they do not exist. For example, if there are no *figures,* the `plot(rand(size([1:10])))` function creates a new *figure* and a set of *axes* with default ***properties,*** and then plots the *line* within those *axes.*

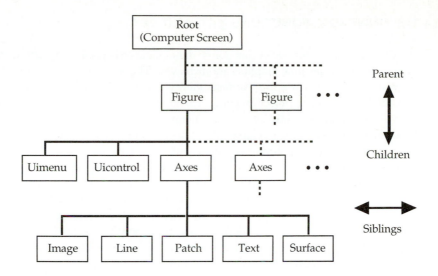

20.3 OBJECT HANDLES

Suppose you have three *figures* open with subplots in two of them and you want to change the color of a line on one of the subplot axes. How would you *identify* the line you wish to change? In MATLAB each object is *identified* by a number, called its **handle.**

Each time an object is created, a unique handle is created for it. The handle of the *root* object, the computer screen, is always zero. The » **Hf_fig=figure** command creates a new *figure* and returns its handle in the **Hf_fig** variable. *Figure* handles are integers that are usually displayed in the *Figure* window title bar. Other object handles are floating-point numbers in full MATLAB precision.

MATLAB commands are available to obtain the handles of *figures, axes,* and other objects. For example, » **Hf_fig=gcf** returns the handle of the current *figure,* and » **Ha_ax=gca** returns the handle of the current *axes* in the current *figure.* These functions and other object-manipulation tools are discussed later in this chapter.

To improve readability, in this text variables that contain object handles are given names beginning with a capital **H**, followed by a letter identifying the object type, then an underscore, and finally one or more descriptive characters. Thus, **Hf_fig** is a handle to a *figure,* **Ha_ax1** is a handle to an *axes* object, and **Ht_title** is a handle to a *text* object. When an object type is unknown, the letter **'x'** is used, such as **Hx_obj**. While handles can be given any name, following this convention makes it easy to spot handle variables in an M-file.

All MATLAB functions that create objects return a handle (or column vector of handles) for each object created. These functions include **plot**, **mesh**, **surf**, and others. Some plots are made up of more than one object. For example, a **mesh** plot consists of a single *surface* object with a single handle, while a **waterfall** plot consists of a number of *line* objects with individual handles associated with each *line.* For example » **Hl_wfall=waterfall(peaks(20))** returns a column vector containing 20 handles to *lines*.

20.4 THE UNIVERSAL FUNCTIONS get AND set

All objects have ***properties*** that define their characteristics. It is by setting these properties that one modifies how graphics are displayed. The list of properties associated with each object type (e.g., *axes, line, surface,* etc.) is unique, although a number of properties are valid for all objects. Object properties can include such things as an object's position, color, type, parent object, child objects, and many others. Each distinct object has properties associated with it that can be changed without affecting other objects of the same type. A complete list of the properties associated with each type of object (*figure, axes, line, text, surface, patch,* and *image*) is presented later in this chapter.

   Object properties consist of property names and their associated values. Property names are character strings. They are typically displayed in mixed case with the initial letter of each word capitalized, e.g., **'LineStyle'**. However, MATLAB recognizes a property regardless of case. In addition, you need only use enough characters to uniquely identify the property name. For example, the position property of an *axes* object can be called **'Position'**, **'position'**, or even **'pos'**.

   When an object is created, it is initialized with a full set of default property values that can be changed in one of two ways. The object creation function can be issued with **(property-name,property-value)** pairs, or properties can be changed after the object is created. An example of the former method is

```
» Hf_1=figure('Color','white')
```

which creates a new *figure* with default properties, except that the background color is set to white rather than the default black.

   Only two functions are necessary to obtain or change Handle Graphics object properties. The function **get** returns the current value of some property of an object. The simplest syntax for using **get** is **get(handle,'PropertyName')**. For example,

```
» p = get(Hf_1,'Position')
```

returns the position vector of the *figure* having the handle **Hf_1**. Similarly,

```
» c = get(Hl_a,'Color')
```

returns the color of an object having the handle **Hl_a.**

   The function **set** changes the values of Handle Graphics object properties and uses the syntax **set(handle,'PropertyName',Value)**. For example,

```
» set(Hf_1,'Position',p_vect)
```

sets the position of the *figure* having the handle **Hf_1** to that specified by the vector **p_vect**. Likewise,

```
» set(Hl_a,'Color','r')
```

sets the color of the object having the handle **Hl_a** to red.

In general, the **set** function can have any number of (**'PropertyName'**, **PropertyValue)** pairs, e.g.,

```
» set(H1_a,'Color','r','LineWidth',2,'LineStyle','--')
```

changes the *line* having the handle **H1_a** to red, its line width to 2 points, and its line style to dashed.

In addition to these primary purposes, the **get** and **set** functions provide help. For example, » **set(handle,'PropertyName')** returns a list of Values that can be assigned to the object described by **handle**, e.g.,

```
» set(Hf_1,'Units')
```

```
[ inches | centimeters | normalized | points | {pixels} ]
```

shows that there are five allowable character string values for the **'Units'** property of the *figure* referenced by **Hf_1**, and that **'pixels'** is the default value.

If you specify a property that does not have a fixed set of values, MATLAB informs you of that fact.

```
» set(Hf_1,'Position')
```

```
A figure's 'Position' property does not have a fixed set of property values.
```

In addition to the **set** command, the Handle Graphics object-creation functions (such as **figure**, **axes**, **line**, etc.) accept multiple pairs of property names and values. For example,

```
» figure('Color','blue','NumberTitle','off','Name','My Figure')
```

creates a new *figure* with a blue background labeled **'My Figure'** rather than the default window title **'Figure No. 1'**.

To illustrate the above concepts, consider the following example:

```
» Hf_fig = figure        % create a figure having an integer handle

Hf_fig =

    1

» H1_line = line         % create a line having a floating-point handle

H1_line =

   59.0002

» set(H1_line);          % list settable properties and potential values
```

```
Color
EraseMode: [ {normal} | background | xor | none ]
LineStyle: [ {-} | -- | : | -. | + | o | * | . | x ]
LineWidth
MarkerSize
Xdata
Ydata
Zdata

ButtonDownFcn
Clipping: [ {on} | off ]
Interruptible: [ {no} | yes ]
Parent
UserData
Visible: [ {on} | off ]
```

```
» get(Hl_line);              % list properties and current property values
```

```
Color = [1 1 1]
EraseMode = normal
LineStyle = -
LineWidth = [0.5]
MarkerSize = [6]
Xdata = [0 1]
Ydata = [0 1]
Zdata = [ ]

ButtonDownFcn =
Children = []
Clipping = on
Interruptible = no
Parent = [58.0002]
Type = line
UserData = []
Visible = on
```

In the above example, the 'Parent' property of the created *line* is the handle of the *axes* object containing the *line*. Moreover, the property lists shown are divided into two groups. The first group, before the blank line, lists properties that are unique to the particular object type. The second group, after the blank line, lists properties common to all object types. Note also that the **set** and **get** functions returned different property lists. **set** lists only those properties that can be changed with the **set** command, while **get** lists all object properties. In the previous example, **get** listed the 'Children' and 'Type' properties while **set** did not. These properties can be read, but not changed, i.e., they are *read-only* properties.

The number of properties associated with each object type is fixed, but the number varies among object types. As shown above, a *line* object lists 16 properties. On the other hand, an *axes* object lists 64 properties. Clearly, it is beyond the scope of this text to thoroughly describe and illustrate all properties of all object types! Many of them, however, are discussed in detail and all known properties are listed later in the text.

As an example of the use of object handles, consider the problem posed earlier, which was to plot a line in a non-standard color. In this case, the line color is specified using an RGB value of [1 .5 0], a medium orange color.

```
» x = -2*pi:pi/40:2*pi;            % create data

» y = sin(x);                      % find the sine of x

» Hl_sin = plot(x,y)               % plot sine and save line handle

Hl_sin =
    59.0002

» set(Hl_sin,'Color',[1 .5 0],'LineWidth',3) % Change the color and width
```

Now add a cosine curve in light blue.

```
» z = cos(x);                      % find the cosine of x

» hold on                          % keep the sine curve

» Hl_cos = plot(x,z);              % plot the cosine and save the handle

» set(Hl_cos,'Color',[.75 .75 1])  % color it light blue

» hold off
```

It's also possible to do the same thing with fewer steps.

```
» Hl_lines = plot(x,y,x,z);        % plot both curves and save both handles

» set(Hl_line(1),'Color',[1 .5 0],'LineWidth',3)

» set(Hl_line(2),'Color',[.75 .75 1])
```

How about adding a title and making the font size larger than normal?

```
» title('Handle Graphics Example') % add a title

» Ht_text = get(gca,'Title')       % get a handle to the title

» set(Ht_text,'FontSize',16)       % customize the font size
```

This last example illustrates an interesting point about *axes* objects. Every object has a `'Parent'` property and a `'Children'` property which contain handles to the descendent objects. A *line* plotted on a set of *axes* has the handle of the *axes* object as its `'Parent'` property value, and the empty matrix as the `'Children'` property value. At the same time, the *axes* object has the handle of its *figure* as the `'Parent'` value, and the handles of *line* objects as `'Children'` property values. The title string and axis labels are not included in the `'Children'` property values of the *axes,* but rather they are kept in the `'Title'`, `'XLabel'`, `'YLabel'`, and `'ZLabel'` properties. These *text* objects are created when the *axes* object is created. The `title` command sets the `'String'` property of the title *text* object within the current *axes.* Finally, the standard MATLAB functions `title`, `xlabel`, `ylabel`, and `zlabel` do not return handles, but do accept property and value arguments. For example, the following command adds a 24-point green title to the current plot:

```
» title('This is a title.','Fontsize',24,'Color','green')
```

In addition to `set` and `get`, MATLAB provides two additional functions to manipulate objects and their properties. Any object and all of its children can be deleted using the `» delete(handle)` function. Similarly, `» reset(handle)` resets all object properties associated with `handle` (except for the `'Position'` property) to the defaults for that object type.

20.5 FINDING OBJECTS

As you have seen, Handle Graphics provides access to the objects in a *figure* and allows you to customize graphics using the `get` and `set` commands. However, what if you forgot to save a handle or handles of objects in a *figure?* Or, maybe your variables were overwritten? How can you change object properties if you don't know their handles? To solve this problem, MATLAB provides tools for finding object handles.

Two of these tools, `gcf` and `gca`, were introduced earlier.

```
» Hf_fig = gcf
```

returns the handle of the current *figure,* and

```
» Ha_ax = gca
```

returns the handle of the current *axes* in the current *figure.*
In addition to the above, there is *get current object,* `gco`.

```
» Hx_obj = gco
```

returns the handle of the current object in the current *figure,* or alternatively

```
» Hx_obj = gco(Hf_fig)
```

returns the handle of the current object in the *figure* associated with handle `Hf_fig`.

The current object is defined as the last object clicked on with the mouse. This object can be any graphics object except the *root* (computer screen). However, if the mouse button is not clicked while the pointer is within a *figure,* **gco** returns an empty matrix. Something must have been selected for the current object to exist.

Once an object handle has been obtained, the object type can be found by querying an object's **'Type'** property, which is a character-string object name such as **'figure'**, **'axes'**, or **'text'**. For example,

> ```
> » x_type = get(Hx_obj,'Type')
> ```

The MATLAB functions **gcf**, **gca**, and **gco** are good examples that illustrate how MATLAB uses Handle Graphics to get information about objects. The function **gcf** gets the **'CurrentFigure'** property value of the *root* object, which is the handle of the current *figure*. The **gcf** M-file contains:

```
function h = gcf()
%GCF    Get current figure handle.
%   H = GCF returns the handle to the current figure. The current
%   figure is the figure (graphics window) that graphics commands
%   like PLOT, TITLE, SURF, etc. draw to if issued.
%
%   Use the commands FIGURE to change the current figure
%   to a different figure, or to create new ones.
%
%   See also FIGURE, CLOSE, CLF, GCA.

%   Copyright (c) 1984-94 by The MathWorks, Inc.

h = get(0,'CurrentFigure');
```

Similarly, the function **gca** returns the **'CurrentAxes'** property of the current *figure* as described by its M-file.

```
function h = gca()
%GCA    Get current axis handle.
%   H = GCA returns the handle to the current axis. The current
%   axis is the axis that graphics commands like PLOT, TITLE,
```

```
%    SURF, etc. draw to if issued.
%
%    Use the commands AXES or SUBPLOT to change the current axis
%    to a different axis, or to create new ones.
%
%    See also AXES,SUBPLOT,DELETE,CLA,HOLD,GCF.

%    Copyright (c) 1984-94 by The MathWorks, Inc.

h =  get(get(0,'CurrentFigure'),'CurrentAxes');
```

The gco function is similar, except that it tests for the existence of a *figure* before attempting to get the current object. **Note that gcf and gca cause the associated object to be created if none existed.** In gco as shown below, the existence of 'Children' is checked first, so that no *figure* object is created if it doesn't already exist.

```
function object = gco(figure)
%GCO    Handle of current object.
%    OBJECT = GCO returns the current object in
%    the current figure.
%
%    OBJECT = GCO(FIGURE) returns the current object
%    in figure FIGURE.
%
%    The current object for a given figure is the last
%    object clicked on with the mouse.

%    Copyright (c) 1984-94 by The MathWorks, Inc.

if isempty (get (0, 'Children')),
    object = [];
    return;
end;

if(nargin == 0)
    figure = get(0,'CurrentFigure');
end

object = get( figure, 'CurrentObject');
```

When something other than the `'CurrentFigure'`, `'CurrentAxes'`, or `'Current Object'` is desired, the function **get** can be used to obtain a vector of handles to the children of an object. For example,

```
» Hx_kids=get(gcf,'Children')
```

returns a vector containing handles of the children of the current *figure*.

This technique of getting `'Children'` handles can be used to search through the Handle Graphics hierarchy to find objects you want. For example, consider the problem of finding the handle of a green *line* object after plotting some data.

```
» x = -pi:pi/20:pi;              % create some data

» y = sin(x); z = cos(x);

» plot(x,y,'r',x,z,'g');         % plot two lines in red and green

» Hl_lines = get(gca,'Children'); % get the line handles

» for k=1:size(Hl_lines)         % find the green line

»    if get(Hl_lines(k),'Color') == [0 1 0]

»      Hl_green = Hl_lines(k)

»    end

» end

Hl_green =

    58.0001
```

Although this technique is effective, it gets complicated if many objects exist. The technique also misses *text* objects in titles and axis labels, unless these objects are tested individually.

Consider the problem of finding the handles of all green *line* objects when there are multiple *figures* with multiple *axes* on each.

```
» Hf_all = get(0,'Children');                    % get all figure handles

» for k = 1:length(Hf_all)

»    Ha_all = [Ha_all; get(Hf_all(k),'Children')];   % get all axes handles
```

```
» end

» for k = 1:length(Ha_all)

»    Hx_all = [Hx_all; get(Ha_all(k),'Children')];   % get axes child handles

» end

» for k = 1:length(Hx_all)

»    if get(Hx_all(k),'Type') == 'line'

»      Hl_all = [Hl_all; Hx_all(k)];                  % get line handles only

»    end

» end

» for k = 1:length(Hl_all)

»    if get(Hl_all(k),'Color') == [0 1 0]

»      Hl_green = [Hl_green; Hl_all(k)];              % find green ones

»    end

» end
```

To simplify the process of finding object handles, MATLAB versions 4.2 and later contain the built-in MATLAB function **findobj**, which returns the handles of objects with specified property values. The on-line help text for **findobj** is

```
» help findobj

FINDOBJ Find objects with specified property values.
     H = FINDOBJ('P1Name',P1Value,...) returns the handles of the
     objects at the root level and below whose property values
     match those passed as param-value pairs to the FINDOBJ
     command.

     H = FINDOBJ(ObjectHandles, 'P1Name', P1Value,...) restricts
     the search to the objects listed in ObjectHandles and their
     descendents.

     H = FINDOBJ(ObjectHandles, 'flat', 'P1Name', P1Value,...)
     restricts the search only to the objects listed in
     ObjectHandles.  Their descendents are not searched.
```

```
H = FINDOBJ returns the handles of the root object and all its
descendents.

H = FINDOBJ(ObjectHandles) returns the handles listed in
ObjectHandles, and the handles of all their descendents.

See also SET, GET, GCF, GCA.
```

`findobj` returns the handles of objects that meet the criteria you select. It checks all `'Children'`, including titles and labels of *axes*. If no objects are found to match the specified criteria, `findobj` returns an empty matrix.

The previous example becomes one line using `findobj`:

```
» Hl_green = findobj(0,'Type','line','Color',[0 1 0]);
```

20.6 SELECTING OBJECTS WITH THE MOUSE

The `gco` command returns the handle of the *current object* which is the last object clicked on with the mouse. However, how does MATLAB know which object was selected? For example, what handle is returned if you click on the intersection of two lines on a plot? Or, how far can the pointer be from a line when the button is clicked and still select the line? The answers are based on the rules MATLAB uses to select objects, and on something called *stacking order.*

Stacking order determines which overlapping object is *on top* of the others. Initially, the stacking order is determined when the object is created, with the object created last at the top of the stack. For example, if you issue two `figure` commands, two *figures* are created. The second *figure* is drawn on top of the first. The resulting stacking order has *figure 2* on top of *figure 1* and the current figure `gcf` is *figure 2*. If the `figure(1)` command is issued or if *figure 1* is clicked on, the stacking order changes. *Figure 1* moves to the top of the stack and becomes the current figure.

In the preceding example, the stacking order was apparent from the window overlap on the computer screen. However, this is not always the case. If two lines are plotted, the second line drawn is on top of the first at points where they intersect. If the first line is clicked on with the mouse at some other point, the first line is now at the top of the stack, and a click on the intersecting point will select the first line. **The current stacking order is given by the order in which `'Children'` handles appear for a given object. That is, » `Hx_kids=get(handle,'Children')` returns handles of children in stacking order. The first element in the vector `Hx_kids` is at the top of the stack and the last element is at the bottom of the stack.**

In addition to stacking order, when you click *near* an object with the mouse, the object is selected. Each type of object has its own *selection region* associated with it. For example, a *line* is selected if the mouse pointer is within 5 pixels of the line. The selection region of a *surface, patch,* or *text* object is the smallest rectangle that contains the object. The selection region of an *axes* object is the axes box itself plus the areas where labels and titles appear. Objects within *axes,* such as *lines* and *surfaces,* are higher in the stacking order, and clicking on them selects the associated object rather than the axes. Selecting an area outside the *axes*-selection region selects the *figure* itself.

The stacking order and object-selection rules discussed here are current as of MATLAB version 4.2c. Be aware, however, that the way MATLAB defines and uses stacking order and object selection regions is likely to change to some extent in version 5.

20.7 POSITION AND UNITS

The **'Position'** property of a *figure* object and many other Handle Graphics objects is a four-element row vector. The values in this vector are **[left, bottom, width, height]** where **[left, bottom]** is the position of the lower-left corner of the object relative to its parent, and **[width, height]** is the width and height of the object.

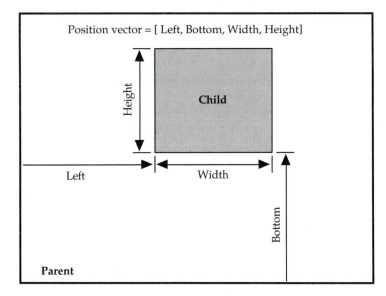

These position-vector values are in units specified by the **'Units'** property of the object. For example,

```
» get(gcf,'Position')

ans =
        360    544    560    420

» get(gcf,'Units')

ans =
        pixels
```

shows that the lower left corner of the current *figure* object is 360 *pixels* to the right and 544 *pixels* above the lower-left corner of the screen, and the *figure* object is 560 *pixels* wide and 420 *pixels* high. **Note that the position vector gives the drawable area within the *figure* object itself and does not include window borders, scroll bars, or title bar of the *Figure* window.**

The **'Units'** property defaults to pixels but can be inches, centimeters, points, or normalized coordinates. Pixels represent screen pixels, the smallest rectangular object that can be represented on a computer screen. For example, a computer display set to a resolution of 800 by 600 is 800 pixels wide by 600 pixels high. Points are a typesetting standard, where one point is equal to 1/72 of an inch. Normalized coordinates are in the range zero to one. In normalized coordinates, the lower-left corner of the screen is at **[0, 0]** and the upper-right corner is at **[1.0, 1.0]**. Inches and centimeters are self-explanatory.

To illustrate various **'Units'** property values, reconsider the above example.

```
» set(gcf,'Units','inches');

» get(gcf,'Position')

ans =
        3.7764    5.7203    5.8907    4.4245

» set(gcf,'Units','centimeters')

» get(gcf,'Position')

ans =
        9.5847    14.5185    14.9511    11.2297

» set(gcf,'Units','points')

» get(gcf,'Position')

ans =
        271.9001    411.8596    424.1339    318.5655

» set(gcf,'Units','normalized')

» get(gcf,'Position')

ans =
        0.2805 0.5303 0.4375 0.4102
```

All of these values represent the same position relative to the computer screen for a particular monitor.

The position of *axes* objects is also a four-element vector having the same form **[left, bottom, width, height]**, but specifying the object position relative to the lower-left corner of the parent *figure*. In general, the **'Position'** property of a child is relative to the position of its parent.

In order to be more descriptive, the computer screen or *root* object position property is not called `'Position'`, but rather `'ScreenSize'`. In this case, `[left, bottom]` is always `[0, 0]`, and `[width, height]` are the dimensions of the computer screen in units specified by the `'Units'` property of the *root* object.

20.8 PRINTING FIGURES

In addition to *figure* placement on the computer screen, MATLAB provides properties that control *figure* placement on the printed page as well as properties that specify attributes of the paper itself. For example, the MATLAB `orient` command uses `get` and `set` to change the values of paper properties. The help text for the `orient` function is

```
»help orient

ORIENT     Hardcopy paper orientation.
   ORIENT LANDSCAPE causes subsequent PRINT operations from the current
   figure window to generate output in full-page landscape orientation
   on the paper.
   ORIENT PORTRAIT returns to the default PORTRAIT orientation with
   the figure window occupying a rectangle with aspect ratio 4/3 in
   the middle of the page.
   ORIENT TALL causes the figure window to map to the whole page
   in portrait orientation.
   ORIENT, by itself, returns a string containing the current paper
   orientation, either PORTRAIT, LANDSCAPE or TALL.
   ORIENT is an M-file that sets the PaperOrientation and PaperPosition
   properties of the current figure window.

   See also PRINT.
```

The *figure* properties that affect printed output are as follows:

Paper Properties of Figures

PaperUnits	[{inches} | centimeters | normalized | points]
PaperOrientation	[{portrait} | landscape]
PaperPosition	A position vector in the form `[left,bottom, width,height]` where `left` and `bottom` are offsets from the lower-left corner of the page and `width` and `height` are *figure* dimensions

PaperSize	A two-element vector containing the paper size (`[8.5 11]`)
PaperType	`[{ usletter}	uslegal	a3	a4letter	a5	b4	tabloid]`

The `'PaperPosition'` and `'PaperSize'` properties are returned in units specified by the `'PaperUnits'` property. Changing the `'PaperPosition'` property
changes the size and location of the plot on a printed page in the same way that changing
the `'Position'` property of a *figure* object changes the size and location of the *figure*
on the screen.

Consider the following examples illustrating the use of paper properties:

```
» set(gcf,'PaperType','a4letter')
```

sets the paper used for the current *figure* to `'a4letter'`,

```
» set(gcf,'PaperOrientation','landscape')
```

sets the *figure* orientation of the current *figure* to `'landscape'`.

As with other *figure* properties, the above paper properties apply to individual
figures. Modifying properties for all *figures* is discussed in the next section on default
properties.

20.9 DEFAULT PROPERTIES

MATLAB assigns **default** properties to each object as it is created. Whenever you want to
deviate from these defaults, you must set them using Handle Graphics tools. In those cases
where you want to change the same properties every time, MATLAB allows you to set your
own default properties. MATLAB lets you change the default properties for individual objects and object types at any point in the object hierarchy.

You can set default values by using a special property-name string consisting of
`'Default'` followed by the object type and the property name. The handle you use in the
`set` command determines the point in the object parent-child hierarchy at which the default is applied. For example,

```
» set(0,'DefaultFigureColor',[.5 .5 .5])
```

sets the default color for all **new** *figure* objects to medium gray rather than the MATLAB
default black. This property applies to the *root* object (whose handle is always zero), so all
new *figures* will have a gray background.

Here are some other examples of defaults you can change.

```
» set(0,'DefaultAxesFontSize',14)          % larger axes fonts - all figures

» set(gcf,'DefaultAxesLineWidth',2)        % thick axis lines - this figure

» set(gcf,'DefaultAxesXColor','y')         % yellow X-axis lines and labels

» set(gcf,'DefaultAxesYGrid','on')         % Y axis grid lines - this figure

» set(0,'DefaultAxesBox','on')             % enclose axes - all figures

» set(gca,'DefaultLineLineStyle',':')      % dotted linestyle - these axes
```

When working with existing objects, it is always a good idea to restore them to their original state after they are used. **If you change default properties of objects in a routine, save the previous settings and restore them when exiting the routine.** For example, consider this function fragment.

```
oldunits = get(0,'DefaultFigureUnits');
set(0,'DefaultFigureUnits','normalized');
    <MATLAB statements>
set(0,'DefaultFigureUnits',oldunits);
return
```

If you want to customize MATLAB to use your default values at all times, simply include the desired **set** commands in your **startup.m** file. For example, if you want grids and axis boxes on all *axes* by default, and you usually print on A4-size paper, add the lines

```
set(0,'DefaultAxesXGrid','on')
set(0,'DefaultAxesYGrid','on')
set(0,'DefaultAxesZGrid','on')
set(0,'DefaultAxesBox','on')
set(0,'DefaultFigurePaperType','a4paper')
```

to your **startup.m** file. For more information on the script M-file **startup.m**, see Chapter 2.

There are three special property-value strings that reverse, override, or get user-defined default properties. They are **'remove'**, **'factory'**, and **'default'**. If you've changed a default property, you can reverse the change, resetting it to the original defaults, using **'remove'** as illustrated below:

```
» set(0,'DefaultFigureColor',[.5 .5 .5])   % set a new default

» set(0,'DefaultFigureColor','remove')     % return to MATLAB defaults
```

To temporarily override a default and use the original MATLAB default value for a particular object, use the special property value **'factory'**. For example,

```
» set(0,'DefaultFigureColor',[.5 .5 .5])    % set a new user default

» figure('Color','factory')    % create a new figure using the MATLAB default
```

The third special property-value string is **'default'**. This value forces MATLAB to search up the object hierarchy until it encounters a default value for the desired property. If found, it uses that default value. If the *root* object is reached and no user-defined default is found, the MATLAB factory default value is used. This concept is useful when you want to set an object property to a default property value after it was created with a different property value. To clarify the use of **'default'**, consider the example

```
» set(0,'DefaultLineColor','r')          % set the default at the root level

» set(gcf,'DefaultLineColor','g')        % current figure level default

» set(gca,'DefaultLineColor','b')        % current axes level default

» Hl_rand = plot(rand(size([1:10])));    % plot a yellow line

» set(Hl_rand,'Color','default')         % the line becomes blue

» set(gca,'DefaultLineColor','remove')   % the axes level default is removed

» set(Hl_rand,'Color','default')         % the line becomes green

» close(gcf)                             % close the window

» Hl_rand = plot(rand(size([1:10])));    % plot a yellow line in a new window

» set(Hl_rand,'Color','default')         % the line becomes red
```

Note that the **plot** command does not use *line* object defaults for the line color. If a color argument is not specified, the **plot** command uses the *axes* **'ColorOrder'** property to specify the color of each *line* it generates.

20.10 UNDOCUMENTED PROPERTIES

The properties listed for each object using the **get** and **set** commands are the *documented* properties. There are also *undocumented* properties used by MATLAB developers. Some of them can be set, but others are read-only.

One useful undocumented property of every object type is the `'Tag'` property. This property is useful to *tag* an object with a user-defined text string. For example,

```
» set(gca,'Tag','My Axes')
```

attaches the string `'My Axes'` to the current *axes* in the current *figure*. This string does not display in the *axes* or in the *figure,* but you can query the `'Tag'` property to identify the object. For example, if there are numerous *axes*, you can find the handle to the above *axes* object by using

```
» Ha_myaxes = findobj(0,'Tag','My Axes');
```

Like the `'UserData'` property, which is discussed in the next chapter, the `'Tag'` property is available for your exclusive use. No MATLAB function or M-file either changes or makes assumptions about the values contained in these properties. However, as will be discussed in the next chapter, some user-contributed M-files and several *Mastering MATLAB Toolbox* functions make use of the `'UserData'` property to store temporary data.

Since undocumented properties have been purposely left undocumented, one must be **very** cautious when using them. They are sometimes less robust than documented properties and are **always** subject to change. Undocumented properties may appear, disappear, change functionality, or even become documented in future versions of MATLAB.

Undocumented properties that should become documented in MATLAB version 5 include the following:

Property	Objects
`'TerminalHideGraphCommand'`	*root*
`'TerminalDimensions'`	*root*
`'TerminalShowGraphCommand'`	*root*
`'Tag'`	all objects
`'Layer'`	*axes*
`'PaletteModel'`	*surface, patch*

20.11 M-FILE EXAMPLES

The *Mastering MATLAB Toolbox* contains numerous utility functions that demonstrate the concepts discussed in this chapter. Some of these functions were illustrated in the 2-D and

3-D graphics chapters. Given the previous Handle Graphics discussion, these functions can now be discussed more thoroughly.

One of the simplest *Mastering MATLAB Toolbox* functions addresses a common problem. The MATLAB function **gcf** returns the handle of the current *figure*. However, it has a side effect. If no *figure* exists, **gcf** creates one and then returns its handle. What if you want to find out if a *figure* exists in the first place, but not create an empty one if there happens to be none? The function **mmgcf** does just that as described by its contents.

```
function Hf=mmgcf
%MMGCF Get Current Figure if it Exists.
% MMGCF returns the handle of the current figure if it exists.
% If no current figure exists, MMGCF returns an empty handle.
%
% Note that the function GCF is different. It creates a figure
% and returns its handle if it does not exist.

% Copyright (c) 1996 by Prentice-Hall, Inc.

Hf=get(0,'Children');  % check for figure children
if isempty(Hf)
    return
else
    Hf=get(0,'CurrentFigure');
end
```

The function **mmgcf** first checks for the existence of *figures,* which are children of the *root* object. If there is at least one *figure* object, the *root* **'CurrentFigure'** property value returns the current figure.

The function **mmgca** does the same thing for *axes* as described in its M-file.

```
function Ha=mmgca
%MMGCA Get Current Axes if it Exists.
% MMGCA returns the handle of the current axes if it exists.
% If no current axes exists, MMGCA returns an empty handle.
%
% Note that the function GCA is different. It creates a figure
% and an axes and returns the axes handle if they do not exist.
```

```
% Copyright (c) 1996 by Prentice-Hall, Inc.

Ha=findobj(0,'Type','axes');
if isempty(Ha)
    return
else
    Ha=get(get(0,'CurrentFigure'),'CurrentAxes');
end
```

There is no need for an **mmgco** function since **gco** already exhibits the behavior of returning an empty matrix when no object exists.

Another function in the *Mastering MATLAB Toolbox* is **mmzap**, which was introduced in the 2-D graphics chapter. As shown in its M-file below, it uses **mmgcf** for error checking, along with **findobj** and **get** to delete a specified graphics object.

```
function mmzap(arg)
%MMZAP Delete graphics object using mouse.
% MMZAP waits for a mouse click on an object in
% a figure window and deletes the object.
% MMZAP or MMZAP text erases text objects.
% MMZAP axes  erases axes objects.
% MMZAP line  erases line objects.
% MMZAP surf  erases surface objects.
% MMZAP patch  erases patch objects.
%
% Clicking on an object other than the selected type or striking
% a key on the keyboard aborts the command.

% Copyright (c) 1996 by Prentice-Hall, Inc.

if nargin<1, arg='text'; end

Hf=mmgcf;
if isempty(Hf), error('No Figure Available.'), end
if length(findobj(0,'Type','figure'))==1
    figure(Hf) % bring only figure forward
end
key=waitforbuttonpress;  % pause until user takes some action
if key  % key on keyboard pressed
    return    % take no action
else    % object selected
```

```
        object=gco;    % get object selected by buttonpress
        type=get(object,'Type');
        if all(type(1:4)==arg(1:4))     % delete only if 'Type' is correct
            delete(object)
        end
    end
```

The **mmzap** function illustrates a technique that is very useful when writing Handle Graphics function M-files. It uses the combination of **waitforbuttonpress** and **gco** to get the handle to an object selected using the mouse. **waitforbuttonpress** is a built-in MATLAB function that waits for a mouse click or key press. Its help text is

```
» help waitforbuttonpress

 WAITFORBUTTONPRESS Wait for key/buttonpress over figure.
     T = WAITFORBUTTONPRESS stops program execution until a key or
     mouse button is pressed over a figure window. Returns 0
     when terminated by a mouse buttonpress, or 1 when terminated
     by a keypress. Additional information about the terminating
     event is available from the current figure.

     See also GINPUT, GCF.
```

After a mouse button is pressed with the mouse pointer over a *figure,* **gco** returns the handle of the object clicked on. Then the handle can be used to manipulate the selected object. Other functions in the *Mastering MATLAB Toolbox* that use this simple selection technique are **mmline** and **mmaxes**. Of these, the M-file description for **mmline** is as follows:

```
function mmline(arg1,arg2,arg3,arg4,arg5,arg6)
%MMLINE Set Line Properties Using Mouse.
% MMLINE waits for a mouse click on a line then
% applies the desired properties to the selected line.
% Properties are given in pairs, e.g., MMLINE name value ...
% Properties:
% NAME          VALUE {default}
% color         [y m c r g b w k] or an rgb in quotes: '[r g b]'
% style         [- -- : -.]
% mark          [o + . * x]
% width         points for linewidth {0.5}
% size          points for marker size {6}
% zap           (n.a.)          delete selected line
```

```
% Examples:
% MMLINE color r width 2    sets color to red and width to 2 points
% MMLINE mark + size 8       sets marker type to + and size to 8 points
%
% Clicking on an object other than a line, or striking
% a key on the keyboard aborts the command.

% Copyright (c) 1996 by Prentice-Hall, Inc.

Hf=mmgcf;
if isempty(Hf), error('No Figure Available.'), end
if length(get(0,'Children'))==1
    figure(Hf) % bring only figure forward
end
key=waitforbuttonpress;
if key % key on keyboard pressed
    return
else % object selected
    Hl=gco;
    if strcmp(get(Hl,'Type'),'line') % line object selected
        for i=1:2:max(nargin-1,1)
            name=eval(sprintf('arg%.0f',i),[]); % get name argument
            if strcmp(name,'zap')
                delete(Hl),return
            end
            value=eval(sprintf('arg%.0f',i+1),[]); % get value
            if strcmp(name,'color')
                set(Hl,'Color',value)
            elseif strcmp(name,'style')
                set(Hl,'Linestyle',value)
            elseif strcmp(name,'mark')
                set(Hl,'Linestyle',value)
            elseif strcmp(name,'width')
                value=abs(eval(value));
                set(Hl,'LineWidth',value)
            elseif strcmp(name,'size')
                value=abs(eval(value));
                set(Hl,'MarkerSize',value)
            else
                disp(['Unknown Property Name: ' name])
            end
        end
    end
end
```

The *Mastering MATLAB Toolbox* function **mmpaper** illustrates the use of paper properties in a simple way. As shown below, **mmpaper** sets the paper properties of the current *figure* as well as sets the default for all future *figures*. In the next chapter, the function **mmpage** is discussed, which is a companion to **mmpaper**. **mmpage** creates a graphical user interface for setting *figure* position on the printed page.

```matlab
function mmpaper(arg1,arg2,arg3,arg4,arg5,arg6)
%MMPAPER Set Default Paper Properties.
% MMPAPER name value ...
% sets default paper properties for the current figure and
% succeeding figures based on name value pairs.
% Properties:
% NAME              VALUE   {default}
% units             [{inches},centimeters,points,normal]
% orient            [{portrait},landscape]
% type              [{usletter},uslegal,a3,a4letter,a5,b4,tabloid]
%
% Examples:
% MMPAPER units inch orient landscape
% MMPAPER type tabloid
%
% MMPAPER with no arguments returns the current paper defaults.

% Copyright (c) 1996 by Prentice-Hall, Inc.

Hf=mmgcf;
flag=0;
if isempty(Hf)
    flag=1;
    Hf=figure('Visible','off');
end
if nargin
    for i=1:2:max(nargin-1,1)
        name=eval(sprintf('arg%.0f',i),[]); % get name argument
        value=eval(sprintf('arg%.0f',i+1),[]); % get value argument
        if name(1)=='o'
            set(0,'DefaultFigurePaperOrientation',value)
            set(Hf,'PaperOrientation',value)
        elseif name(1)=='t'
            set(0,'DefaultFigurePaperType',value)
            set(Hf,'PaperType',value)
        elseif name(1)=='u'
            set(0,'DefaultFigurePaperUnits',value)
            set(Hf,'PaperUnits',value)
```

```
        else
            disp(['Unknown Property Name: ' name])
        end
    end
end
disp(['Paper Orientation: ' get(Hf,'PaperOrientation')])
disp(['Paper Type: ' get(Hf,'PaperType')])
disp(['Paper Units: ' get(Hf,'PaperUnits')])
if flag, delete(Hf); end
```

When placing objects in a particular position, it is sometimes useful to convert between pixels and normalized coordinates. The *Mastering MATLAB Toolbox* contains two functions to do these conversions. The first, **mmpx2n** converts from pixels to normalized; the other, **mmn2px** converts the opposite way. These functions demonstrate how to obtain **'Position'** in a desired set of units. First, save the present **'Units'** property of the object, then set the **'Units'** property to the desired value and get the **'Position'** property value desired, and finally reset the **'Units'** property to the original value. The **mmpx2n** M-file is shown below.

```
function y=mmpx2n(x,Hf)
%MMPX2N Pixel to Normalized Coordinate Transformation.
% MMPX2N(X) converts the position vector X from
% pixel coordinates to normalized coordinates w.r.t.
% the computer screen.
%
% MMPX2N(X,H) converts the position vector X from
% pixel coordinates to normalized coordinates w.r.t.
% the figure window having handle H.
%
% X=[left bottom width height] or X=[width height]

% Copyright (c) 1996 by Prentice-Hall, Inc.

msg='Input is not a pixel position vector.';
lx=length(x);

sz='Position';
if nargin==1,Hf=0;sz='ScreenSize';end

if any(x<1)|(lx~=4&lx~=2)
    error(msg)
end
```

```
      if lx==2,x=[1 1 x(:)'];end % [width height] input format
      u=get(Hf,'Units'); % get units
      set(Hf,'Units','pixels'); % set units to pixels
      s=get(Hf,sz);
      y=(x-1)./([s(3:4) s(3:4)]-1); % convert
      set(Hf,'Units',u); % reset units
      if any(y>1)
          error(msg)
      end
      if lx==2,y=y(3:4);end % [width height] output format
```

Both *Mastering MATLAB Toolbox* functions **mmcont2** and **mmcont3** plot contours with user-specified color maps. Each function parses the input arguments and builds a string containing a color specification. Once this string is created, the **'ColorOrder'** property of the current *axes* is set. Finally, they call the **contour** and **contour3** functions respectively with appropriate arguments to render the plots. The content of the **mmcont2** M-file is shown below.

```
      function [cs,h]=mmcont2(arg1,arg2,arg3,arg4,arg5)
      %MMCONT2 2-D contour plot using a colormap.
      % MMCONT2(X,Y,Z,N,C) plots N contours of Z in 2-D using the color
      % specified in C. C can be a linestyle and color as used in plot,
      % e.g., 'r-', or C can be the string name of a colormap. X and Y
      % define the axis limits.
      % If not given default argument values are: N = 10, C = 'hot',
      % X and Y = row and column indices of Z. Examples:
      % MMCONT2(Z)                  10 lines with hot colormap
      % MMCONT2(Z,20)               20 lines with hot colormap
      % MMCONT2(Z,'copper')         10 lines with copper colormap
      % MMCONT2(Z,20,'gray')        20 lines with gray colormap
      % MMCONT2(X,Y,Z,'jet')        10 lines with jet colormap
      % MMCONT2(Z,'c-')             10 dashed lines in cyan
      % MMCONT2(X,Y,Z,25,'pink')    25 lines in pink colormap
      %
      % CS=MMCONT2(...) returns the contour matrix CS as described in
      % CONTOURC.
      % [CS,H]=MMCONT2(...) returns a column vector H of handles to
      % line objects.

      % Copyright (c) 1996 by Prentice-Hall, Inc.
```

```
n=10;     c='hot';     % default values
nargs=nargin; cflag=1;
if nargin<1,error('Not enough input arguments.'),end


for i=2:nargin     % check input arguments for N and C
    argi=eval(sprintf('arg%.0f',i));
    if ~isstr(argi)&length(argi)==1 % must be N, grab it
        n=argi;
        nargs=i;                    % # args to pass to contour2
    elseif isstr(argi)              % must be C
        if exist(argi)==2           % is colormap, so grab it
            c=argi;
            nargs=i-1;
        else                        % is single color/linestyle
            cflag=0;
            nargs=i;
        end
    end
end
if cflag  % a colormap has been chosen
    clf               % clear figure
    view(2)           % make it 2-D
    hold on           % hold it
    mapstr=sprintf([c '(%.0f)'],n);
    set(gca,'ColorOrder',eval(mapstr));
end
evalstr='[CS,H]=contour(';
for i=1:nargs
    evalstr=[evalstr sprintf('arg%.0f',i) ','];
end
lstr=length(evalstr);
evalstr(lstr:lstr+1)=');';
eval(evalstr)
hold off
if nargout==1,          cs=CS;
elseif nargout==2,      cs=CS; h=H;
end
```

The last *Mastering MATLAB Toolbox* function to be discussed here is **mmtile**. As described in the 2-D graphics chapter, this function arranges up to four existing *Figure* windows in a tile pattern on the computer screen. The content of **mmtile.m** is shown below.

```
function h=mmtile(n)
%MMTILE Tile Figure Windows.
```

```
% MMTILE with no arguments, tiles the current figure windows
% and brings them to the foreground.
% Figure size is adjusted so that 4 figure windows fit on the screen.
% Figures are arranged in a clockwise fashion starting in the
% upper-left corner of the display.
%
% MMTILE(N) makes tile N the current figure if it exists.
% Otherwise, the next tile is created for subsequent plotting.
%
% Tiled figure windows are titled TILE #1, TILE #2, TILE #3, TILE #4.

% Copyright (c) 1996 by Prentice-Hall, Inc.

HT=40;      % tile height fudge in pixels
WD=20;      % tile width fudge
% adjust the above as necessary to eliminate tile overlaps
% bigger fudge numbers increase gaps between tiles

Hf=sort(get(0,'Children'));  % get handles of current figures
nHf=length(Hf);
set(0,'Units','Pixels')      % set screen dimensions to pixels
sz=get(0,'Screensize');      % get screen size in pixels
tsz=0.9*sz(3:4);             % default tile area is almost whole monitor
if sz(4)>sz(3),              % if portrait monitor
    tsz(2)=.75*tsz(1);       % take a landscape chunk
end
tsz=min(tsz,[920 690]);% hold tile area on large screens to 920 by 690
tl(1,1)=sz(3)-tsz(1)+1;     % left side of left tiles
tl(2,1)=tl(1,1)+tsz(1)/2;   % left side of right tiles
tb(1,1)=sz(4)-tsz(2)+1;     % bottom of bottom tiles
tb(2,1)=tb(1,1)+tsz(2)/2;   % bottom of top tiles

tpos=zeros(4);               % matrix holding tile position vectors
tpos(:,1)=tl([1 2 2 1],1);        % left sides
tpos(:,2)=tb([2 2 1 1],1);        % bottoms
tpos(:,3)=(tsz(1)/2-WD)*ones(4,1);  % widths
tpos(:,4)=(tsz(2)/2-HT)*ones(4,1);  % heights
tpos=fix(tpos);              % make sure pixel positions are integers
if nargin==0                 % tile figures as needed
    for i=1:min(nHf,4)
        set(Hf(i),'Units','pixels')
        if any(get(Hf(i),'Position')~=tpos(i,:))
            set(Hf(i), 'Position',tpos(i,:),...
                   'NumberTitle','off',...
```

```
                        'Name',['TILE #' int2str(i)])
            end
            figure(Hf(i))
        end
    else                        % go to tile N or create it
        n=rem(abs(n)-1,4)+1;    % N must be between 1 and 4
        if n<=nHf               % tile N exists, make it current
            figure(Hf(n))
        else                    % tile N does not exist, create next one
            n=nHf+1;
            figure( 'Position',tpos(n,:),...
                    'NumberTitle','off',...
                    'Name',['TILE #' int2str(n)])
        end
    end
end
```

As described above, **mmtile** gets the handles of all *figure* objects and the screen size from the *root* object, calculates the new positions and sizes for the *figures,* and then sets the **'Units'**, **'Position'**, **'NumberTitle'**, and **'Name'** properties of each *figure.* This has the effect of placing and sizing the *figures,* and changing the name string in the title bar of each window. The **HT** and **WD** fudge numbers are computer-platform dependent. They compensate for the fact that the **'Position'** of a *figure* describes the drawable area within a *Figure* window rather than its external dimensions.

20.12 PROPERTY NAMES AND VALUES

The following tables list the object properties and values found in MATLAB version 4.2c. Properties with an asterisk (*****) are undocumented. Property values enclosed in set brackets **{}** are default values.

Root Object Properties

BlackAndWhite	Automatic hardware checking flag
on:	Assume display is monochrome; do not check
{off}:	Check the display type
* BlackOutUnusedSlots	Values are [{no} | yes]
* CaptureMap	

`CaptureMatrix`	Read-only matrix of image data of the region enclosed by the `CaptureRect` rectangle. Use `image` to display
`CaptureRect`	Size and position of rectangle to capture. A four-element vector `[left,bottom,width, height]` in units specified by the `Units` property
* `CaseSen`	Values are `[ {on} | off ]`
`CurrentFigure`	The handle of the current *figure* window
`Diary`	Session logging
`on:`	Copy all keyboard input and most output to a file
`{off}:`	Do not save input and output to a file
`DiaryFile`	A string containing the file name of the `Diary` file. The default file name is `diary`
`Echo`	Script echoing mode
`on:`	Display each line of a script file as it executes
`{off}:`	Do not echo unless `echo on` is specified
`Format`	Number display format
`{short}:`	Fixed-point format with 5 digits
`shortE:`	Floating-point format with 5 digits
`long:`	Scaled fixed-point format with 15 digits
`longE:`	Floating-point format with 15 digits
`hex:`	Hexadecimal format
`bank:`	Fixed-format of dollars and cents
`+:`	Displays + and − symbols
`rat:`	Approximation by ratio of small integers
`FormatSpacing`	Output spacing
`{loose}:`	Display extra line feeds
`compact:`	Suppress extra line feeds

* `HideUndocumented`	Control display of undocumented properties
`no:`	Display undocumented properties
`{yes}:`	Do not display undocumented properties
`PointerLocation`	Read-only vector `[left, bottom]` or `[x, y]` of pointer location relative to the lower-left corner of the screen in units specified by the `Units` property
`PointerWindow`	Handle of the *Figure* window containing the mouse pointer. If not in a *Figure* window, contains 0
`ScreenDepth`	Integer specifying depth of the screen in bits, e.g., 1 for monochrome, 8 for 256 colors or grayscales
`ScreenSize`	Position vector `[left, bottom, width, height]` where `[left, bottom]` is always `[0 0]` and `[width, height]` are the screen dimensions in units specified by the `Units` property
* `StatusTable`	Vector
* `TerminalHideGraphCommand`	Text string
`TerminalOneWindow`	Used by the terminal graphics driver
`no:`	Terminal has multiple windows
`yes:`	Terminal has only one window
* `TerminalDimensions`	Vector `[width, height]` of terminal dimensions
`TerminalProtocol`	Terminal type set at startup, then read-only
`none:`	Not in terminal mode, not connected to X server
`x:`	X display server found; X Windows mode
`tek401x:`	Tektronix 4010/4014 emulation mode

	`tek410x:`	Tektronix 4100/4105 emulation mode
*	`TerminalShowGraphCommand`	Text string
	`Units`	Unit of measurement for `position` property values
	`inches:`	Inches
	`centimeters:`	Centimeters
	`normalized:`	Normalized coordinates, where the lower-left corner of the screen maps to `[0  0]` and the upper-right corner maps to `[1  1]`
	`points:`	Typesetting points; equal to 1/72 of an inch
	`{pixels}:`	Screen pixels; the smallest unit of resolution of the computer screen
*	`UsageTable`	Vector
	`ButtonDownFcn`	MATLAB callback string, passed to the `eval` function whenever the object is selected; initially an empty matrix
	`Children`	Read-only vector of handles to all *figure* objects
	`Clipping`	Data clipping
	`{on}:`	No effect for *root* objects
	`off:`	No effect for *root* objects
	`Interruptible`	`ButtonDownFcn` callback string interruptibility
	`{no}:`	Not interruptible by other callbacks
	`yes:`	Interruptible by other callbacks
	`Parent`	Handle of parent object, always the empty matrix
*	`Selected`	Values are `[on	off]`
*	`Tag`	Text string
	`Type`	Read-only object identification string; always `root`
	`UserData`	User-specified data. Can be matrix, string, etc
	`Visible`	Object visibility
	`{on}:`	No effect for *root* objects
	`off:`	No effect for *root* objects

Figure Object Properties

`BackingStore`	Store a copy of the *Figure* window for fast redraw
`{on}:`	Copy from backing store when previously covered portions of a *figure* are exposed. Faster window refresh, but uses more memory
`off:`	Redraw previously covered portions of the *figure*. Slower window refresh, but saves memory
* `CaptureMap`	Matrix
* `Clint`	Matrix
`Color`	*figure* background color. A 3-element RGB vector or one of MATLAB's predefined color names. The default color is `black`
`Colormap`	An m-by-3 matrix of RGB vectors. See the `colormap` function
* `Colortable`	Matrix; probably contains a copy of the system color table
`CurrentAxes`	The handle of the *figure's* current *axes*
`CurrentCharacter`	The most recent key pressed on the keyboard when the mouse pointer was in the *Figure* window
`CurrentMenu`	The handle of the most-recently selected menu item
`CurrentObject`	The handle of the most-recently selected object within the *Figure* window. This is the handle returned by the function `gco`
`CurrentPoint`	A position vector `[left, bottom]` or `[x, y]` of the point within the *Figure* window where the mouse pointer was located when the mouse button was last pressed or released
`FixedColors`	An n-by-3 matrix of RGB values that define the colors using slots in the system color table. The initial fixed colors are `black` and `white`
* `Flint`	
`InvertHardcopy`	Change *figure* element colors for printing

`{on}:`	Change black *figure* background to white, and lines, text, and axes to black for printing
`off:`	Color of printed output exactly matches display
`KeyPressFcn`	MATLAB callback string passed to the **eval** function whenever a key is pressed while the mouse pointer is within the *Figure* window
`MenuBar`	Display MATLAB menus at the top of the *Figure* window or at the top of the screen on some systems
`{figure}:`	Display default MATLAB menus
`none:`	Do not display default MATLAB menus
`MinColormap`	Minimum number of color table entries to use. This affects the system color table and, if set too low, can cause unselected *figures* to display in false colors
`Name`	*Figure* window title (**not** axis title). By default, the empty string. If set to **string**, the window title becomes: **Figure No. n: string**
`NextPlot`	Determines drawing action for new plots
`new:`	Create a new *figure* before drawing
`{add}:`	Add new objects to the current *figure*
`replace:`	Reset all *figure* object properties except **Position** to defaults and delete all children before drawing
`NumberTitle`	Prepend the *figure* number to the *figure* title
`{on}:`	The window title is **Figure No. N**, with **: string** appended if the **Name** property is set to **string**
`off:`	The window title is the **Name** property string only
`PaperUnits`	Unit of measurement for **Paper ...** properties
`{inches}:`	Inches
`centemeters:`	Centimeters
`normalized:`	Normalized coordinates
`points:`	Points. One point is 1/72 of an inch

PaperOrientation	Paper orientation for printing
{portrait}:	Portrait orientation; longest page dimension is vertical
landscape:	Landscape orientation; longest page dimension is horizontal
PaperPosition	Position vector **[left, bottom, width, height]** representing the location of the *figure* on the printed page. **[left, bottom]** represents the location of the lower-left corner of the *figure* with respect to the lower-left corner of the printed page **[width, height]** are the dimensions of the printed *figure*. Units are specified by the **PaperUnits** property
PaperSize	Vector **[width, height]** representing the dimensions of the paper to be used for printing. Units are specified by the **PaperUnits** property. The default **PaperSize** is **[8.5, 11]**
PaperType	Paper type for printed figures. When **PaperUnits** is set to normalized, MATLAB uses **PaperType** to scale figures for printing
{usletter}:	Standard U.S. letter paper
uslegal:	Standard U.S. legal paper
a3:	European A3 paper
a4letter:	European A4 letter paper
a5:	European A5 paper
b4:	European B4 paper
tabloid:	Standard U.S. tabloid paper
Pointer	Mouse pointer shape
crosshair:	Crosshair pointer
{arrow}:	Arrow pointer
watch:	Watch pointer
topl:	Arrow pointing top-left
topr:	Arrow pointing top-right
botl:	Arrow pointing bottom-left
botr:	Arrow pointing bottom-right

`circle:`	Circle
`cross:`	Double-line cross
`fleur:`	Four-headed arrow or compass
`Position`	Position vector `[left, bottom, width, height]` where `[left, bottom]` represents the location of the lower-left corner of the *figure* with respect to the lower-left corner of the computer screen and `[width, height]` are the screen dimensions. Units are specified by the `Units` property
`Resize`	Allow/disallow interactive *Figure* window resizing
`{on}:`	The window may be resized using the mouse
`off:`	The window cannot be resized using the mouse
`ResizeFcn`	MATLAB callback string passed to the `eval` function whenever the window is resized using the mouse
* `Scrolled`	Values are `[{on}	off]`
`SelectionType`	A read-only string providing information about the method used for the last mouse button selection. The actual key and/or button press is platform-dependent
`{normal}:`	Click (press and release) the left, or only, mouse button
`extended:`	Hold down the **shift** key and make multiple normal selections; click both buttons of a two-button mouse; or click the middle button of a three-button mouse
`alt:`	Hold down the **Control** key and make a normal selection; or click the right button of a two or three-button mouse
`open:`	Double-click any mouse button
`ShareColors`	Share color table slots
`no:`	Do not share color table slots with other windows
`{yes}:`	Reuse existing color table slots whenever possible
* `StatusTable`	Vector

Units	Unit of measurement for **position** property values
inches:	Inches
centimeters:	Centimeters
normalized:	Normalized coordinates, where the lower-left corner of the screen maps to **[0 0]** and the upper right corner maps to **[1 1]**
points:	Typesetting points; equal to 1/72 of an inch
{pixels}:	Screen pixels; the smallest unit of resolution of the computer screen
* **UsageTable**	Vector
WindowButtonDownFcn	MATLAB callback string passed to the **eval** function whenever a mouse button is pressed while the mouse pointer is within a *figure*
WindowButtonMotionFcn	MATLAB callback string passed to the **eval** function whenever a mouse pointer is moved while the mouse pointer is within a *figure*
WindowButtonUpFcn	MATLAB callback string passed to the **eval** function whenever a mouse button is released while the mouse pointer is within a *figure*
* **WindowID**	Large integer
ButtonDownFcn	MATLAB callback string, passed to the **eval** function whenever the *figure* is selected; initially an empty matrix
Children	Read-only vector of handles to all children of the *figure; axes, uicontrol,* and *uimenu* objects
Clipping	Data clipping
{on}:	No effect for *figure* objects
off:	No effect for *figure* objects
Interruptible	Specifies if *figure* callback strings are interruptible
{no}:	Not interruptible by other callbacks
yes:	Callback strings are interruptible by other callbacks
Parent	Handle of the *figure's* parent object, always **0**

* `Selected`	Values are [`on` | `off`]
* `Tag`	Text string
`Type`	Read-only object identification string; always `figure`
`UserData`	User-specified data. Can be matrix, string, etc.
`Visible`	Visibility of *Figure* window
`{on}:`	Window is visible on the screen
`off:`	Window is not visible

Axes Object Properties

`AspectRatio`	Aspect ratio vector [`axis_ratio`, `data_ratio`] where `axis_ratio` is the aspect ratio of the *axes* object (width/height), and `data_ratio` is the ratio of the lengths of data units along the horizontal and vertical axes. If set, MATLAB creates the largest *axes* that will fit within the rectangle defined by `Position` while preserving these ratios. The default value for `AspectRatio` is [`NaN`, `NaN`]
`Box`	*Axes* bounding box
`on:`	Enclose *axes* in a box or cube
`{off}:`	Do not enclose *axes*
`CLim`	Color limit vector [`cmin cmax`] that determines the mapping of data to the colormap. `cmin` is the data value that maps to the first colormap entry, and `cmax` maps to the last. See the `caxis` function
`CLimMode`	Color-limits mode
`{auto}:`	Color limits span the full range of data of the *axes* children
`manual:`	Color limits do not automatically change. Setting `Clim` sets `ClimMode` to manual
`Color`	*Axes* background color. A 3-element RGB vector or one of MATLAB's predefined color

		names. The default color is **none**, which uses the *figure* background color
	`ColorOrder`	An m-by-3 matrix of RGB values. If a *line* color is not specified with **plot** and **plot3**, these colors will be used. The default **ColorOrder** is yellow, magenta, cyan, red, green, blue
	`CurrentPoint`	Coordinate matrix containing a pair of points in the *axes* data space that define a line in 3-D that extends from the *front* to the *back* of the *axes* volume. The form is [xb yb zb; xf yf zf]. Units are specified by the **Units** property. The points [xf yf zf] are the coordinates of the last mouse click on the *axes* object
	`DrawMode`	Object rendering order
	`{normal}:`	Sort objects and draw them from back to front based on the current **view**
	`fast:`	Draw objects in the order created; do not sort first
*	`ExpFontAngle`	Values are [`{normal}` | `italic` | `oblique`]
*	`ExpFontName`	Default is **Helvetica**
*	`ExpFontSize`	Default is 8 points
*	`ExpFontStrikeThrough`	Values are [`on` | `{off}`]
*	`ExpFontUnderline`	Values are [`on` | `{off}`]
*	`ExpFontWeight`	Values are [`light` | `{normal}` | `demi` | `bold`]
	`FontAngle`	Italics for *axes* text
	`{normal}:`	Regular font angle
	`italic:`	Italics
	`oblique:`	Italics on some systems
	`FontName`	Name of the font used for *axes* tick labels. *Axes* labels do not change fonts until they are redisplayed by setting the **XLabel**, **YLabel**, and **ZLabel** properties. The default font is **Helvetica**
	`FontSize`	Size in points used for *axes* labels and titles. Default is 12 points
*	`FontStrikeThrough`	Values are [`on` | `{off}`]

*	`FontUnderline`	Values are [`on` | `{off}`]
	`FontWeight`	Bolding for *axes* text
	`light`:	Light font weight
	`{normal}`:	Normal font weight
	`demi`:	Medium or bold font weight
	`bold`:	Bold font weight
	`GridLineStyle`	Line style of grid lines
	`-`:	Solid lines
	`--`:	Dashed lines
	`{:}`:	Dotted lines
	`-.`:	Dash-dot lines
*	`Layer`	Values are [`top` | `{bottom}`]
	`LineStyleOrder`	A text string specifying the order of line styles to use to plot multiple lines on the axes. For example, `'.-|:|--|-'` will cycle through dot-dash, dotted, dashed, and solid lines. The default `LineStyleOrder` is `'-'`; solid lines only
	`LineWidth`	Width of X, Y, and Z axis lines. Default is 0.5 points
*	`MinorGridLineStyle`	Values are [`-`|`--`| `{:}` | `-.`]
	`NextPlot`	Action to be taken for new plots
	`new`:	Create new *axes* before drawing
	`add`:	Add new objects to the current *axes*. See `hold`
	`{replace}`:	Delete the current *axes* and its children, and replace it with a new *axes* object before drawing
	`Position`	Position vector [`left, bottom, width, height]` where [`left, bottom]` represents the location of the lower-left corner of the *axes* with respect to the lower-left corner of the *figure* object and [`width, height]` are the *axes* dimensions. Units are specified by the `Units` property
	`TickLength`	Vector [`2Dlength 3Dlength]` representing the length of *axes* tick marks in 2-D and 3-D views. The length is relative to the axis length. Default is [0.01 0.025] representing 1/100 of the axis

		length in 2-D views and 25/1000 of the axis length in 3-D views
TickDir		[{in} | out]
	in:	Tick marks point inward from the axis line. Default for 2-D view
	out:	Tick marks point outward from the axis line. Default for 3-D view
Title		The handle of the *axes* title *text* object
Units		Unit of measurement for position property values
	inches:	Inches
	centimeters:	Centimeters
	{normalized}:	Normalized coordinates, where the lower-left corner of the object maps to [0 0] and the upper-right corner maps to [1 1]
	points:	Typesetters points; equal to 1/72 of an inch
	pixels:	Screen pixels; the smallest unit of resolution of the computer screen
View		A vector [az el] representing the viewpoint of the observer. az is the azimuth or rotation of the viewpoint in degrees to the right of the negative Y axis. el is the elevation in degrees above the plane of the X-Y axis. See the 3-D graphics chapter for details
XColor		An RGB vector or predefined MATLAB color string that specifies the color of the X axis line, labels, tick marks, and grid lines. Default is white
XDir		Direction of increasing X values
	{normal}:	X values increase from left to right
	reverse:	X values increase from right to left
XForm		A 4-by-4 view transformation matrix. Setting view affects XForm
XGrid		X-axis grid lines
	on:	Grid lines are drawn at each tick mark on the X axis
	{off}:	No grid lines are drawn

XLabel	The handle of the **X**-axis label *text* object
XLim	Vector [**xmin xmax**] specifying the minimum and maximum **X**-axis values
XLimMode	**X**-axis limits mode
{auto}:	**XLim** is automatically calculated to include all **XData** of all *axes* children
manual:	**X**-axis limits are taken from **XLim**
* XMinorGrid	Values are [**on** | {**off**}]
* XMinorTicks	Values are [**on** | {**off**}]
XScale	**X**-axis scaling
{linear}:	Linear scaling
log:	Logarithmic scaling
XTick	Vector of data values at which tick marks will be drawn on the **X** axis. Setting **XTick** to the empty matrix will suppress tick marks
XTickLabels	A matrix of text strings to be used to label tick marks on the **X** axis. If empty, MATLAB will use the data values at the tick marks
XTickLabelMode	**X**-axis-tick mark labeling mode
{auto}:	**X**-axis tick-labels span the **XData**
manual:	Take **X**-axis tick labels from **XTickLabels**
XTickMode	**X**-axis tick-mark-spacing mode
{auto}:	**X**-axis tick-mark spacing to span **XData**
manual:	**X**-axis tick-mark spacing from **XTick**
YColor	An RGB vector or predefined MATLAB color string that specifies the color of the **Y** axis line, labels, tick marks, and grid lines. Default is **white**
YDir	Direction of increasing Y values
{normal}:	**Y** values increase from bottom to top
reverse:	**Y** values increase from top to bottom
YGrid	**Y**-axis grid lines
on:	Grid lines are drawn at each tick mark on the **Y** axis
{off}:	No grid lines are drawn
YLabel	The handle of the **Y**-axis label *text* object

`YLim`	Vector `[ymin ymax]` specifying the minimum and maximum `Y`-axis values
`YLimMode`	Y-axis limits mode
`{auto}:`	`YLim` is automatically calculated to include all `YData` of all *axes* children
`manual:`	Y-axis limits are taken from `YLim`
* `YMinorGrid`	Values are [`on` | `{off}`]
* `YMinorTicks`	Values are [`on` | `{off}`]
`YScale`	Y-axis scaling
`{linear}:`	Linear scaling
`log:`	Logarithmic scaling
`YTick`	Vector of data values at which tick marks will be drawn on the `Y` axis. Setting `YTick` to the empty matrix will suppress tick marks
`YTickLabels`	A matrix of text strings to be used to label tick marks on the `Y` axis. If empty, MATLAB will use the data values at the tick marks
`YTickLabelMode`	Y-axis tick-mark labeling mode
`{auto}:`	Y-axis tick labels span the `YData`
`manual:`	Take Y-axis tick labels from `YTickLabels`
`XTickMode`	Y-axis tick-mark spacing mode
`{auto}:`	Y-axis tick-mark spacing to span `YData`
`manual:`	Y-axis tick-mark spacing from `YTick`
`ZColor`	An RGB vector or predefined MATLAB color string that specifies the color of the `Z` axis line, labels, tick marks, and grid lines. Default is `white`
`ZDir`	Direction of increasing `Z` values
`{normal}:`	`Z` values increase from bottom to top (3-D) or pointing out of the screen (2-D)
`reverse:`	`Z` values increase from top to bottom (3-D) or pointing into the screen (2-D)
`ZGrid`	Z-axis grid lines
`on:`	Grid lines are drawn at each tick mark on the `Z` axis
`{off}:`	No grid lines are drawn
`ZLabel`	The handle of the `Z`-axis label *text* object

`ZLim`	Vector [`zmin zmax`] specifying the minimum and maximum `Z`-axis values
`ZLimMode`	Z-axis limits mode
{`auto`}:	`ZLim` is automatically calculated to include all `ZData` of all *axes* children
`manual`:	Z-axis limits are taken from `ZLim`
* `ZMinorGrid`	Values are [`on` | {`off`}]
* `ZMinorTicks`	Values are [`on` | {`off`}]
`ZScale`	Z-axis scaling
{`linear`}:	Linear scaling
`log`:	Logarithmic scaling
`ZTick`	Vector of data values at which tick marks will be drawn on the `Z` axis. Setting `ZTick` to the empty matrix will suppress tick marks
`ZTickLabels`	A matrix of text strings to be used to label tick marks on the `Z` axis. If empty, MATLAB will use the data values at the tick marks
`ZTickLabelMode`	Z-axis tick-mark labeling mode
{`auto`}:	Z-axis tick labels span the `ZData`
`manual`:	Take `Z`-axis tick labels from `ZTickLabels`
`XTickMode`	Z-axis tick-mark spacing mode
{`auto`}:	Z-axis tick-mark spacing to span `ZData`
`manual`:	Z-axis tick-mark spacing from `ZTick`
`ButtonDownFcn`	MATLAB callback string, passed to the **eval** function whenever the *axes* is selected; initially an empty matrix
`Children`	Read-only vector of handles to all children of the *axes* except *axes* labels and titles: *line, surface, image, patch,* and *text* objects
`Clipping`	Data clipping
{`on`}:	No effect for *axes* objects
`off`:	No effect for *axes* objects
`Interruptible`	Specifies if **ButtonDownFcn** callback string is interruptible
{`no`}:	Not interruptible by other callbacks

	yes:	ButtonDownFcn callback string is interruptible
Parent		Handle of the *figure* containing the *axes* object
* Selected		Values are [on | off]
* Tag		Text string
Type		Read-only object identification string; always **axes**
UserData		User-specified data. Can be matrix, string, etc.
Visible		Visibility of *axes* lines, tick marks, and labels
	{on}:	Axes are visible on the screen
	off:	Axes are not visible

Line Object Properties

Color		*Line* color. A 3-element RGB vector or one of MATLAB's predefined color names. The default color is **white**
EraseMode		Erase and redraw mode
	{normal}:	Redraws the affected region of the display ensuring that all objects are rendered correctly. This is the most accurate mode but is also the slowest
	background:	The *line* is erased by drawing it in the figure's background color. This can damage objects behind the erased *line*
	xor:	The *line* is drawn and erased by performing an exclusive OR (XOR) with the color of the screen beneath it. Incorrect color can be used when drawing over other objects
	none:	The *line* is not erased when moved or deleted
LineStyle		Line style control
	{-}:	Solid *line* is drawn through all data points
	--:	Dashed *line* is drawn through all data points

	`: :`	Dotted *line* is drawn through all data points
	`-.:`	Dash-dot *line* is drawn through all data points
	`+:`	Plus symbol is used as a marker at all data points
	`o:`	Circle is used as a marker at all data points
	`*:`	Star symbol is used as a marker at all data points
	`.:`	A solid dot is used as a marker at all data points
	`x:`	An X-mark is used as a marker at all data points
`LineWidth`		*Line* width in points; defaults to `0.5` points
`MarkerSize`		Marker size in points; defaults to `6` points
`Xdata`		Vector of `X` coordinates for the *line*
`Ydata`		Vector of `Y` coordinates for the *line*
`Zdata`		Vector of `Z` coordinates for the *line*
`ButtonDownFcn`		MATLAB callback string, passed to the `eval` function whenever the *line* is selected; initially an empty matrix
`Children`		The empty matrix. *Line* objects have no children
`Clipping`		Data clipping mode
	`{on}:`	Any portion of the *line* outside the *axes* limits is not displayed
	`off:`	*Line* data is not clipped
`Interruptible`		Specifies if `ButtonDownFcn` callback string is interruptible
	`{no}:`	Not interruptible by other callbacks
	`yes:`	`ButtonDownFcn` callback string is interruptible
`Parent`		Handle of the *axes* containing the *line* object
* `Selected`		Values are `[on	off]`
* `Tag`		Text string
`Type`		Read-only object identification string; always `line`
`UserData`		User-specified data. Can be matrix, string, etc
`Visible`		Visibility of *line*
	`{on}:`	*Line* is visible on the screen
	`off:`	*Line* is not visible

Text Object Properties

`Color`	Line color. A 3-element RGB vector or one of MATLAB's predefined color names. The default color is `white`
`EraseMode`	Erase and redraw mode
`{normal}:`	Redraws the affected region of the display ensuring that all objects are rendered correctly. This is the most accurate mode but is also the slowest
`background:`	The *text* is erased by drawing it in the *figure's* background color. This can damage objects behind the erased *text*
`xor:`	The *text* is drawn and erased by performing an exclusive OR (XOR) with the color of the screen beneath it. Incorrect color can be used when drawing over other objects
`none:`	The *text* is not erased when moved or deleted
`Extent`	*Text* position vector `[left, bottom, width, height]` where `[left, bottom]` represents the location of the lower-left corner of the *text* with respect to the lower-left corner of the *axes* object and `[width, height]` are the dimensions of a rectangle enclosing the *text* string. Units are specified by the `Units` property
`FontAngle`	Italics for *text* object
`{normal}:`	Regular font angle
`italic:`	Italics
`oblique:`	Italics on some systems
`FontName`	Name of the font used for *text* objects. The default font is `Helvetica`
`FontSize`	Size in points of *text* objects. Default is `12` points
* `FontStrikeThrough`	Values are `[on	{off}]`
* `FontUnderline`	Values are `[on	{off}]`
`FontWeight`	Bolding for *text* object
`light:`	Light font weight
`{normal}:`	Normal font weight

`demi`:	Medium or bold font weight
`bold`:	Bold font weight
`HorizontalAlignment`	Horizontal text alignment
`{left}`:	Text is left-justified with respect to its `Position`
`center`:	Text is centered with respect to its `Position`
`right`:	Text is right-justified with respect to its `Position`
`Position`	Two- or three-element vector [x y (z)] specifying the location of the *text* object in three dimensions. Units are specified by the `Units` property
`Rotation`	Text orientation in degrees of rotation
`{0}`:	Horizontal orientation
`±90`:	Rotate text ± 90 degrees
`±180`:	Rotate text ± 180 degrees
`±270`:	Rotate text ± 270 degrees
`String`	The text string that is displayed
`Units`	Unit of measurement for position property values
`inches`:	Inches
`centimeters`:	Centimeters
`normalized`:	Normalized coordinates, where the lower-left corner of the axes maps to `[0  0]` and the upper-right corner maps to `[1  1]`
`points`:	Typesetters points; equal to 1/72 of an inch
`pixels`:	Screen pixels; the smallest unit of resolution of the computer screen
`{data}`:	Data units of the parent axes
`VerticalAlignment`	Vertical text alignment
`top`:	String is placed at the top of the specified `Y`-position
`cap`:	Font's capital letter height is placed at the specified `Y`-position
`{middle}`:	String is placed at the middle of the specified `Y`-position
`baseline`:	Font's baseline is placed at the specified `Y`-position
`bottom`:	String is placed at the bottom of the specified `Y`-position

ButtonDownFcn	MATLAB callback string, passed to the **eval** function whenever the *text* is selected; initially an empty matrix
Children	The empty matrix. *Text* objects have no children
Clipping	Data clipping mode
{on}:	Any portion of the *text* outside the *axes* is not displayed
off:	Text data is not clipped
Interruptible	Specifies if **ButtonDownFcn** callback string is interruptible
{no}:	Not interruptible by other callbacks
yes:	**ButtonDownFcn** callback string is interruptible
Parent	Handle of the *axes* object containing the *text* object
* Selected	Values are [on | off]
* Tag	Text string
Type	Read-only object identification string; always **text**
UserData	User-specified data. Can be matrix, string, etc.
Visible	Visibility of *text*
{on}:	*Text* is visible on the screen
off:	*Text* is not visible

Surface Object Properties

CData	Matrix of values that specify the color at every point in **ZData**. If **CData** is not the same size as **ZData**, the image contained in **CData** is mapped to the surface defined by **ZData**
EdgeColor	Surface edge color control
none:	Edges are not drawn
{flat}:	Edges are a single color determined by the first **CData** entry for that face. The default is **black**
interp:	Each edge color is determined by linear interpolation through the values at the vertices

A ColorSpec: A 3-element RGB vector or one of MATLAB's predefined color names specifying a single color for edges. The default color is **black**

EraseMode   Erase-and-redraw mode

{normal}:   Redraws the affected region of the display ensuring that all objects are rendered correctly. This is the most accurate mode but is also the slowest

background: The *surface* is erased by drawing it in the figure's background color. This can damage objects behind the erased *surface*

xor: The *surface* is drawn and erased by performing an exclusive OR (XOR) with the color of the screen beneath it. Incorrect color can be used when drawing over other objects

none: The *surface* is not erased when moved or deleted

FaceColor   *Surface* face color control

none: Faces are not drawn, but edges may be drawn

{flat}:   The first **CData** entry determines face color

interp: Each face color is determined by linear interpolation through the mesh points on the *surface*

A ColorSpec: An RGB vector or one of MATLAB's predefined color names specifying a single color for faces

LineStyle   Edge line style control

{-}:   Solid line is drawn through all mesh points

--:   Dashed line is drawn through all mesh points

::   Dotted line is drawn through all mesh points

-.:   Dash-dot line is drawn through all mesh points

+:   Plus symbol is used as a marker at all mesh points

o: Circle is used as a marker at all mesh points

*:   Star symbol is used as a marker at all mesh points

.:   A solid dot is used as a marker at all mesh points

x: An X-mark is used as a marker at all mesh points

`LineWidth`	Edge line width in points; defaults to `0.5` points
`MarkerSize`	Edge line marker size in points; defaults to `6` points
`MeshStyle`	Draw row and/or column lines
`{both}:`	Draw all edges
`row:`	Draw row edges only
`column:`	Draw column edges only
* `PaletteModel`	Values are
	`[{scaled}	direct	bypass]`
`XData`	X-coordinates of *surface* points
`YData`	Y-coordinates of *surface* points
`ZData`	Z-coordinates of *surface* points
`ButtonDownFcn`	MATLAB callback string, passed to the `eval` function whenever the *surface* is selected; initially an empty matrix
`Children`	The empty matrix. *Surface* objects have no children
`Clipping`	Data clipping mode
`{on}:`	Any portion of the *surface* outside the *axes* is not displayed
`off:`	*Surface* data is not clipped
`Interruptible`	Specifies if `ButtonDownFcn` callback string is interruptible
`{no}:`	Not interruptible by other callbacks
`yes:`	`ButtonDownFcn` callback string is interruptible
`Parent`	Handle of the *axes* object containing the *surface* object
* `Selected`	Values are `[on	off]`
* `Tag`	Text string
`Type`	Read-only object identification string; always `surface`
`UserData`	User-specified data. Can be matrix, string, etc.
`Visible`	Visibility of *surface*
`{on}:`	*Surface* is visible on the screen
`off:`	*Surface* is not visible

Patch Object Properties

`CData`	Matrix of values that specify the color at every point along the edge of the *patch*. Only used if `EdgeColor` or `FaceColor` is set to `interp` or `flat`
`EdgeColor`	*Patch* edge color control
`none`:	Edges are not drawn
`{flat}`:	Edges are a single color determined by the average of the color data for the *patch*. The default color is `black`
`interp`:	Edge color is determined by linear interpolation through the values at the *patch* vertices
A *ColorSpec*:	A 3-element RGB vector or one of MATLAB's predefined color names specifying a single color for edges. The default color is `black`
`EraseMode`	Erase-and-redraw mode
`{normal}`:	Redraws the affected region of the display ensuring that all objects are rendered correctly. This is the most accurate mode but is also the slowest
`background`:	The *patch* is erased by drawing it in the *figure's* background color. This can damage objects behind the erased *patch*
`xor`:	The *patch* is drawn and erased by performing an exclusive OR (XOR) with the color of the screen beneath it. Incorrect color can be used when drawing over other objects
`none`:	The *patch* is not erased when moved or deleted
`FaceColor`	*Patch* face color control
`none`:	Faces are not drawn, but edges may be drawn
`{flat}`:	The values in the color argument `c` determine face color for each *patch*
`interp`:	Each face color is determined by linear interpolation through the values specified in the `CData` property
A *ColorSpec*:	An RGB vector or one of MATLAB's predefined color names specifying a single color for faces

	LineWidth	Edge line width in points; defaults to `0.5` points
*	PaletteModel	Values are `[ {scaled} | direct | bypass ]`
	XData	X-coordinates of points along the edge of the *patch*
	YData	Y-coordinates of points along the edge of the *patch*
	ZData	Z-coordinates of points along the edge of the *patch*
	ButtonDownFcn	MATLAB callback string, passed to the `eval` function whenever the *patch* is selected; initially an empty matrix
	Children	The empty matrix. *Patch* objects have no children
	Clipping	Data clipping mode
	`{on}:`	Any portion of the *patch* outside the *axes* is not displayed
	`off:`	*Patch* data is not clipped
	Interruptible	Specifies if `ButtonDownFcn` callback string is interruptible
	`{no}:`	Not interruptible by other callbacks
	`yes:`	`ButtonDownFcn` callback string is interruptible
	Parent	Handle of the *axes* object containing the *patch* object
*	Selected	Values are `[ on | off ]`
*	Tag	Text string
	Type	Read-only object identification string; always `patch`
	UserData	User-specified data. Can be matrix, string, etc.
	Visible	Visibility of *patch*
	`{on}:`	*Patch* is visible on the screen
	`off:`	*Patch* is not visible

Image Object Properties

`CData`	A matrix of values that specifies the color of each element of the *image*. `image(C)` assigns `C` to `CData` The elements of `CData` are indices into the current colormap
`XData`	Image `X`-data; specifies the position of the *image* rows. If omitted, the row indices of `CData` are used
`YData`	Image `Y`-data; specifies the position of the *image* columns. If omitted, the column indices of `CData` are used
`ButtonDownFcn`	MATLAB callback string, passed to the `eval` function whenever the *image* is selected; initially an empty matrix
`Children`	The empty matrix. *Image* objects have no children
`Clipping`	Data clipping mode
`{on}:`	Any portion of the *image* outside the *axes* is not displayed
`off:`	*Image* data is not clipped
`Interruptible`	Specifies if `ButtonDownFcn` callback string is interruptible
`{no}:`	Not interruptible by other callbacks
`yes:`	`ButtonDownFcn` callback string is interruptible
`Parent`	Handle of *axes* object containing the *image* object
* `Selected`	Values are [on | off]
* `Tag`	Text string
`Type`	Read-only object identification string; always `image`
`UserData`	User-specified data. Can be matrix, string, etc.
`Visible`	Visibility of *image*
`{on}:`	*Image* is visible on the screen
`off:`	*Image* is not visible

20.13 SUMMARY

Handle Graphics functions let you fine-tune the graphs and displays that you create. Each graphics object has a handle associated with it that can be used to manipulate the object. Object properties can be changed using **get** and **set** to customize your plots and graphs. The functions discussed in this chapter are summarized in following tables.

Handle Graphics Functions

`set(handle,'PropertyName',Value)`	Set object properties
`get(handle,'PropertyName')`	Get object properties
`reset(handle)`	Reset object properties to defaults
`delete(handle)`	Delete object and all its children
`gcf`	Get handle to the current *figure*
`gca`	Get handle to the current *axes*
`gco`	Get handle to the current object
`findobj('PropertyName',Value)`	Get handles to objects with specified property values
`waitforbuttonpress`	Wait for key or button press over a *figure*
`figure('PropertyName',Value)`	Create *figure* object
`axes('PropertyName',Value)`	Create *axes* object
`line(X,Y,'PropertyName',Value)`	Create *line* object
`text(X,Y,S,'PropertyName',Value )`	Create *text* object
`patch(X,Y,C,'PropertyName',Value)`	Create *patch* object
`surface(X,Y,Z,'PropertyName',Value)`	Create *surface* object
`image(C,'PropertyName',Value)`	Create *image* object

These are the *Mastering MATLAB Toolbox* functions mentioned in this chapter.

Mastering MATLAB Handle Graphics Functions

`mmgcf`	Get handle to the current *figure* if it exists
`mmgca`	Get handle to the current *axes* if it exists
`mmzap(T)`	Delete graphics object of type `T` using the mouse
`mmpx2n(X)`	Pixel to normalized coordinate transformation
`mmn2px(X)`	Normalized to pixel coordinate transformation
`mmline Name Value...`	Set *line* attributes using the mouse
`mmaxes Name Value...`	Set *axes* attributes using the mouse
`mmcont2(X,Y,Z,N,C)`	2-D contour plot with user-defined colors
`mmcont3(X,Y,Z,N,C)`	3-D contour plot with user-defined colors
`mmtile`	Arrange *Figure* windows in a tile pattern
`mmpaper Name Value...`	Set default paper attributes for printing

21

Creating Graphical User Interfaces

A *user interface* is the point of contact or method of interaction between a person, i.e., the user, and a computer or computer program. It is the method used by the computer and the user to communicate information. The computer displays text and graphics on the computer screen and may generate sounds with a speaker. The user communicates with the computer through input devices such as a keyboard, mouse, trackball, drawing pad, or microphone. The user interface defines the *look* and *feel* of the computer, operating system, or application. Often a computer or program is chosen on the basis of pleasing design and functional efficiency of its user interface.

A *graphical user interface* or GUI (pronounced *goo'ey*) is a user interface incorporating graphical objects such as windows, icons, buttons, menus, and text. Selecting or activating these objects in some way usually causes an action or change to occur. The most common activation method is to use a mouse or other pointing device to control the movement of a pointer on the screen, and to press a mouse button to signal object selection or some other action.

In the same way that the Handle Graphics capabilities of MATLAB discussed in the previous chapter let you customize the way MATLAB displays information, the Handle

Graphics user-interface functions described in this chapter let you customize the way you can interact with MATLAB. The *Command* window is not the only way to interact with MATLAB.

This chapter illustrates the use of Handle Graphics *uicontrol* and *uimenu* objects to add graphical interfaces to MATLAB functions and M-files. *uimenu* objects create drop-down menus and submenus in *Figure* windows. *uicontrol* objects create objects such as buttons, sliders, popup menus, and text boxes.

MATLAB includes an excellent example of its GUI capabilities in the `demo` command.

```
» demo
```

Explore this command to see how *uimenus* and *uicontrols* provide interactive input to MATLAB functions.

21.1 WHO SHOULD CREATE GUIs—AND WHY?

After running the demo, you are likely to be asking yourself *"Why would I want to create a GUI in MATLAB?"* Good question! The short answer is that you probably won't. Most people who use MATLAB to analyze data, solve problems, and plot results will not find GUI tools very useful.

On the other hand, GUIs can be used to create very effective tools and utilities in MATLAB, or to build interactive demonstrations of your work. The most common reasons to create a graphical user interface are

- you are writing a utility function that you will use over and over again and menus, buttons, or text boxes make sense as input methods; or
- you are writing a function or developing an application for others to use; or
- you want to create an interactive demonstration of a process, technique, or analysis method; or
- you think GUIs are neat and want to experiment with them.

A number of GUI-based tools and utility functions are included in the *Mastering MATLAB Toolbox* and are discussed later in this chapter. Other tools and utilities have been written by MATLAB users incorporating MATLAB's GUI functions. Most of these tools are available on the *Mathworks Anonymous FTP Site* and from other sources. The Internet Resources chapter explains how to access the FTP site as well as a number of other Internet resources.

Before we begin, remember that a basic understanding of Handle Graphics is a prerequisite to designing and implementing a GUI in MATLAB. If you skipped the previous chapter, you should go back and read it now. Otherwise, press on.

21.2 GUI OBJECT HIERARCHY

As was demonstrated in the previous chapter, everything created by a graphics command is a graphics object. These include *uimenu* and *uicontrol* objects as well as *figures*, *axes*, and their children. Let's take another look at the object hierarchy. The computer screen itself is the *root* object and *figures* are children of the *root; axes, uimenus*, and *uicontrols* are children of *figures*. This hierarchy is diagrammed below.

The *root* can contain one or more *figures,* and each *figure* can contain one or more sets of *axes* and their children. Each *figure* can also contain one or more *uimenus* and *uicontrols* that are independent of *axes*. While *uicontrol* objects have no children, they do have a variety of styles. However, *uimenu* objects often have other *uimenu* objects as children.

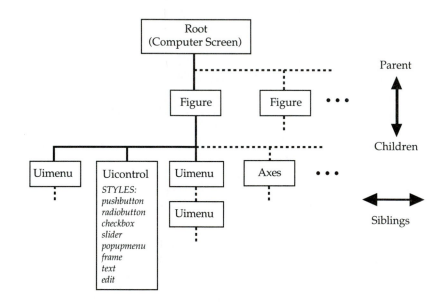

Graphic displays are generated differently on each type of computer, or *platform,* that runs MATLAB. Unix workstations use the X Window System having a number of *window managers,* such as **mwm** or **twm**, to control the layout of the display. PCs rely on Microsoft Windows or Windows NT for window management. Macintosh computers use code in the Macintosh Toolbox for window management. Although displays may be visually different on each platform, in most cases Handle Graphics' code is identical. MATLAB takes care of platform and window system differences internally. Functions incorporating Handle Graphics routines, including those using *uimenu* and *uicontrol* objects, usually run on all platforms. Where known differences exist, they are pointed out later in this chapter.

21.3 MENUS

Menus are used in each windowing system to let users select commands and options. Commonly, there is a **menu bar** across the top of the window or display. When one of these *top-level* menus is selected by moving the pointer over the menu label and pressing the mouse button, a list of *menu items* drops down from the menu label. This type of menu is often called a *pull-down* menu. Menu items are selected by holding down the mouse button, moving the pointer over a menu item, and releasing the mouse button. MS-Windows and some X Window System platforms provide an additional method of selecting a menu item. Pressing and releasing the mouse button, or *clicking,* on a top-level menu opens the pull-down menu. Menu items can then be selected by moving the pointer over the menu item and clicking again. Selecting one of the menu items in the pull-down menu causes some action to occur.

A menu item can also act as a *submenu* with its own list of menu items. Submenu items display a small triangle or arrowhead to the right of the submenu label to indicate that there are more menu items available for this choice. When a submenu item is selected, another menu with more menu items is displayed in a pull-down menu to the right of the submenu item. This is sometimes called a *walking menu*. Selecting one of these items also causes some action to occur.

Submenus may be nested, with the number of levels limited only by the windowing system used and available system resources.

Menu Placement

The Macintosh uses a menu bar containing the titles of each pull-down menu at the top of the display. When a *Figure* window is selected, the menu bar changes to reflect the options available for the active *Figure* window. The standard Macintosh menu titles include **Apple, File, Edit, Window,** and **Help** or **Balloon Help.** Menu titles added by *uimenu* objects are positioned between the **Window** and **Help** titles. If you want to remove the **File, Edit,** and **Window** menu titles from the menu bar, you can issue the following **set** command.

```
» set(Hf_fig1,'MenuBar','none')
```

The **Apple** and **Help** menu titles cannot be removed from the menu bar. Likewise, the standard menus are restored by the command

```
» set(Hf_fig1,'MenuBar','figure')
```

Under Microsoft Windows, the menu bar is located at the top of the *Figure* window. Each *Figure* window has its own menu bar containing **File, Edit, Window,** and **Help** titles. Menu titles added by *uimenu* objects are positioned after the **Help** title. You can remove or restore all standard menus from the *Figure* window by issuing the same **set** commands shown above.

There are no standard menu titles for MATLAB *Figure* windows on X Window System workstations. The window manager may place its own menu bar above each window on the screen, but this menu bar is independent of the MATLAB menus. MATLAB creates its own menu bar at the top edge of the *Figure* window when the first *uimenu* object is created.

Creating Menus and Submenus

Menu items are created with the `uimenu` function. The general syntax is similar to other object-creation functions. For example,

```
» Hm_1 = uimenu(Hx_parent,'PropertyName',PropertyValue,...)
```

where `Hm_1` is the handle of the menu item created by `uimenu`, and `'PropertyName'`, `PropertyValue` pairs define the characteristics of the menu by setting *uimenu* object property values. `Hx_parent` is the handle of the parent object, which must be either a *figure* or another *uimenu*.

The most important properties of a *uimenu* object are the `'Label'` and `'CallBack'` properties. The `'Label'` property value is the text string that is placed on the menu bar or in the pull-down menu to identify the menu item. The `'CallBack'` property value is a MATLAB string that is passed to `eval` for execution when the menu item is selected. These and other properties are discussed in detail later in this chapter.

Menu Example

The following example uses the `uimenu` function to add simple menus to the current *Figure* window. It is presented here to illustrate how working menus can be created with just a few MATLAB commands. Later examples will discuss `uimenu` commands and properties in more detail.

This example adds a menu bar with two pull-down menus to the current *Figure* window. First a top-level menu having the name **Example** is created by entering

```
» Hm_ex = uimenu(gcf,'Label','Example');
```

Under this menu, two menu items are added. The first item is labeled **Grid** and toggles the state of the axis grid on and off.

```
» Hm_exgrid = uimenu(Hm_ex,'Label','Grid','CallBack','grid');
```

Note that the handle `Hm_ex` is associated with the top-level menu. This *uimenu* will appear under the top-level menu. Note also that the value for the `'CallBack'` property is a quoted string, as required by `eval`.

The second item under **Example** is a menu item labeled **View** with a submenu.

```
» Hm_exview = uimenu(Hm_ex,'Label','View');
```

The **View** submenu contains two items to select a 2-D or a 3-D view.

```
» Hm_exv2d = uimenu(Hm_exview,'Label','2-D','CallBack','view(2)');
```

```
» Hm_exv3d = uimenu(Hm_exview,'Label','3-D','CallBack','view(3)');
```

Note here that these are submenus to the **View** menu because they specify **Hm_exview** as their parent.

Now add a second top-level menu in the menu bar titled **Close.**

```
» Hm_close = uimenu(gcf,'Label','Close');
```

From this top-level menu add two menu items. The first item closes the *Figure* window, and the second leaves the *Figure* window open, but removes the user menus.

```
» Hm_clfig = uimenu(Hm_close,'Label','Close Figure','CallBack','close');
```

```
» Hm_clmenu = uimenu(Hm_close,'Label','Remove Menus',...
               'CallBack','delete(Hm_ex); delete(Hm_close); drawnow');
```

The preceding example is contained in the *Mastering MATLAB Toolbox* script M-file **mmenu1.m**. Therefore, you can run **mmenu1.m** to experiment with this example.

Menu Properties

As with all Handle Graphics functions, *uimenu* properties can be defined at the time the object is created, as shown above, or changed with the **set** command. All settable properties can be changed by **set**, including the label text, menu colors, and even the callback string. This ability is very useful for customizing menus and attributes *on-the-fly*.

The following table lists the object properties and values found in MATLAB version 4.2c for *uimenu* objects. Properties with an asterisk (*****) are undocumented and should be used with caution. Property values enclosed in set brackets **{}** are default values.

Uimenu Object Properties	
Accelerator	A character specifying the keyboard equivalent or *shortcut* key for the menu item. For X-Windows, the key sequence is **Control-Char**; for Macintosh systems the sequence is **Command-Char** or ⌘ **-Char**

BackgroundColor		*Uimenu* background color. A 3-element RGB vector or one of MATLAB's predefined color names. The default background color is **light gray**
CallBack		MATLAB callback string, passed to the **eval** function whenever you select the menu item; initially the empty matrix
Checked		Checkmark for selected items
	{on}:	Checkmark appears next to selected items
	off:	Checkmarks are not displayed
Enable		Menu-enable mode
	{on}:	The menu item is enabled. Selecting the menu item will send the **CallBack** string to **eval**
	off:	The menu item is disabled and the label string is dimmed. Selecting the menu item has no effect
ForegroundColor		*Uimenu* foreground (text) color. A 3-element RGB vector or one of MATLAB's predefined color names. The default text color is **black**
Label		A text string containing the label of the menu item. On PC systems, a **&** preceding a character in the label defines a shortcut key that is activated with the key sequence **Alt-Char**
Position		Relative position of the *uimenu* object. Top-level menus are numbered from left to right, and submenus are numbered top to bottom
Separator		Separator-line mode
	on:	A dividing line is drawn above the menu item
	{off}:	No dividing line is drawn
* **Visible**		Visibility of **vimenu** object
	{on}:	*Uimenu* is visible on the screen
	off:	*Uimenu* is not visible
ButtonDownFcn		MATLAB callback string, passed to the **eval** function whenever the object is selected; initially the empty matrix
Children		Object handles of other *uimenu* objects (submenus)
Clipping		Clipping mode

{on}:	No effect for *uimenu* objects
off:	No effect for *uimenu* objects
DestroyFcn	Macintosh version 4.2c only. No documentation available
Interruptible	Specifies if **ButtonDownFcn** and **CallBack** strings are interruptible
{no}:	Callbacks are not interruptible by other callbacks
yes:	Callback strings are interruptible
Parent	Handle of the parent object; either a *figure* handle if the *uimenu* object is a top-level menu, or the parent *uimenu* object handle for submenus
* **Selected**	Values are [on | off]
* **Tag**	Text string
Type	Read-only object identification string; always **uimenu**
UserData	User-specified data. Can be matrix, string, etc.
Visible	Visibility of *uimenu* object
{on}:	*Uimenu* is visible on the screen
off:	*Uimenu* is not visible

Property values simply define the attributes of *uimenu* objects and control how the menus are displayed. They also determine actions that result from selecting menu items. Some of these properties are discussed in more detail below.

Menu Shortcut Keys

The **'Label'** property defines the label that appears on the menu or menu item. It can also be used to define a *shortcut* key for Microsoft Windows systems by preceding the desired character in the label string with the ampersand (**&**) character. For example,

```
» Hm_top = uimenu('Label','Example');

» uimenu(Hm_top,'Label','&Grid','CallBack','grid');
```

defines the shortcut key as the letter **G** on the keyboard. The menu item label will appear as **<u>G</u>rid** on the menu. To invoke the shortcut key, hold down the **Alt** key on the keyboard and press the **G** key while the *Figure* window is selected. The shortcut key need not be the first character in the string. The following example defines the **R** key as the shortcut key:

» uimenu(Hm_top,'Label','G&rid','CallBack','grid');

and the label will appear as **Gr̲id** on the menu.

Macintosh platforms use the **'Accelerator'** property rather than the **'Label'** property to define shortcut keys. On the Macintosh,

» uimenu(Hm_top,'Label','Grid','Accelerator','G','CallBack','grid');

defines the shortcut key as the letter **G.** The menu-item label will appear as **Grid ⌘G.** To invoke the shortcut key, hold down the **Command** or ⌘ key on the keyboard and press the **G** key while the *Figure* window is selected.

A shortcut key cannot be defined for top-level menus on the Macintosh. In addition, any shortcut keys already defined on standard Macintosh menus, such as ⌘ **C** or **Command C,** cannot be assigned without removing the standard Macintosh menu.

Defining and using shortcut keys on X Window Systems is similar to the Macintosh, but there are some differences. Like the Macintosh, the **'Accelerator'** property is used rather than the **'Label'** property to define shortcut keys. However, the shortcut key is displayed differently on the menus. For example,

» uimenu(Hm_top,'Label','Grid','Accelerator','G','CallBack','grid');

defines the shortcut key as the letter **G.** The menu-item label will usually appear as **Grid** <**Ctrl**>**-G** on the menu. To utilize the shortcut key, hold down the **Control** key on the keyboard and press the **G** key. Like the Macintosh, a shortcut key cannot be defined for top-level menus.

While we have not been able to verify this, the MATLAB manuals indicate that some workstations using X act differently when given the same command. If a top-level menu label contains an underlined character, that menu can be selected by holding down the **Meta** key and pressing the character key. If a pull-down menu-item label contains an underlined character, press that character key to select the menu item. Consult your keyboard documentation to determine the appropriate **Meta** key for your system.

Menu Appearance

Three properties that affect the placement and appearance of your menus are **'Position'**, **'Checked'**, and **'Separator'**. The **'Position'** property value for *uimenu* objects is an integer that specifies the menu location relative to other menus or menu items. The **'Position'** property value is set when the object is created. The left-most menu in a menu bar and the top menu item in a pull-down menu are at position 1.

Menus can be reordered by setting the **'Position'** property. Consider the following example:

```
» Hm_1 = uimenu('Label','First');              % Create two menus

» Hm_2 = uimenu('Label','Second');

» get(Hm_1,'Position')                         % Check the locations

ans =
    1

» get(Hm_2,'Position')

ans =
    2

» set(Hm_2,'Position',1)                       % Change menu order

» get(Hm_1,'Position')                         % Check the locations

ans =
    2

» get(Hm_2,'Position')

ans =
    1
```

Note that when the **'Position'** property of one *uimenu* is changed, the other *uimenus* are shifted to accommodate the change, and their **'Position'** property values are updated. Menu items in submenus can be reordered in the same way.

The **'Checked'** property value controls the appearance of a checkmark to the left of the menu item label. The default value is **'off'**. The command

```
» set(Hm_item,'Checked','on')
```

causes a checkmark to appear next to the label of the **Hm_item** *uimenu*. This property is useful for creating menu items that represent attributes. For example,

```
» Hm_top = uimenu('Label','Example');

» Hm_box = uimenu(Hm_top,'Label','Axis Box',...
              'CallBack',[...
                'if strcmp(get(gca,''Box''),''on''),',...
                  'set(gca,''Box'',''off''),',...
                  'set(Hm_box,''Checked'',''off''),',...
                'else,',...
                  'set(gca,''Box'',''on''),',...
                  'set(Hm_box,''Checked'',''on''),',...
                'end']);
```

creates a pull-down menu item labeled **Axis Box.** When selected, the callback string is evaluated. The callback string determines the present value of the current *axes* **'Box'** property, and sets the *axes* **'Box'** property and the *uimenu* **'Checked'** property appropriately. This example is available in the script M-file **mmenu2.m** of the *Mastering MATLAB Toolbox.*

The *uimenu* **'Label'** property can also be changed to reflect the current status of a menu item. The following example (**mmenu3.m**) changes the menu-item label itself rather than adding a checkmark:

```
» Hm_top = uimenu('Label','Example');

» Hm_box = uimenu(Hm_top,'Label','Axis Box',...
            'CallBack',[...
              'if strcmp(get(gca,''Box''),''on''),',...
                'set(gca,''Box'',''off''),',...
                'set(Hm_box,''Label'',''Set Box On''),',...
              'else,',...
                'set(gca,''Box'',''on''),',...
                'set(Hm_box,''Label'',''Set Box Off''),',...
              'end']);
```

Pull-down menu items can be separated into logical groups by using the **'Separator'** property. If **'Separator'** is set to **'on'** for a *uimenu* item, that item will appear in the pull-down menu with a horizontal line drawn above it to separate it visually from previous items. The default value is **'off'**, but can be changed when the object is created.

```
» Hm_box = uimenu(Hm_top,'Label','Box','Seperator','on');
```

or later using the **set** command.

```
» set(Hm_box,'Separator','on');
```

Top-level menus ignore the value of the **'Separator'** property.

Color Control

Two color properties can be set for *uimenu* objects. The **'BackGroundColor'** property controls the color used to fill the menu *background,* or the rectangular area defined by the *uimenu* object. The default background color is a light gray. The other color property is **'ForeGroundColor'**. This property determines the color of the text of the menu item label. The default text color is black.

Color properties apply equally well to top-level menus and pull-down menu items. Colors can be used to indicate status information or simply to add interest to your menus. For example, the background of each menu item in a submenu for selecting *line* color properties can be filled with the appropriate color.

```
» Hm_green = uimenu(Hm_lcolor,'Label','Green','BackGroundColor','g',...
          'CallBack','set(Hl_line,''Color'',''g'')');
```

Disabling Menu Items

A menu item may be temporarily disabled by changing the value of the `'Enable'` or the `'Visible'` property of a *uimenu* object. The `'Enable'` property is normally set to `'on'`. When `'Enable'` is set to `'off'`, the label string is dimmed and the menu item is disabled. In this state, the menu item remains visible, but cannot be selected. This property can be used to disable an inappropriate menu choice.

The following example (`mmenu4.m`) illustrates another way to set the *axes* `'Box'` property using two menu items and the `'Enable'` property:

```
» Hm_top = uimenu('Label','Example');

» Hm_boxon = uimenu(Hm_top,'Label','Set Box On',...
              'CallBack',[...
                'set(gca,''Box'',''on''),',...
                'set(Hm_boxon,''Enable'',''off''),',...
                'set(Hm_boxoff,''Enable'',''on'')']);

» Hm_boxoff = uimenu(Hm_top,'Label','Set Box Off',...
              'Enable','off',...
              'CallBack',[...
                'set(gca,''Box'',''off''),',...
                'set(Hm_boxon,''Enable'',''on''),',...
                'set(Hm_boxoff,''Enable'',''off'')']);
```

A menu item can be completely hidden by setting the `'Visible'` property to `'off'`. The menu item seems to disappear from the menu, and the other menu items change position on the display to fill the gap caused by the now-invisible menu item. However, the invisible menu item still exists, and the `'Position'` property values of the *uimenu* objects do not change. The menu item reappears in its normal position when the `'Visible'` property is set to `'on'` again.

This property can be used to temporarily remove a menu. The following example (`mmenu5.m`) creates two top-level menus and two menu items:

```
» Hm_control = uimenu('Label','Control');

» Hm_extra = uimenu('Label','Extra');

» Hm_limit = uimenu(Hm_control,'Label','Limited Menus',...
              'CallBack','set(Hm_extra,''Visible'',''off'')');

» Hm_full = uimenu(Hm_control,'Label','Full Menus',...
              'CallBack','set(Hm_extra,''Visible'',''on'')');
```

When the **Limited Menus** item is selected, the **Extra** menu disappears from the menu bar. When the **Full Menus** menu item is selected, the **Extra** menu reappears on the menu bar in its original location.

The CallBack Property

The `'CallBack'` property value is a MATLAB string that is passed to the function **eval** to be executed **in the *Command* window workspace.** This has important implications for function M-files and is addressed later in this chapter.

  Since the `'CallBack'` property value must be a string, multiple MATLAB commands, continuation lines, and strings within strings can make the necessary syntax quite complicated. If there is more than a single command to be executed, the commands must be separated appropriately. For example,

```
» uimenu('Label','Test','CallBack','grid on; set(gca,''Box'',''on'')');
```

passes a single string to **eval** and causes the command

```
» grid on; set(gca,'Box','on')
```

to be executed in the *Command* window workspace. This is legal syntax, since multiple commands can be entered on the same command line if they are separated by a comma or semicolon. The MATLAB convention of using two single quotemarks to represent a single quote within a quoted string is also followed when defining the callback string.

  Strings can be concatenated to create a legal MATLAB string by enclosing them in square brackets.

```
» uimenu('Label','Test','CallBack',['grid on,','set(gca,''Box'',''on'')']);
```

Note that the string `'grid on,'` contains the required comma to separate the two commands.

  If continuation lines are used, the preceding command becomes

```
» uimenu('Label','Test',...
          'CallBack',[...
               'grid on,',...
               'set(gca,''Box'',''on'')'...
                   ]);
```

Here the lines are broken appropriately and three periods are added to the end of each of the lines to show that the statement is continued. Note that all elements of the single-line example above are preserved, including the comma within the strings to separate the commands. The comma after the final quote in the line `'grid on,',`... is optional; the spaces at the beginning of the next line serve the same purpose. Refer to the discussion in an earlier chapter about creating row vectors for more details.

MATLAB will complain if the quotes, commas, and parentheses are not entered correctly, but an error in a complicated callback string can be very difficult to find. To minimize errors, remember these guidelines for callback strings containing multiple MATLAB statements:

- Enclose the entire callback in square brackets, and don't forget the final close-parentheses `')'`.
- Enclose each statement in `'single quotes'`.
- Quotes within a quoted string double up: `'quoted'`; `'a   ''quoted''` `string'`;`'Quote ''a '''quoted''' string'' now'`.
- Each statement except the last ends with a comma or semicolon **within the quotes** and a comma or space **after the quotes.**
- Each line that is to be continued ends with three periods (`...`).

One of the earlier examples, **mmenu4.m**, is a good illustration of the syntax of more involved callback strings.

```
» Hm_top = uimenu('Label','Example');

» Hm_boxon = uimenu(Hm_top,...
              'Label','Set Box On',...
              'CallBack',[...
                 'set(gca,''Box'',''on''),',...
                 'set(Hm_boxon,''Enable'',''off''),',...
                 'set(Hm_boxoff,''Enable'',''on'')']);

» Hm_boxoff = uimenu(Hm_top,...
              'Label','Set Box Off',...
              'Enable','off',...
              'CallBack',[...
                 'set(gca,''Box'',''off''),',...
                 'set(Hm_boxon,''Enable'',''on''),',...
                 'set(Hm_boxoff,''Enable'',''off'')']);
```

The preceding example brings up another important point about callback strings. **Hm_boxoff** is used in the callback string for **Hm_boxon** before the variable **Hm_boxoff** is defined. MATLAB does not complain because the callback string is simply a string and is not executed by MATLAB until the *uimenu* object is activated and the string is passed to **eval**. This has implications for function M-file design and testing that will be discussed later in this chapter.

Example M-file

The following example demonstrates the creation of a simple set of menus. This example is contained in the *Mastering MATLAB Toolbox* function M-file **mmenus**. As illustrated below, this function is broken into fragments so that various aspects can be discussed separately.

First, define the function and create a menu bar with a top-level **Line** menu in the current *figure*, containing three submenus for **Line Style, Line Width,** and **Line Color.**

```
function mmenus()
%MMENUS Simple menu example.
% MMENUS uses waitforbuttonpress and gco in the callback strings
% to let the user make a menu selection and then select an object
% by clicking on it with the mouse. The callback strings then use
% the set function to apply the property value to the selected
% object.

% Copyright (c) 1996 by Prentice-Hall, Inc.

Hm_line = uimenu(gcf,'label','Line');
Hm_lstyle = uimenu(Hm_line,'label','Line Style');
Hm_lwidth = uimenu(Hm_line,'label','Line Width');
Hm_lcolor = uimenu(Hm_line,'label','Line Color');
```

Next, use **waitforbuttonpress** and **gco** to get a handle to the *current object,* make sure it is a *line* object, and apply the appropriate **'LineStyle'** value. Note that the handles to these menu items are never used, so they are not saved.

```
uimenu(Hm_lstyle,'Label','Solid',...
    'CallBack',['waitforbuttonpress;',...
        'if get(gco,''Type'') == ''line'',',...
            'set(gco,''LineStyle'',''-''),',...
        'end']);

uimenu(Hm_lstyle,'Label','Dotted',...
    'CallBack',['waitforbuttonpress;',...
        'if get(gco,''Type'') == ''line'',',...
            'set(gco,''LineStyle'','':''),',...
        'end']);
```

```
    uimenu(Hm_lstyle,'Label','Dashed',...
        'CallBack',['waitforbuttonpress;',...
            'if get(gco,''Type'') == ''line'',',...
                'set(gco,''LineStyle'',''-''),',...
            'end']);

    uimenu(Hm_lstyle,'Label','DashDot',...
        'CallBack',['waitforbuttonpress;',...
            'if get(gco,''Type'') == ''line'',',...
                'set(gco,''LineStyle'',''-.''),',...
            'end']);
```

Now do the same for the **Line Width** submenu items.

```
        uimenu(Hm_lwidth,'Label','Default',...
            'CallBack',['waitforbuttonpress;',...
                'if get(gco,''Type'') == ''line'',',...
                    'set(gco,''LineWidth'',0.5),',...
                'end']);

        uimenu(Hm_lwidth,'Label','Thick',...
            'CallBack',['waitforbuttonpress;',...
                'if get(gco,''Type'') == ''line'',',...
                    'set(gco,''LineWidth'',2.0),',...
                'end']);

        uimenu(Hm_lwidth,'Label','Thicker',...
            'CallBack',['waitforbuttonpress;',...
                'if get(gco,''Type'') == ''line'',',...
                    'set(gco,''LineWidth'',3.0),',...
                'end']);

        uimenu(Hm_lwidth,'Label','Thickest',...
            'CallBack',['waitforbuttonpress;',...
                'if get(gco,''Type'') == ''line'',',...
                    'set(gco,''LineWidth'',4.0),',...
                'end']);
```

and the **Line Color** submenu items. Color the background of the menu items and change the text color when appropriate.

```
uimenu(Hm_lcolor,'Label','Yellow',...
    'BackgroundColor','y',...
    'CallBack',['waitforbuttonpress;',...
        'if get(gco,''Type'') == ''line'',',...
            'set(gco,''Color'',''y''),',...
        'end']);

uimenu(Hm_lcolor,'Label','Magenta',...
    'BackgroundColor','m','ForegroundColor','w',...
    'CallBack',['waitforbuttonpress;',...
        'if get(gco,''Type'') == ''line'',',...
            'set(gco,''Color'',''m''),',...
        'end']);

uimenu(Hm_lcolor,'Label','Cyan',...
    'BackgroundColor','c',...
    'CallBack',['waitforbuttonpress;',...
        'if get(gco,''Type'') == ''line'',',...
            'set(gco,''Color'',''c''),',...
        'end']);

uimenu(Hm_lcolor,'Label','Red',...
    'BackgroundColor','r','ForegroundColor','w',...
    'CallBack',['waitforbuttonpress;',...
        'if get(gco,''Type'') == ''line'',',...
            'set(gco,''Color'',''r''),',...
        'end']);

uimenu(Hm_lcolor,'Label','Green',...
    'BackgroundColor','g',...
    'CallBack',['waitforbuttonpress;',...
        'if get(gco,''Type'') == ''line'',',...
            'set(gco,''Color'',''g''),',...
        'end']);

uimenu(Hm_lcolor,'Label','Blue',...
    'BackgroundColor','b','ForegroundColor','w',...
    'CallBack',['waitforbuttonpress;',...
        'if get(gco,''Type'') == ''line'',',...
            'set(gco,''Color'',''b''),',...
        'end']);

uimenu(Hm_lcolor,'Label','White',...
    'BackgroundColor','w',...
```

```
                    'CallBack',['waitforbuttonpress;',...
                        'if get(gco,''Type'') == ''line'',',...
                            'set(gco,''Color'',''w''),',...
                        'end']);

          uimenu(Hm_lcolor,'Label','Black',...
                'BackgroundColor','k','ForegroundColor','w',...
                'CallBack',['waitforbuttonpress;',...
                    'if get(gco,''Type'') == ''line'',',...
                        'set(gco,''Color'',''k''),',...
                    'end']);
```

To make this function more complete, additional menus can be added in the same way to change *axes, surface, patch,* and *figure* properties. For example, the following adds a **Color Map** menu for changing the *figure* color map:

```
Hm_cmap = uimenu(gcf,'Label','Color Map');

uimenu(Hm_cmap,'Label','Lighter','CallBack','brighten(.3)');
uimenu(Hm_cmap,'Label','Darker','CallBack','brighten(-.3)');
uimenu(Hm_cmap,'Label','Default','CallBack','colormap(''default'')');
uimenu(Hm_cmap,'Label','Gray','CallBack','colormap(gray)');
uimenu(Hm_cmap,'Label','Hot','CallBack','colormap(hot)');
uimenu(Hm_cmap,'Label','Cool','CallBack','colormap(cool)');
uimenu(Hm_cmap,'Label','Bone','CallBack','colormap(bone)');
uimenu(Hm_cmap,'Label','Copper','CallBack','colormap(copper)');
uimenu(Hm_cmap,'Label','Pink','CallBack','colormap(pink)');
uimenu(Hm_cmap,'Label','Prism','CallBack','colormap(prism)');
uimenu(Hm_cmap,'Label','Jet','CallBack','colormap(jet)');
uimenu(Hm_cmap,'Label','Flag','CallBack','colormap(flag)');
uimenu(Hm_cmap,'Label','HSV','CallBack','colormap(hsv)');
```

And finally, a **Quit** menu is added with two items. **Close Figure** closes the *figure*. **Remove Menus** leaves the *figure* open but removes the top-level user menus and all of their children. The **drawnow** command removes the menus immediately.

```
Hm_quit = uimenu(gcf,'Label','Quit');

uimenu(Hm_quit,'Label','Close Figure','CallBack','close; return');

uimenu(Hm_quit,'Label','Remove Menus',...
    'CallBack',[...
        'delete(findobj(gcf,''Type'',''uimenu'',''Parent'',gcf)),',...
        'drawnow']);
```

The *Mastering MATLAB Toolbox* contains a number of additional functions that use these techniques and others to create useful object-based tools. Many of them will be discussed in more detail later in this chapter.

21.4 CONTROLS

Controls and menus are used by windowing systems on each computer platform to let users perform some action or to set an option or attribute. Controls are graphic objects, such as icons, text boxes, and scroll bars, that are used along with menus to create the graphical user interface known as the windowing system and/or window manager on your computer.

MATLAB controls, called *uicontrols,* are very similar to those used by your window manager. They are graphical objects that can be placed anywhere in a MATLAB *Figure* window and activated with the mouse. MATLAB *uicontrols* include buttons, sliders, text boxes, and popup menus.

Uicontrols created by MATLAB have a slightly different appearance on Macintosh, MS-Windows, and X Window System computer platforms, due to differences in the way the windowing systems render graphical objects. However, the functionality is essentially the same, so the same MATLAB code will create the same object that performs the same function across platforms.

Uicontrols are created by the function **uicontrol**. The general syntax is similar to **uimenu** discussed earlier.

> » Hc_1 = uicontrol(Hf_fig,'PropertyName',PropertyValue,...)

Here **Hc_1** is the handle of the *uicontrol* object created by **uicontrol**, and **'PropertyName',PropertyValue** pairs define the characteristics of the *uicontrol* by setting *uicontrol* object property values. **Hf_fig** is the handle of the parent object, which must be a *figure*. If the *figure* handle is omitted, the current *figure* is used.

Creating Different Types of Controls

There are eight different types, or *styles,* of MATLAB controls. They are all created using the **uicontrol** function. The **'Style'** property value determines the type of control that is created. The **'CallBack'** property value is the MATLAB string passed to **eval** to execute in the *Command* window workspace when the control is activated.

In the following, each of the eight types of *uicontrol* objects is discussed individually and illustrated with an example below. A more thorough discussion of the properties of *uicontrol* objects and more complete examples of their use will follow later.

Push Buttons.   Push buttons, sometimes called *command buttons* or just *buttons,* are small rectangular screen objects that are usually labeled with text on the object itself. Selecting the push button *uicontrol* with the mouse by moving the pointer over the object and clicking the mouse button causes MATLAB to perform the action defined by the object's callback string. The **'Style'** property value of a push button is **'pushbutton'**.

Push buttons are typically used to perform an action rather than change a state or set an attribute. The following example (**mmct11.m**) creates a push button *uicontrol* labeled **Close** that closes the current *figure* when activated. The **'Position'** property defines the size and location of the push button in pixels, which is the default **'Units'** property value. The **'String'** property specifies the button's label.

```
» Hc_close = uicontrol(gcf,'Style','push',...
            'Position',[10 10 100 25],...
            'String','Close',...
            'CallBack','close');
```

Radio Buttons.   Radio buttons, sometimes called *option buttons* or *toggle buttons,* consist of a button containing a label and a small circle or diamond to the left of the label text. When *selected,* the circle or diamond is filled and the **'Value'** property is set to 1; when *unselected,* the indicator is cleared and the **'Value'** property is set to 0. The **'Style'** property value of a radio button is **'radiobutton'**.

Radio buttons are typically used to select one of a group of mutually-exclusive options. To enforce this exclusivity, however, the callback string for each radio button *uicontrol* must *unselect* all other buttons in the group by setting the **'Value'** of each of them to 0. This is only a convention, however. Radio buttons can be used interchangeably with check boxes, if desired.

The following example (**mmct12.m**) creates two mutually-exclusive radio buttons that set the *axes* **'Box'** property on or off.

```
» Hc_boxon = uicontrol(gcf,'Style','radio',...
            'Position',[20 45 100 20],...
            'String','Set Box On',...
            'Value',0,...
            'CallBack',[...
```

```
                    'set(Hc_boxon,''Value'',1),'...
                    'set(Hc_boxoff,''Value'',0),'...
                    'set(gca,''Box'',''on'')']);

    » Hc_boxoff = uicontrol(gcf,'Style','radio',...
                    'Position',[20 20 100 20],...
                    'String','Set Box Off',...
                    'Value',1,...
                    'CallBack',[...
                        'set(Hc_boxon,''Value'',0),'...
                        'set(Hc_boxoff,''Value'',1),'...
                        'set(gca,''Box'',''off'')']);
```

Check Boxes.   Check boxes, sometimes also called *toggle buttons,* consist of a button with a label and a small square box to the left of the label text. When activated, the *uicontrol* is toggled between *checked* and *cleared.* When checked, the box is filled or contains an **X** depending on the platform, and the **'Value'** property is set to 1; when cleared, the box becomes empty and the **'Value'** property is set to 0. The **'Style'** property value of a check box is **'checkbox'**.

Check boxes are typically used to indicate the state of an option or attribute. Check boxes are usually independent objects, but can be used interchangeably with radio buttons, if desired.

The following example (**mmct13.m**) creates a check box *uicontrol* to set the *axes* **'Box'** property. When activated, the **'Value'** property of the *uicontrol* is tested to determine if the check box had just been checked or cleared, and the *axes* **'Box'** property is set appropriately.

```
        » Hc_box = uicontrol(gcf,'Style','check',...
                    'Position',[100 50 100 20],...
                    'String','Axis Box',...
                    'CallBack',[...
                        'if get(Hc_box,''Value'')==1,'...
                            'set(gca,''Box'',''on''),'...
                        'else,'...
                            'set(gca,''Box'',''off''),'...
                        'end']);
```

Static Text Boxes.   Static text boxes are *uicontrols* that simply display a text string as determined by the **'String'** property. The **'Style'** property value of a static text box is **'text'**. Static text boxes are typically used to display labels, user information, or current values.

Static text boxes are *static* in the sense that the user cannot dynamically change the text displayed. The text can only be changed by changing the **'String'** property.

On X Window Systems, a static text box can contain only one line of text; if the text box is too narrow to contain the text string, only a portion of the text is displayed. On Macintosh and MS-Windows platforms, however, text strings that are longer than the width of the text box will *word wrap,* that is, multiple lines will be displayed with the lines broken between words where possible. If the height of the text box is too small for the text string, some of the text will not be visible.

This example (**mmct14.m**) creates a text box containing the MATLAB version number.

```
» Hc_ver = uicontrol(gcf,'Style','text',...
        'Position',[10 10 150 20],...
        'String',['MATLAB Version ',version]);
```

Unlike other *text* objects, e.g., axis titles and labels, the function writer has no explicit control over the font used in *uicontrol* text strings. The fonts used in the *Command* window and in *uicontrols* are identical and can be set by the user.

X Window Systems users can give a command line argument when invoking MATLAB, such as

```
matlab -fn 9x14bold
```

This command uses the **9x14bold** font in both the *Command* window and *uicontrol* text strings. The Macintosh uses the same font for the *Command* window and *uicontrol* strings as well, but the font can be set by the user from the **Text Style** menu item in the *Command* window **Options** menu. PC users have the option of setting the *Command* window font from the **Command Window Font** menu item of the *Command* window **Options** menu, and setting a different font for *uicontrol* text strings from the **Uicontrols Font** item in the same menu. We hope that future versions of MATLAB will add font control properties to *uimenu* and *uicontrol* text strings.

Editable Text Boxes.   Editable text boxes, like static text boxes, display text on the screen. Unlike static text, however, editable text boxes allow the user to modify or replace the text string dynamically just as you would using a text editor or word processor. That information is then available in the **'String'** property. The **'Style'** property value of an editable text box *uicontrol* is **'edit'**. Editable text boxes are typically used to let the user enter a text string or a specific value.

Editable text boxes may contain one or more lines of text. A single-line editable text box will accept one line of text from the user, while a multi-line text box will accept more than one line of text. Single-line text entry is terminated by pressing the **Return** key. Multi-line text entry is terminated with **Control-Return** on X Window Systems and MS-Windows systems, and by **Command-Return** on Macintoshes.

This example (**mmct15.m**) creates a static text label and a single-line editable text box. The user enters a color map name string in the text box and the callback string applies it to the *figure.*

```
» Hc_label = uicontrol(gcf,'Style','text',...
           'Position',[10 10 70 20],...
           'String','Colormap:');

» Hc_map = uicontrol(gcf,'Style','edit',...
           'Position',[80 10 60 20],...
           'String','hsv',...
           'callback','colormap(eval(get(Hc_map,''String'')))');
```

Multi-line text boxes are created by setting the 'Max' or 'Min' property to a number such that Max-Min>1. The 'Max' property does not specify the maximum number of lines. Multi-line text boxes can have an unlimited number of lines. A multi-line editable text box is shown below:

```
» Hc_multi = uicontrol(gcf,'Style','edit',...
           'Position',[20 50 75 75],...
           'String','Line 1|Line 2|Line 3',...
           'Max',2);
```

Multi-line strings are specified as a single quoted string using the vertical bar character '|' to designate where the lines are broken.

Sliders. Sliders, or *scroll bars,* consist of three distinct parts. These parts are the *trough,* or rectangular area representing the range of valid object values, the indicator within the trough representing the current value of the slider, and arrows at each end of the trough. The 'Style' property value of a slider *uicontrol* is 'slider'.

Sliders are typically used to select a value from a range of values. Slider values can be set in three ways. First, the indicator can be moved by positioning the mouse pointer over the indicator, holding down the mouse button while moving the mouse, and releasing the mouse button when the indicator is in the desired location. The second method is to click the mouse button while the pointer is in the trough, but to one side of the indicator. The indicator moves in that direction over a distance equal to about 10% of the total range of the slider. Finally, clicking on one of the arrows at either end of the slider moves the indicator about 1% of the slider range in the direction of the arrow. Sliders are often accompanied by separate text *uicontrol* objects used to display labels, the current slider value, and range limits.

The following example (**mmct16.m**) implements a slider that can be used to set the azimuth of a viewpoint. Three text boxes are used to indicate the maximum, minimum, and current value of the slider.

```
» vw = get(gca,'View');

» Hc_az = uicontrol(gcf,'Style','slider',...
           'Position',[10 5 140 20],...
           'Min',-90,'Max',90,'Value',vw(1),...
```

```
              'CallBack',[...
                 'set(Hc_cur,''String'',num2str(get(Hc_az,''Value''))),'...
                 'set(gca,''View'',[get(Hc_az,''Value'') vw(2)])']);


    » Hc_min = uicontrol(gcf,'Style','text',...
              'Position',[10 25 40 20],...
              'String',num2str(get(Hc_az,'Min')));


    » Hc_max = uicontrol(gcf,'Style','text',...
              'Position',[110 25 40 20],...
              'String',num2str(get(Hc_az,'Max')));


    » Hc_cur = uicontrol(gcf,'Style','text',...
              'Position',[60 25 40 20],...
              'String',num2str(get(Hc_az,'Value')));
```

The **'Position'** property of sliders contains the familiar **[left bottom width height]** vector in units designated by the **'Units'** property. The orientation of the slider depends on the aspect ratio of **width** to **height**. A horizontal slider will be drawn if **width > height**, and a vertical slider will be drawn if **width < height**. On X Window System platforms only, the arrows will not appear if one dimension is less than four times the other. All sliders have arrows on other platforms.

Popup Menus.   Popup menus are typically used to present a list of mutually exclusive options to the user, who can then make a selection. Popup menus, unlike the pull-down menus discussed previously, are not restricted to a menu bar. A popup menu can be positioned anywhere in the *Figure* window. The **'Style'** property value of a popup menu is **'popupmenu'**.

When closed, a popup menu appears as a rectangle or button containing the label of the current selection with a small raised rectangle or downward-pointing arrow to the right of the label to show that the *uicontrol* object is a popup menu. When the pointer is positioned over a popup *uicontrol* and the mouse button is pressed, other choices appear. Moving the pointer to a different choice and releasing the mouse button closes the popup menu and displays the new selection. MS-Windows and some X Window System platforms allow the user to click on a popup to open it, and then click on another choice to select it.

When a popup item is selected, the **'Value'** property is set to the index of the selected element of the vector of choice. Choice labels are specified as a single string separated by vertical bar (|) characters, similar to the way multi-line text strings are specified. The following example (**mmct17.m**) creates a popup menu of *figure* colors. The callback sets the *figure's* **'Color'** property to the chosen value. The RGB values associated with each color are stored in the **'UserData'** property of the popup control. **The 'UserData' property of all Handle Graphics objects simply provides isolated storage for a single matrix.**

```
» Hc_fcolor = uicontrol(gcf,'Style','popupmenu',...
      'Position',[20 20 80 20],...
      'String','Black|Red|Yellow|Green|Cyan|Blue|Magenta|White',...
      'Value',1,...
      'UserData',[[0 0 0];...
                  [1 0 0];...
                  [1 1 0];...
                  [0 1 0];...
                  [0 1 1];...
                  [0 0 1];...
                  [1 0 1];...
                  [1 1 1]],...
      'CallBack',[...
          'UD=get(Hc_fcolor,''UserData'');',...
          'set(gcf,''Color'',UD(get(Hc_fcolor,''Value''),:))']);
```

The **'Position'** property of a popup menu contains the familiar **[left bottom width height]** vector, where the **width** and **height** values determine the dimensions of the popup object. On X Window Systems and Macintosh systems, these are the dimensions of the *closed* popup menu. When opened, the menu expands to display all of the choices that can fit on the screen. On MS-Windows systems, the **height** value is essentially ignored. These platforms create a popup that is tall enough to display one line of text regardless of the **height** value.

Frames.   Frame *uicontrol* objects are simply shaded rectangular regions. Frames are analogous to the **'Separator'** property of *uimenu* objects in the sense that they provide visual separation. Frames are typically used to form groups of radio buttons or other *uicontrol* objects. Frames should be defined before other objects are placed within the frames. Otherwise, the frame may cover the controls making them invisible.

The following example (**mmct18.m**) creates a frame and places two push buttons and a label within it.

```
» Hc_frame = uicontrol(gcf,'Style','frame','Position',[250 200 95 65]);

» Hc_pb1 = uicontrol(gcf,'Style','pushbutton',...
        'Position',[255 205 40 40],'String','OK');

» Hc_pb2 = uicontrol(gcf,'Style','pushbutton',...
        'Position',[300 205 40 40],'String','NOT');

» Hc_lbl = uicontrol(gcf,'Style','text',...
        'Position',[255 250 85 10],'Str','Push Me');
```

Control Properties

As with all Handle Graphics object-creation functions, *uicontrol* properties can be defined at the time the object is created, as shown above, or changed with the **set** command. All settable properties can be changed by `set`, including the string text, callback string, and even the type of control. Some examples appear later in this chapter.

The following table lists the object properties and values found in MATLAB version 4.2c for *uicontrol* objects. Properties with an asterisk (`*`) are undocumented and should be used with caution. Property values enclosed in set brackets `{}` are default values.

Uicontrol Object Properties

`BackgroundColor`	*Uicontrol* background color. A 3-element RGB vector or one of MATLAB's predefined color names. The default background color is light gray
`CallBack`	MATLAB callback string, passed to the `eval` function whenever the *uicontrol* object is *activated;* initially an empty matrix
`ForegroundColor`	*Uicontrol* foreground (text) color. A 3-element RGB vector or one of MATLAB's predefined color names. The default text color is black
`HorizontalAlignment`	Horizontal alignment of label string
`left:`	Text is left-justified with respect to the *uicontrol*
`{center}:`	Text is centered with respect to the *uicontrol*
`right:`	Text is right-justified with respect to the *uicontrol*
`Max`	The largest value allowed for the `Value` property. The value depends on the *uicontrol* `Type`. Radio buttons and check boxes set `Value` to `Max` while the *uicontrol* is in the `on` state. This value defines the maximum index value of a popup menu or the maximum value of a slider. Editable text boxes are multi-line text when `Max-Min>1`. The default value is 1
`Min`	The smallest value allowed for the `Value` property. The value depends on the *uicontrol* `Type`. Radio buttons and check boxes set `Value` to `Min` while the *uicontrol* is in the `off` state. This value defines the minimum index value of a popup menu or minimum value of a slider. Editable text boxes are multi-line text when `Max-Min>1`. The default is 0

Position	Position vector [left, bottom, width, height] where [left, bottom] represents the location of the lower-left corner of the *uicontrol* with respect to the lower-left corner of the *figure* object and [width, height] are the *uicontrol* dimensions. Units are specified by the Units property
* Enable	Control-enable mode
{on}:	The *uicontrol* is enabled. Activating the *uicontrol* will send the CallBack string to eval
off:	The *uicontrol* is disabled and the label string is dimmed. Activating the *uicontrol* has no effect
String	Text string specifying the *uicontrol* label on push buttons, radio buttons, check boxes, and popup menus. For editable text boxes, this property is set to the string typed by the user. For multiple items in a popup menu or an editable text box, separate each item with a vertical bar character (	), and quote the entire strings. Not used for frames or sliders
Style	Defines the type of *uicontrol* object
{pushbutton}:	Push button: performs an action when selected
radiobutton:	Radio button: when used alone, toggles between two states; when used in a group, lets the user choose one option
checkbox:	Check box: when used alone, toggles between two states; when used in a group, lets the user choose one option
edit:	Editable text box: displays a string and lets the user change it
text:	Static text box: displays a string
slider:	Slider: lets the user choose a value within a range of values
frame:	Frame: displays a border around one or more *uicontrols,* forming a logical group
popupmenu:	Popup menu containing a number of mutually-exclusive choices
Units	Unit of measurement for position property values
inches:	Inches
centimeters:	Centimeters

`{normalized}:`	Normalized coordinates, where the lower-left corner of the *figure* maps to `[0  0]` and the upper-right corner maps to `[1  1]`
`points:`	Typesetters points; equal to 1/72 of an inch
`pixels:`	Screen pixels; the smallest unit of resolution of the computer screen
`Value`	Current value of the *uicontrol*. Radio buttons and check boxes set **Value** to **Max** when on and **Min** when **off**. Sliders set **Value** to the number set by the slider (**Min≤Value≤Max**). Popup menus set **Value** to the index of the item selected (**1≤Value≤Max**). Text objects and push buttons do not set this property
`{on}:`	*Uicontrol* is visible on the screen
`off:`	*Uicontrol* is not visible, but still exists
`ButtonDownFcn`	MATLAB callback string, passed to the **eval** function whenever the *uicontrol* is selected; initially the empty matrix
`Children`	*Uicontrol* objects have no children; always returns the empty matrix
`Clipping`	Clipping mode
`{on}:`	No effect for *uicontrol* objects
`off:`	No effect for *uicontrol* objects
`DestroyFcn`	Macintosh version 4.2c only. No documentation available
`Interruptible`	Specifies if **ButtonDownFcn** and **CallBack** strings are interruptible
`{no}:`	Callbacks are not interruptible by other callbacks
`yes:`	Callback strings are interruptible
`Parent`	Handle of the *figure* containing the *uicontrol* object
* `Selected`	Values are [on | off]
* `Tag`	Text string
`Type`	Read-only object identification string; always **uicontrol**
`UserData`	User-specified data. Can be matrix, string, etc.
`Visible`	Visibility of *uicontrol* object
`{on}:`	*Uicontrol* is visible on the screen
`off:`	*Uicontrol* is not visible, but still exists

Control Placement Considerations. The `'Position'` and `'Units'` properties of *uicontrols* are used to locate control objects in the *Figure* window. The default `'Position'` vector for *uicontrols* is `[20 20 60 20]` represented in pixels, which is the default `'Units'` value. This is a `60-by-20` pixel rectangle with the lower-left corner of the *uicontrol* positioned 20 pixels to the right and 20 pixels above the lower-left corner of the parent *figure*. The default *figure* size is around `560-by-420` pixels, placed in the upper middle of the display. Using this information, placing *uicontrols* becomes a layout problem in 2-D geometry.

Some restrictions apply. For example, MS-Windows popup menus ignore the `height` value of the position vector, and use just enough vertical space to display one line of text. In all other situations, one must make sure that the control is large enough to contain the control-label string. Since the font that is used to display the `'String'` property value of controls is the same font that is used in the *Command* window, one has no control over the properties of the font used to label each control. Different fonts will be used on different platforms and can be changed globally by the user.

Often sizing and placing controls is a process of trial and error. Even when you are satisfied with the result, the appearance of the *figure* on another platform may be sufficiently different to require more adjustments. Often it is desirable to make the control larger than appears necessary to be sure that the label is readable on all platforms.

Just because a *figure* has a default size, there is no guarantee that all *figures* are that size. If you add *uicontrols* or *uimenus* to an existing *figure*, it may be smaller or larger than the default. In addition, the user can resize any *figure* at any time unless prevented from doing so by setting the *figure's* `'Resize'` property to `'off'`.

Two things to consider when adding *uicontrols* to a *figure* that may be resized are the `'Units'` property and the restrictions imposed by fixed-font label strings. If the position of each *uicontrol* is specified in absolute units such as `pixels`, `inches`, `centimeters`, or `points`, resizing the window will not change the size or placement of the *uicontrols*. The *uicontrols* will maintain the same position relative to the bottom and left sides of the *figure*. If the *figure* is made smaller, some of the *uicontrols* may no longer be visible.

If the position is specified in `normalized` units, the *uicontrols* will maintain their relationship to each other and to the *figure* itself when the *figure* is resized. There is one drawback to using `normalized` units, however. If the *figure* is made smaller and the *uicontrols* are resized as well, *uicontrol* labels may become unreadable since the font size is fixed. Any portion of the label that is outside the dimensions of the resized *uicontrol* is clipped. It is hoped that the next release of MATLAB will give the programmer more control over the font properties of *uicontrol-* and *uimenu*-label strings.

M-file Examples

The following example illustrates the use of some of the controls discussed in this chapter. The *Mastering MATLAB Toolbox* function `mmclock` creates a digital clock on the display screen with optional arguments to set the clock position. Frame, text, radio button, check box, and push button *uicontrols* are used.

To run this example on a PC, select the **Enable Background Process** menu item in the **Options** menu of the *Command* window. Failing to do so will cause MATLAB to go into an infinite loop and hang. For best results on a Macintosh, turn off the **Background Operation** checkmark in the **Options** menu. X Window System versions need no adjustments.

First, the `function` statement and help text is created.

```
        function T=mmclock(X,Y)
        %MMCLOCK Place a digital clock on the screen.
        % MMCLOCK places a digital clock at the upper-right corner
        % of the display screen.
        % MMCLOCK(X,Y) places a digital clock at position X pixels
        % to the right and Y pixels above the bottom of the screen.
        % T=MMCLOCK returns the current date and time as a string
        % when 'Close' is pressed.
```

Define some initial values and parse any input arguments.

```
        fsize = [200 150]; sec = 1; mil = 0;
        mstr = ['Jan';'Feb';'Mar';'Apr';'May';'Jun'
                'Jul';'Aug';'Sep';'Oct';'Nov';'Dec'];
        scr = get(0,'ScreenSize');

        if nargin == 0
            figpos = [scr(3)-fsize(1)-20 scr(4)-fsize(2)-5 fsize(1:2)];
        elseif nargin == 2
            figpos = [X Y fsize(1:2)];
        else
            error('Invalid Arguments');
        end
```

Create the *figure,* and set some defaults for *uicontrols* in this *figure.*

```
      Hf_clock = figure('Position',figpos',...
              'Color',[.7 .7 .7],...
              'NumberTitle','off',...
              'Name','Digital Clock');
      set(Hf_clock,'DefaultUicontrolUnits','normalized',...
              'DefaultUicontrolBackgroundColor',get(Hf_clock,'Color'));
```

Create the **Close** push button, radio button for seconds, and check box for 24-hour style.

```
            Hc_close = uicontrol('Style','push',...
                    'Position',[.65 .05 .30 .30],...
                    'BackgroundColor',[.8 .8 .9],...
                    'String','Close',...
                    'CallBack','close(gcf)');

            Hc_sec = uicontrol('Style','radiobutton',...
                    'Position',[.05 .05 .50 .13],...
                    'Value',sec,...
                    'String','Seconds');

            Hc_mil = uicontrol('Style','checkbox',...
                    'Position',[.05 .22 .50 .13],...
                    'Value',mil,...
                    'String','24-Hour');
```

Create frames and text strings for the date and time displays.

```
   Hc_dframe = uicontrol('Style','frame','Position',[.04 .71 .92 .24]);
   Hc_date   = uicontrol('Style','text', 'Position',[.05 .72 .90 .22]);
   Hc_tframe = uicontrol('Style','frame','Position',[.04 .41 .92 .24]);
   Hc_time   = uicontrol('Style','text', 'Position',[.05 .42 .90 .22]);
```

Loop until the *figure* is closed by the push button callback, updating the display once every second.

```
   while find(get(0,'Children') == Hf_clock) % Loop while clock exists
       sec = get(Hc_sec,'Value');
       mil = get(Hc_mil,'Value');
```

```
        now = fix(clock);
        datestr = sprintf('%s %2d, %4d',mstr(now(2),:),now(3),now(1));
        timestr = [num2str(now(4)) ':' sprintf('%02d',now(5))];
        if sec
             timestr = [timestr ':' sprintf('%02d',now(6))];
        end
        if mil
             suffix = '';
        else
             if now(4) > 12
                  suffix = ' PM';
                  now(4) = rem(now(4),12);
             else
                  suffix = ' AM';
             end
        end
        timestr = [timestr suffix];
        set(Hc_date,'String',datestr);
        set(Hc_time,'String',timestr);
        pause(1)
   end
```

And finally, return a date and time string if requested.

```
             if nargout
                  T = [datestr ' - ' timestr];
             end
```

While this example illustrates how a complete GUI function can be created using *uicontrols,* it is not very practical. Because the function keeps looping until the **Close** button is pressed, the *Command* window is not available until the clock *figure* is closed.

Notice that this example contains only one callback string: the `close` statement in the push button callback. The other button values are obtained within the While Loop rather than having the buttons themselves trigger an action using callbacks. More complex M-files require more complex callbacks.

21.5 PROGRAMMING AND CALLBACK CONSIDERATIONS

Now that the basics of Handle Graphics User Interface functions have been covered, it is time to apply them. As you have seen, creating *uimenus* and *uicontrols* by typing them in on the command line is not very efficient. Script or function M-files are a lot easier to use. Suppose you have an idea for an M-file that you want to implement. The first decision is whether to write a script or a function.

Scripts vs. Functions

Scripts seem to be the obvious choice. In scripts, everything executes in the *Command* window workspace, so all MATLAB functions and all object handles are available at all times. There is no difficulty passing information to callback strings. There are a number of trade-offs, however. The first is that while all variables are available, the workspace becomes cluttered with variable names and values even after they are no longer useful. On the other hand, if the user issues the **clear** command, important object handles may be lost. Another disadvantage is that defining callback strings can become very complicated using scripts. For example, here is a slider-definition fragment from a script M-file in the *Mastering MATLAB Toolbox* called **mmsetclr**.

```
Hc_rsli = uicontrol(Hf_fig,'Style','slider',...
  'Position',[.10 .55 .35 .05],...
  'Min',0,'Max',1,'Value',initrgb(1),...
  'CallBack',[...
    'set(Hc_nfr,''BackgroundColor'',',...
     '[get(Hc_rsli,''Val''),get(Hc_gsli,''Val''),get(Hc_bsli,''Val'')]),',...
    'set(Hc_ncur,''String'',',...
    'sprintf(''[%.2f %.2f %.2f]'',get(Hc_nfr,''BackgroundColor''))),',...
    'hv=rgb2hsv(get(Hc_nfr,''BackgroundColor''));',...
    'set(Hc_hsli,''Val'',hv(1)),',...
    'set(Hc_hcur,''String'',sprintf(''%.2f'',hv(1))),',...
    'set(Hc_ssli,''Val'',hv(2)),',...
    'set(Hc_scur,''String'',sprintf(''%.2f'',hv(2))),',...
    'set(Hc_vsli,''Val'',hv(3)),',...
    'set(Hc_vcur,''String'',sprintf(''%.2f'',hv(3))),',...
    'set(Hc_rcur,''String'',sprintf(''%.2f'',get(Hc_rsli,''Val'')))']);
```

Another problem is that scripts execute more slowly than functions, which are compiled the first time they are run. Finally, scripts are less flexible than functions. Functions can accept input arguments and can return values. Functions, therefore, can be used as arguments to other functions.

Functions keep the *Command* window workspace uncluttered, execute rapidly when called repeatedly, accept input arguments and return values, and callback strings are less complicated to write. Therefore, in many situations, function M-files are the best choice.

Consider the previous example of a slider definition from a script file. Here is the equivalent fragment taken from a function M-file in the *Mastering MATLAB Toolbox* called **mmsetc**.

```
Hc_rsli = uicontrol(Hf_fig,'Style','slider',...
        'Position',[.10 .55 .35 .05],...
        'Min',0,'Max',1,'Value',initrgb(1),...
        'CallBack','mmsetc(0,''rgb2new'')');
```

Note that the callback string calls **mmsetc** again with different arguments. This is an example of using recursive function calls in callbacks. Functions have their own problems, however. The major difficulty with functions results from the fact that callback strings are passed to **eval** and **executed in the *Command* window workspace**, while the rest of the code executes within the function workspace. The scope rules for variables and functions discussed in an earlier chapter apply here. Variables defined within the function are not available in the *Command* window workspace, and therefore cannot be used in callback strings. At the same time, variables used in the *Command* window workspace are not available inside the function itself.

There are a number of ways to work around this problem while gaining the advantages of using functions. Global variables, **'UserData'** properties, special function M-files used only for callbacks, and recursive function calls can all be useful techniques for creating GUI function M-files.

Separate Callback Functions

One effective technique for creating GUI functions is to write separate M-file functions designed specifically to execute one or more callbacks. Object handles and other variables used by the callback function can be passed as arguments, and the callback function can return values if necessary.

Consider an earlier example that creates an azimuth slider implemented as a script.

```
% setview.m script file

vw = get(gca,'View');

Hc_az = uicontrol(gcf,'Style','slider',...
        'Position',[10 5 140 20],...
        'Min',-90,'Max',90,'Value',vw(1),...
        'CallBack',[...
            'set(Hc_cur,''String'',num2str(get(Hc_az,''Value''))),'...
            'set(gca,''View'',[get(Hc_az,''Value'') vw(2)])']);
```

```
Hc_min = uicontrol(gcf,'Style','text',...
         'Position',[10 25 40 20],...
         'String',num2str(get(Hc_az,'Min')));

Hc_max = uicontrol(gcf,'Style','text',...
         'Position',[110 25 40 20],...
         'String',num2str(get(Hc_az,'Max')));

Hc_cur = uicontrol(gcf,'Style','text',...
         'Position',[60 25 40 20],...
         'String',num2str(get(Hc_az,'Value')));
```

Here is the same example as a function using the **'Tag'** property to identify controls and use a separate function M-file to execute the callback.

```
function setview()

vw=get(gca,'View');

Hc_az = uicontrol(gcf,'Style','slider',...
        'Position',[10 5 140 20],...
        'Min',-90,'Max',90,'Value',vw(1),...
        'Tag','AZslider',...
        'CallBack','svcback');

Hc_min = uicontrol(gcf,'Style','text',...
        'Position',[10 25 40 20],...
        'String',num2str(get(Hc_az,'Min')));

Hc_max = uicontrol(gcf,'Style','text',...
        'Position',[110 25 40 20],...
        'String',num2str(get(Hc_az,'Max')));

Hc_cur = uicontrol(gcf,'Style','text',...
        'Position',[60 25 40 20],...
        'Tag','AZcur',...
        'String',num2str(get(Hc_az,'Value')));
```

And the callback function itself is given by

```
function svcback()

vw = get(gca,'View');

Hc_az = findobj(gcf,'Tag','AZslider');
Hc_cur = findobj(gcf,'Tag','AZcur');

str = num2str(get(Hc_az,'Value'));
newview = [get(Hc_az,'Value') vw(2)];
set(Hc_cur,'String',str)
set(gca,'View',newview)
```

The previous example does not save much coding, but gains all the advantages of using functions rather than scripts. The callback function can use temporary variables without cluttering up the *Command* window workspace, and the syntax of the commands in call-back functions becomes much simpler without all those quotes and strings required for **eval**. More complicated callbacks are much simpler using this technique.

The disadvantage of separate callback functions is the potentially large number of M-files required to implement a single GUI function that uses a number of controls or menu items. All of the M-files must be available in the MATLAB path, and each must have a distinctive name. On platforms that have filename size limits and are not case-sensitive, e.g., MS-Windows, the chances of a filename conflict are increased. Also, the callback functions should only be called by the GUI function itself and not by the user.

Recursive Function Calls

You can avoid the complexity of multiple M-files while maintaining the advantages of functions by using a single M-file and calling it recursively to execute callbacks. The callback functions can be incorporated into the calling function by using *switches* or **if** and **elseif** statements. The generic structure of such a function call is

<code>function guifunc(switch)</code>

where **switch** is the argument that determines which function switch is executed. This could be a string such as **'startup'**, **'close'**, **'setcolor'**, etc. It could also be a code or number. If **switch** is a string, the switch can be coded as shown in the following M-file fragment:

```
if nargin < 1, switch = 'startup'; end;
if ~isstr(switch), error('Invalid argument'), end;
if strcmp(switch,'startup'),
```

```
        <statements to create controls or menus>
        <statements to implement the GUI function>
elseif strcmp(switch,'setcolor'),
        <statements to perform the callback associated with setcolor>
elseif strcmp(switch,'close'),
        <statements to perform the callback associated with close>
end
```

If codes or numbers are used, the switch can be coded in a similar manner.

```
if nargin < 1, switch = 0; end;
if isstr(switch), error('Invalid argument'), end;
if switch == 0,
        <statements to create controls or menus>
        <statements to implement the GUI function>
elseif switch == 1,
        <statements to perform the callback associated with setcolor>
elseif switch == 2,
        <statements to perform the callback associated with close>
end
```

The following example shows how the azimuth-slider example could be implemented as a single function M-file:

```
function setview(switch)

if nargin < 1, switch = 'startup'; end;
if ~isstr(switch), error('Invalid argument.'); end;

vw = get(gca,'View'); % This information is needed in both sections

if strcmp(switch,'startup') % Define the controls and tag them

    Hc_az = uicontrol(gcf,'Style','slider',...
        'Position',[10 5 140 20],...
        'Min',-90,'Max',90,'Value',vw(1),...
        'Tag','AZslider',...
        'CallBack','setview(''set'')');

    Hc_min = uicontrol(gcf,'Style','text',...
        'Position',[10 25 40 20],...
        'String',num2str(get(Hc_az,'Min')));
```

```
        Hc_max = uicontrol(gcf,'Style','text',...
            'Position',[110 25 40 20],...
            'String',num2str(get(Hc_az,'Max')));

        Hc_cur = uicontrol(gcf,'Style','text',...
            'Position',[60 25 40 20],...
            'Tag','AZcur',...
            'String',num2str(get(Hc_az,'Value')));

    elseif strcmp(switch,'set') % Execute the callback

        Hc_az=findobj(gcf,'Tag','AZslider');
        Hc_cur=findobj(gcf,'Tag','AZcur');

        str = num2str(get(Hc_az,'Value'));
        newview = [get(Hc_az,'Value') vw(2)];

        set(Hc_cur,'String',str)
        set(gca,'View',newview)
    end
```

Both of the preceding examples set the `'Tag'` property and use this property with the `findobj` function to find handles to desired objects for the callbacks. Two alternative methods are described in the next subsections.

Global Variables

Global variables can be used in a function to make certain variables available to all parts of the GUI function. The global variables are declared in the common area of the function and are therefore available throughout the function and all of its recursive calls. This example shows how the azimuth-slider function can be coded using global variables.

```
    function setview(switch)

    global HC_AZ HC_CUR % Create global variables

    if nargin < 1, switch = 'startup'; end;
    if ~isstr(switch), error('Invalid argument.'); end;

    vw = get(gca,'View'); % This information is needed in both sections
```

```
if strcmp(switch,'startup') % Define the controls

    HC_AZ = uicontrol(gcf,'Style','slider',...
        'Position',[10 5 140 20],...
        'Min',-90,'Max',90,'Value',vw(1),...
        'CallBack','setview(''set'')');

    Hc_min = uicontrol(gcf,'Style','text',...
        'Position',[10 25 40 20],...
        'String',num2str(get(HC_AZ,'Min')));

    Hc_max = uicontrol(gcf,'Style','text',...
        'Position',[110 25 40 20],...
        'String',num2str(get(HC_AZ,'Max')));

    HC_CUR = uicontrol(gcf,'Style','text',...
        'Position',[60 25 40 20],...
        'String',num2str(get(HC_AZ,'Value')));

elseif strcmp(switch,'set') % Execute the callback

    str = num2str(get(HC_AZ,'Value'));
    newview = [get(HC_AZ,'Value') vw(2)];

    set(HC_CUR,'String',str)
    set(gca,'View',newview)

end
```

The global variables follow the MATLAB convention and use uppercase variable names. The `'Tag'` property is not needed and is not used. In addition, the callback code is simpler because the object handles are available without using `findobj` to obtain them. Global variables generally make a function more efficient.

One word of caution, however. Even though a variable is declared `global` within the function, the variable is not automatically available in the *Command* window workspace and cannot be used within a callback string. However, if the user issues the command » `clear global`, *all* global variables are destroyed, including those within the function.

The use of global variables and recursive-function calls are effective techniques when there is a single *figure* or a limited number of variables that need to be available to all callbacks. For more complex functions involving multiple *figures*, or implementations using separate callback functions, the `'UserData'` property may be more appropriate. In addition, the `'UserData'` property value of an object is available in the *Command* window workspace as long as the object handle can be obtained.

UserData Properties

As is the case with the **'Tag'** property, the **'UserData'** property is available to pass information between functions or between different parts of a recursive function. If many variables are needed, they can be passed in the **'UserData'** property of an easily-identified object. **As described earlier, 'UserData' provides storage for a single matrix of data that stays with a Handle Graphics object.** The following code implements the azimuth slider using the **'UserData'** property of the current *figure:*

```
function setview(switch)

if nargin < 1, switch = 'startup'; end;
if ~isstr(switch), error('Invalid argument.'); end;

vw = get(gca,'View'); % This information is needed in both sections

if strcmp(switch,'startup') % Define the controls

    Hc_az = uicontrol(gcf,'Style','slider',...
        'Position',[10 5 140 20],...
        'Min',-90,'Max',90,'Value',vw(1),...
        'CallBack','setview(''set'')');

    Hc_min = uicontrol(gcf,'Style','text',...
        'Position',[10 25 40 20],...
        'String',num2str(get(Hc_az,'Min')));

    Hc_max = uicontrol(gcf,'Style','text',...
        'Position',[110 25 40 20],...
        'String',num2str(get(Hc_az,'Max')));

    Hc_cur = uicontrol(gcf,'Style','text',...
        'Position',[60 25 40 20],...
        'String',num2str(get(Hc_az,'Value')));

    set(gcf,'UserData',[Hc_az Hc_cur]); % Store the object handles

elseif strcmp(switch,'set') % Execute the callback

    Hc_all = get(gcf,'UserData'); % retrieve the object handles
    Hc_az  = Hc_all(1);
    Hc_cur = Hc_all(2);

    str = num2str(get(Hc_az,'Value'));
    newview = [get(Hc_az,'Value') vw(2)];
```

```
      set(Hc_cur,'String',str)
      set(gca,'View',newview)
end
```

The handles are stored in the *figure's* **'UserData'** property at the end of the **'startup'** section and are retrieved before the callback is executed. If there are many callbacks, the **'UserData'** value needs to be retrieved only once as shown in the following code fragments:

```
if strcmp(switch,'startup') % Define the controls and tag them

    % <The 'startup' code is here>

    set(gcf,'UserData',[Hc_az Hc_cur]); % Store the object handles

else    % This must be a callback

    Hc_all = get(gcf,'UserData'); % Retrieve the object handles
    Hc_az  = Hc_all(1);
    Hc_cur = Hc_all(2);

    if strcmp(switch,'set')

        % <The 'set' callback code is here>

    elseif strcmp(switch,'close')

        % <The 'close' callback code is here>

    % <Any other callback code uses additional elseif clauses>
    end
end
```

Debugging GUI M-files

The fact that the callback strings are evaluated and executed in the *Command* window workspace has certain implications for writing and debugging GUI functions and scripts. Callback strings can be very complex, especially in scripts, introducing many opportunities for syntax errors. Keeping track of single quotes, commas, and parentheses can be a daunting task. MATLAB complains if the syntax is not correct, but as long as the value of the `'CallBack'` property of an object is an actual text string, MATLAB is satisfied. The string is not checked for internal syntax errors until the object is activated and the callback string is passed to **eval**.

This lets you define callback strings that reference variables and object handles that have not yet been defined, making it much easier to write MATLAB code that cross-references other objects. However, each callback must be tested separately to make sure that the callback string is a legal MATLAB command and that all variables referenced in the callback string are available in the *Command* window workspace.

Coding callbacks as function M-files or as switches within the GUI function itself lets you change and test individual callbacks without running the entire GUI function.

Because callback strings are evaluated in the *Command* window workspace rather than within the function itself, passing data between the function and each callback can become complex. For example, if a function **test** contains the following code:

```
function test()

tpos1 = [20 20 50 20];

tpos2 = [20 80 50 20];

Hc_text = uicontrol('Style','text','String','Hello','Position',tpos1);

Hc_push = uicontrol('Style','push','String','Move Text',...
    'Position',[15 50 100 25],...
    'CallBack','set(Hc_text,''Position'',tpos2)');
```

all of the statements are valid MATLAB commands, and the callback string evaluates to a valid MATLAB statement as well. The text object and the push button appear on the *figure,* but when the push button is activated, MATLAB complains.

```
» test
»
??? Undefined function or variable Hc_text.

??? Error while evaluating callback string.
```

If **test** were a script, there would be no problem since all variables would be available in the *Command* window workspace. Since **test** is a function, neither **Hc_text** nor **tpos2** is defined in the *Command* window workspace, and the callback string execution fails.

One solution is to create the callback string using individual string elements created from values rather than from variables. For example, changing the callback string as follows:

```
'CallBack',['set(',...
            sprintf('%.15g',Hc_text),...
            ',''Position'','',...
            sprintf('[ %.15g %.15g %.15g %.15g ]',tpos2),...
            ')']);
```

creates a string that includes the value of the **Hc_text** object handle converted to a string with up to 15 digits of precision, and the value of the **tpos2** variable converted to the string representation of a matrix. The **sprintf** statements are evaluated within the function, and the resulting strings are then used in the callback. The actual command that is executed in the *Command* window workspace looks like this:

```
eval('set( 87.000244140625 ,''Position'', [ 20 80 50 100 ] )')
```

Full precision must be maintained when converting an object handle into a string, The conversion above uses up to 15 digits after the decimal point. This is the precision that should be used for object-handle conversions in MATLAB.

Remember, however, that subsequent changes to the variables will not change the callback string. In the previous example, changing the value of **tpos2** after the control is defined has no effect. For example, adding the command

```
tpos2 = [20 200 50 20];
```

to the end of the function has no effect since the callback string is created by evaluating **tpos2** before **tpos2** is redefined.

21.6 POINTER AND MOUSE BUTTON EVENTS

GUI functions use the location of the mouse pointer and the status of mouse buttons to control MATLAB actions. This section discusses the interaction between pointer and object locations and mouse button actions, and how MATLAB responds to changes or *events,* such as a button press, button release, or pointer movement.

Callback Properties, Selection Regions, and Stacking Order

All Handle Graphics objects have a **'ButtonDownFcn'** property that has not yet been addressed, but will be discussed here. Both *uimenus* and *uicontrols* have a **'CallBack'**

property that is central to the use of menus and controls. In addition, *figures* have `'Window-`
`ButtonDownFcn'`, `'WindowButtonUpFcn'`, and `'WindowButtonMotionFcn'`
properties as well as `'KeyPressFcn'` and `'ResizeFcn'` properties. The value associated with each of these properties is a *callback* string, a MATLAB string that is passed to
`eval` when the property is invoked. The pointer location determines which callbacks are involved and the order in which they are invoked when an event occurs.

The previous chapter contained a discussion of stacking order and object-selection
regions that is relevant to this discussion. MATLAB determines which callback will be invoked based on three regions within a *figure*. If the pointer is within a Handle Graphics object as determined by its `'Position'` property, the pointer is **on** the object. If the pointer
is not **on** an object but is within an object's **selection region,** the pointer is **near** the object.
Finally, if the pointer is within the *figure* but not **on** or **near** another object, the pointer can
be considered **off** of other objects. When objects or their selection regions overlap, the
stacking order determines which object is selected.

The selection region of Handle Graphics' *line, surface, patch, text,* and *axes* objects
were discussed in the previous chapter. *Uimenu* objects have no external selection region.
The pointer is either **on** a *uimenu* object or it is not. *Uicontrols* have a selection region that
extends about 5 pixels beyond the control's position in all directions. The pointer can be either **on** or **near** a control.

Remember that the stacking order and selection regions discussed here are current as of MATLAB version 4.2c, but may change to some extent in future versions
of MATLAB.

Button Click

A button click can be defined as the press and subsequent release of a mouse button while
the mouse pointer is over the same object. If the mouse pointer is over a *uicontrol* or
uimenu item, a button click triggers the execution of the object's `'CallBack'` property
string. The button press prepares the control and often changes the *uicontrol* or *uimenu*
visually, and the button release triggers the callback. If the mouse pointer is not **on** a
uicontrol or *uimenu,* both **Button Press** and **Button Release** events are triggered as ex-
plained below.

Button Press

When the mouse button is pressed with the pointer located within a *Figure window,* a number of different actions can occur, based on the location of the pointer and the proximity of
Handle Graphics objects. If an object is selected, it becomes the current object and rises to
the top of the stacking order. If no object is selected, the *figure* itself becomes the current
object. The *figure's* `'CurrentPoint'` and `'SelectionType'` properties are also updated. The appropriate callbacks are then invoked.

The following table lists the pointer location options and the callbacks that are invoked for a button press event.

Pointer Location	Property Invoked
on a *uicontrol* or *uimenu* item	Prepare for a release event
on or **near** a Handle Graphics object, or **near** a control	*Figure's* `WindowButtonDownFcn` and then the object's `ButtonDownFcn`
within the *figure*, but not **on** or **near** any other object (**off**)	*Figure's* `WindowButtonDownFcn` and then the *figure's* `ButtonDownFcn`

Note that a button press event always invokes the *figure's* `'WindowButtonDownFcn'` callback before the selected object's `'ButtonDownFcn'` callback except when the pointer is **on** a *uicontrol* or *uimenu* object. When the pointer is **near** a control, the control's `'ButtonDownFcn'` callback is invoked rather than the `'CallBack'` property callback after the *figure's* `'WindowButtonDownFcn'` callback has finished.

Button Release

When the mouse button is released, the *figure's* `'CurrentPoint'` property is updated, and the *figure's* `'WindowButtonUpFcn'` callback is invoked. If the `'WindowButtonUpFcn'` callback is not defined, the `'CurrentPoint'` property is not updated when the button is released.

Pointer Movement

When the pointer is moved within a *figure*, the *figure's* `'CurrentPoint'` property is updated, and the *figure's* `'WindowButtonMotionFcn'` callback is invoked. If the `'WindowButtonMotionFcn'` callback is not defined, the `'CurrentPoint'` property is not updated when the pointer moves.

Combinations of callbacks can produce many interesting effects. Try the `sigdemo1` and `sigdemo2` function M-files included in the MATLAB `demo` directory to see examples of some of these effects. Another example that will be discussed later is the *Mastering MATLAB Toolbox* function `mmdraw`.

21.7 RULES FOR INTERRUPTING CALLBACKS

Once a callback begins executing, it normally runs to completion before the next callback event is processed. You can change this default behavior by setting the object's `'Interruptible'` property to `'yes'`, allowing pending callback events to be processed when the executing callback reaches a `drawnow`, `figure`, `getframe`, or `pause` command.

Event Queue

MATLAB processes commands that perform computations or set object properties at the time they are issued. Commands that involve *Figure* window input or output generate events. Events include pointer movements and mouse button actions that generate callbacks, and commands that redraw graphics.

CallBack Processing

A callback will execute until it reaches a **drawnow**, **getframe**, **pause**, or **figure** command. Note that **gcf** and **gca** spawn **figure** commands, while the *Mastering MATLAB Toolbox* functions **mmgcf** and **mmgca** do not. Callbacks that do not contain any of these special commands will not be interrupted at all.

When one of these special commands is reached, MATLAB will stop execution of the callback, *suspending* it, and examine each of the pending events in the event queue. If the **'Interruptible'** property of the object that generated the suspended callback is set to **'yes'**, all pending events are processed in order before the suspended callback is resumed. If the **'Interruptible'** property is set to **'no'**, the default value, then only pending redraw events are processed. Callback events are discarded.

Preventing Interruptions

Even if the executing callback is not interruptible, pending redraw events are still processed when the callback reaches a **drawnow**, **figure**, **getframe**, or **pause** command. These events can be suppressed by avoiding the use of all of these special commands in your callback. If they are needed in your callback, but you do not want *any* pending events, even refresh events, to interrupt your callback, you can use a special form of the **drawnow** command, which is discussed next, to prevent interruptions.

Drawnow

The **drawnow** command forces MATLAB to update the screen. Screen updates occur whenever MATLAB returns to the *Command* prompt », or whenever a **drawnow**, **figure**, **getframe**, or **pause** command is executed. The special form **drawnow('discard')** causes *all* events in the event queue to be discarded. Including a **drawnow('discard')** command in your callback prior to one of these special commands will have the effect of flushing the event queue, preventing refresh events as well as callback events from interrupting your callback.

21.8 M-FILE EXAMPLES

A number of functions from the *Mastering MATLAB Toolbox* illustrate some of the techniques discussed previously. The first example, **mmview3d**, uses global variables and recursive function calls to add azimuth and elevation sliders to a *figure*. There are a large number of objects, but the function is straightforward. Since **mmview3d** is fairly long, it is presented in fragments. The first fragment defines the function, the help text, and the global variables.

```
function mmview3d(cmd)
%MMVIEW3D GUI controlled Azimuth and Elevation adjustment.
% MMVIEW3D adds sliders and text boxes to the current figure window
% for adjusting azimuth and elevation using the mouse.
%
% The 'Revert' pushbutton reverts to the original view.
% The 'Done' pushbutton removes all GUIs.

% The 'cmd' argument executes the callbacks.

% Copyright (c) 1996 by Prentice-Hall, Inc.

global Hc_asli Hc_acur Hc_esli Hc_ecur CVIEW
```

The next fragment handles the initial user call, creating the necessary *uicontrol* objects and defining the callbacks as recursive function calls.

```
if nargin == 0

   % - - - - - - - - - - - - - - - - - - - - - - - - - - - - - - - - -
   % Assign a handle to the current figure window.
   % Get the current view for slider initial values.
   % If the view is out of range, adjust as best we can.
   % Use normalized uicontrol units rather than the default 'pixels'.
   % - - - - - - - - - - - - - - - - - - - - - - - - - - - - - - - - -

   Hf_fig = gcf;
   CVIEW = get(gca,'View');
   if abs(CVIEW(1))>180, CVIEW(1)=CVIEW(1)-(360*sign(CVIEW(1))); end
   set(Hf_fig,'DefaultUicontrolUnits','normalized');

   % - - - - - - - - - - - - - - - - - - - - - - - - - - - - - - - - -
   % Define azimuth and elevation sliders.
```

```
% The position is in normalized units (0-1).
% Maximum, minimum, and initial values are set.
% - - - - - - - - - - - - - - - - - - - - - - - - - - - - - -

Hc_asli = uicontrol( Hf_fig,'style','slider',...
   'position',[.09 .02 .3 .05],...
   'min',-180,'max',180,'value',CVIEW(1),...
   'callback','mmview3d(991)');

Hc_esli = uicontrol( Hf_fig,'style','slider',...
   'position',[.92 .5 .04 .42],...
   'min',-90,'max',90,'val',CVIEW(2),...
   'callback','mmview3d(992)');

% - - - - - - - - - - - - - - - - - - - - - - - - - - - - - -
% Place the text boxes showing the minimum and maximum values at the
% ends of each slider. These are text displays, and cannot be edited.
% - - - - - - - - - - - - - - - - - - - - - - - - - - - - - -

uicontrol(Hf_fig,'style','text',...
   'pos',[.02 .02 .07 .05],...
   'string',num2str(get(Hc_asli,'min')));

uicontrol(Hf_fig,'style','text',...
   'pos',[.39 .02 .07 .05],...
   'string',num2str(get(Hc_asli,'max')));

uicontrol(Hf_fig,'style','text',...
   'pos',[.915 .45 .05 .05],...
   'string',num2str(get(Hc_esli,'min')));

uicontrol(Hf_fig,'style','text',...
   'pos',[.915 .92 .05 .05],...
   'string',num2str(get(Hc_esli,'max')));

% - - - - - - - - - - - - - - - - - - - - - - - - - - - - - -
% Place labels for each slider
% - - - - - - - - - - - - - - - - - - - - - - - - - - - - - -

uicontrol(Hf_fig,'style','text',...
     'pos',[.095 .08 .15 .05],...
     'string','Azimuth');

uicontrol(Hf_fig,'style','text',...
   'pos',[.885 .39 .11 .05],...
   'string','Elevation');
```

```
% - - - - - - - - - - - - - - - - - - - - - - - - - - - - - - -
% Define the current value text displays for each slider.
% - - - - - - - - - - - - - - - - - - - - - - - - - - - - - - -
% These are editable text displays to display the current value
% of the slider and at the same time allow the user to enter
% a value using the keyboard.
%
% Note that the text is centered on X Window Systems, but is
% left-justified on MS-Windows and Macintosh machines.
%
% The initial value is found from the value of the sliders.
% When text is entered into the text area and the return key is
% pressed, the callback string is evaluated.
% - - - - - - - - - - - - - - - - - - - - - - - - - - - - - - -

Hc_acur = uicontrol(Hf_fig,'style','edit',...
   'pos',[.25 .08 .13 .053],...
   'string',num2str(get(Hc_asli,'val')),...
   'callback','mmview3d(993)');

Hc_ecur = uicontrol(Hf_fig,'style','edit',...
   'pos',[.885 .333 .11 .053],...
   'string',num2str(get(Hc_esli,'val')),...
   'callback','mmview3d(994)');

% - - - - - - - - - - - - - - - - - - - - - - - - - - - - - - -
% Place a 'Done' button in the lower right corner.
% When clicked, all of the uicontrols will be erased.
% - - - - - - - - - - - - - - - - - - - - - - - - - - - - - - -

uicontrol(Hf_fig,'style','push',...
   'pos',[.88 .02 .10 .08],...
   'backgroundcolor',[.7 .7 .8],...
   'string','Done',...
   'callback','delete(findobj(gcf,''Type'',''uicontrol''))');

% - - - - - - - - - - - - - - - - - - - - - - - - - - - - - - -
% Place a 'Revert' button next to the 'Done' button.
% When clicked, the view reverts to the original view.
% - - - - - - - - - - - - - - - - - - - - - - - - - - - - - - -

uicontrol(Hf_fig,'style','push',...
   'pos',[.77 .02 .10 .08],...
   'backgroundcolor',[.7 .7 .8],...
   'string','Revert',...
   'callback','mmview3d(995)');
```

Now that the controls have been created, the callbacks are defined.

```
else

  % - - - - - - - - - - - - - - - - - - - - - - - - - - - - - - - -
  % The callbacks for the azimuth and elevation sliders:
  % - - - - - - - - - - - - - - - - - -
  %    1) get the value of the slider and display it in the text window
  %    2) set the 'View' property to the current values of the azimuth
  %          and elevation sliders.
  % - - - - - - - - - - - - - - - - - - - - - - - - - - - - - - - -

  if cmd == 991
     set(Hc_acur,'string',num2str(get(Hc_asli,'val')));
     set(gca,'View',[get(Hc_asli,'val'),get(Hc_esli,'val')]);

  elseif cmd == 992
     set(Hc_ecur,'string',num2str(get(Hc_esli,'val')));
     set(gca,'View',[get(Hc_asli,'val'),get(Hc_esli,'val')]);

  % - - - - - - - - - - - - - - - - - - - - - - - - - - - - - - - -
  % The 'slider current value' text display callbacks:
  % - - - - - - - - - - - - - - - - - - - - - - - - - - - - - - - -
  % When text is entered into the text area and the return key is
  % pressed, the entered value is compared to the limits.
  %
  % If the limits have been exceeded, the display is reset to the
  % value of the slider and an error message is displayed.
  %
  % If the value is within the limits, the slider is set to the
  % new value, and the view is updated.
  % - - - - - - - - - - - - - - - - - - - - - - - - - - - - - - - -

  elseif cmd == 993
     if str2num(get(Hc_acur,'string')) < get(Hc_asli,'min')...
       | str2num(get(Hc_acur,'string')) > get(Hc_asli,'max')
         set(Hc_acur,'string',num2str(get(Hc_asli,'val')));
         disp('ERROR - Value out of range');
     else
       set(Hc_asli,'val',str2num(get(Hc_acur,'string')));
       set(gca,'View',[get(Hc_asli,'val'),get(Hc_esli,'val')]);
     end
```

```
  elseif cmd == 994
    if str2num(get(Hc_ecur,'string')) < get(Hc_esli,'min')...
      | str2num(get(Hc_ecur,'string')) > get(Hc_esli,'max')
        set(Hc_ecur,'string',num2str(get(Hc_esli,'val')));
        disp('ERROR - Value out of range');
    else
      set(Hc_esli,'val',str2num(get(Hc_ecur,'string')));
      set(gca,'View',[get(Hc_asli,'val'),get(Hc_esli,'val')]);
    end

  % - - - - - - - - - - - - - - - - - - - - - - - - - - - - - - -
  % Revert to the original view.
  % - - - - - - - - - - - - - - - - - - - - - - - - - - - - - - -

  elseif cmd == 995
    set(Hc_asli,'val',CVIEW(1));
    set(Hc_esli,'val',CVIEW(2));
    set(Hc_acur,'string',num2str(get(Hc_asli,'val')));
    set(Hc_ecur,'string',num2str(get(Hc_esli,'val')));
    set(gca,'View',[get(Hc_asli,'val'),get(Hc_esli,'val')]);

  %- - - - - - - - - - - - - - - - - - - - - - - - - - - - - - -
  % Must be bad arguments.
  %- - - - - - - - - - - - - - - - - - - - - - - - - - - - - - -

  else
    disp('mmview3d: Illegal argument.')
  end
end
```

The next example, **mmcxy**, creates a small text box at the lower-left corner of a *figure* to display the **[X,Y]** coordinates of the mouse pointer while the pointer is in the *figure*. Clicking the mouse button erases the coordinate display.

Although **mmcxy** is a short function, it still uses many of the elements of effective GUI functions including recursive function calls, global variables, and **'UserData'** properties. It also illustrates the use of the *figure* **'WindowButtonDownFcn'** and **'WindowButtonMotionFcn'** properties to initiate callbacks.

```
function out=mmcxy(arg)
%MMCXY Show x-y Coordinates Using Mouse.
% MMCXY places the x-y coordinates of the mouse in the
% lower left hand corner of the current 2-D figure window.
% When the mouse is clicked, the coordinates are erased.
% XY=MMCXY returns XY=[x,y] coordinates where mouse was clicked.
```

```
% XY=MMCXY returns XY=[] if a key press was used.

% Copyright (c) 1996 by Prentice-Hall, Inc.

global MMCXY_OUT

if ~nargin
    Hf=mmgcf;
    if isempty(Hf), error('No Figure Available.'),end
    Ha=findobj(Hf,'Type','axes');
    if isempty(Ha), error('No Axes in Current Figure.'),end

    Hu=uicontrol(Hf,'Style','text',...
                    'units','pixels',...
                    'Position',[1 1 140 15],...
                    'HorizontalAlignment','left');
    set(Hf,'Pointer','crossh',...
            'WindowButtonMotionFcn','mmcxy(''move'')',...
            'WindowButtonDownFcn','mmcxy(''end'')',...
            'Userdata',Hu)
    figure(Hf) % bring figure forward
    if nargout % must return x-y data
        key=waitforbuttonpress; % pause until mouse is pressed
        if key,
            out=[];          % return empty if aborted
            mmcxy('end')     % clean things up
        else
            out=MMCXY_OUT; % now that move is complete return point
        end
        return
    end

elseif strcmp(arg,'move') % mouse is moving in figure window
    cp=get(gca,'CurrentPoint'); % get current mouse position
    MMCXY_OUT=cp(1,1:2);
    xystr=sprintf('[%.3g, %.3g]',MMCXY_OUT);
    Hu=get(gcf,'Userdata');
    set(Hu,'String',xystr) % put x-y coordinates in text box

elseif strcmp(arg,'end') % mouse click occurred, clean things up
    Hu=get(gcf,'Userdata');
    set(Hu,'visible','off') % make sure text box disappears
    delete(Hu)
    set(gcf,'Pointer','arrow',...
            'WindowButtonMotionFcn','',...
            'WindowButtonDownFcn','',...
            'Userdata',[])
end
```

When called the first time, **mmcxy** creates the text *uicontrol*, changes the pointer shape, sets up the **'WindowButtonMotionFcn'** and **'WindowButtonDownFcn'** callbacks, and then waits for a key or button press. If a key is pressed, the cleanup routine is called to delete the text box, restore the mouse pointer, and clean up the *figure* callbacks and **'UserData'** property. If a mouse button is clicked, the **'WindowButtonDownFcn'** callback takes care of the cleanup. While waiting, pointer movement in the *figure* triggers the **'WindowButtonMotionFcn'** callback which updates the text string in the *uicontrol*.

The next *Mastering MATLAB Toolbox* function is **mmtext**, another short function that uses **'WindowButtonDownFcn'**, **WindowButtonMotionFcn'**, and **WindowButtonUpFcn'** callbacks to place and drag text.

```
function mmtext(arg)
%MMTEXT Place and drag text with mouse.
% MMTEXT waits for a mouse click on a text object
% in the current figure then allows it to be dragged
% while the mouse button remains down.
% MMTEXT('whatever') places the string 'whatever' on
% the current axes and allows it to be dragged.
%
% MMTEXT becomes inactive after the move is complete or
% no text object is selected.

% Copyright (c) 1996 by Prentice-Hall, Inc.

if ~nargin,arg=0;end

if isstr(arg) % user entered text to be placed
    Ht=text('Units','normalized',...
            'Position',[.05 .05],...
            'String',arg,...
            'HorizontalAlignment','left',...
            'VerticalAlignment','baseline');
    mmtext(0) % call mmtext again to drag it

elseif arg==0 % initial call, select text for dragging
    Hf=mmgcf;
    if isempty(Hf), error('No Figure Available.'), end
    set(Hf,'BackingStore','off',...
            'WindowButtonDownFcn','mmtext(1)')
    figure(Hf) % bring figure forward

elseif arg==1 & strcmp(get(gco,'Type'),'text') % text object selected
    set(gco,'Units','data',...
            'HorizontalAlignment','left',...
```

```
                    'VerticalAlignment','baseline',...
                    'EraseMode','xor');
        set(gcf,'Pointer','topr',...
                    'WindowButtonMotionFcn','mmtext(2)',...
                    'WindowButtonUpFcn','mmtext(99)')

    elseif arg==2 % dragging text object
        cp=get(gca,'CurrentPoint');
        set(gco,'Position',cp(1,1:3))

    elseif arg==99 % mouse button up, reset everything
        set(gco,'Erasemode','normal')
        set(gcf,'WindowButtonDownFcn','',...
                    'WindowButtonMotionFcn','',...
                    'WindowButtonUpFcn','',...
                    'Pointer','arrow',...
                    'Units','pixels',...
                    'BackingStore','on')

    else        % incorrect object selected
        mmdrag(99) % reset everything
    end
```

mmdraw is another useful GUI function from the *Mastering MATLAB Toolbox* that is very similar to **mmtext,** but is a bit more complex. This function lets you draw a *line* on the current *axes* with the mouse and set *line* properties as well.

```
    function mmdraw(arg1,arg2,arg3,arg4,arg5,arg6,arg7)
    %MMDRAW Draw a Line and Set Properties Using Mouse.
    % MMDRAW draws a line in the current axes using the mouse.
    % Click at the starting point and drag to the end point.
    % In addition, properties can be given to the line.
    % Properties are given in pairs, e.g., MMDRAW name value ...
    % Properties:
    % NAME         VALUE        {default}
    % color        [y m c r g b {w} k] or an rgb in quotes: '[r g b]'
    % style        [- -- {:} -.]
    % mark         [o + . * x)]
    % width        points for linewidth {0.5}
    % size         points for marker size {6}
    % Examples:
    % MMDRAW color r width 2  sets color to red and width to 2 points
```

```
% MMDRAW mark + size 8    sets marker type to + and size to 8 points
% MMDRAW color '[1 .5 0]' sets color to orange

% Copyright (c) 1996 by Prentice-Hall, Inc.

global MMDRAW_HL MMDRAW_EVAL

if nargin==0
    arg1='color';arg2='w';arg3='style';arg4=':';nargin=4;
end

if isstr(arg1) % initial call, set things up
    Hf=mmgcf;
    if isempty(Hf), error('No Figure Available.'), end
    Ha=findobj(Hf,'Type','axes');
    if isempty(Ha), error('No Axes in Current Figure.'),end
    set(Hf, 'Pointer','crossh',... % set up callback for line start
            'BackingStore','off',...
            'WindowButtonDownFcn','mmdraw(1)')
    figure(Hf)
    MMDRAW_EVAL='mmdraw(99'; % set up string to set attributes
    for i=1:nargin
        argi=eval(sprintf('arg%.0f',i));
        MMDRAW_EVAL=[MMDRAW_EVAL ',''' argi ''''];
    end
    MMDRAW_EVAL=[MMDRAW_EVAL ')'];

elseif arg1==1 % callback is line start point
    fp=get(gca,'CurrentPoint');    % start of line point
    set(gca,'Userdata',fp(1,1:2))  % store in axes userdata
    set(gcf,'WindowButtonMotionFcn','mmdraw(2)',...
            'WindowButtonUpFcn','mmdraw(3)')

elseif arg1==2 % callback is mouse motion
    cp=get(gca,'CurrentPoint');cp=cp(1,1:2);
    fp=get(gca,'Userdata');
    Hl=line('Xdata',[fp(1);cp(1)],'Ydata',[fp(2);cp(2)],...
            'EraseMode','xor',...
            'Color','w','LineStyle',':',...
            'Clipping','off');
    if ~isempty(MMDRAW_HL) % delete prior line if it exists
        delete(MMDRAW_HL)
    end
    MMDRAW_HL=Hl; % store current line handle

elseif arg1==3 % callback is line end point, finish up
    set(gcf,'Pointer','arrow',...
            'BackingStore','on',...
```

```
                    'WindowButtonDownFcn','',...
                    'WindowButtonMotionFcn','',...
                    'WindowButtonUpFcn','')
       set(gca,'Userdata',[])
       set(MMDRAW_HL,'EraseMode','normal') % render line better
       eval(MMDRAW_EVAL)
       MMDRAW_EVAL=[];

   elseif arg1==99 % process line properties
       for i=2:2:nargin-1
           name=eval(sprintf('arg%.0f',i),[]); % get name argument
           value=eval(sprintf('arg%.0f',i+1),[]); % get value argument
           if strcmp(name,'color')
               if value(1)=='[', value=eval(value);end
               set(MMDRAW_HL,'Color',value)
           elseif strcmp(name,'style')
               set(MMDRAW_HL,'Linestyle',value)
           elseif strcmp(name,'mark')
               set(MMDRAW_HL,'Linestyle',value)
           elseif strcmp(name,'width')
               value=abs(eval(value));
               set(MMDRAW_HL,'LineWidth',value)
           elseif strcmp(name,'size')
               value=abs(eval(value));
               set(MMDRAW_HL,'MarkerSize',value)
           else
               disp(['Unknown Property Name: ' name])
           end
       end
       MMDRAW_HL=[];
   end
```

Though too long to illustrate here, the *Mastering MATLAB Toolbox* functions **mmsetc** and **mmsetf** are straightforward examples of GUI functions that use recursion, global variables, and **'UserData'** properties. You might also like to look at the **mmsetclr** M-file. **mmsetclr** is a *script M-file* version of the **mmsetc** *function M-file*. If you compare these files, you can see some of the tradeoffs that were made to implement each type of GUI M-file.

21.9 DIALOG BOXES AND REQUESTERS

MATLAB contains several useful GUI tools for creating *dialog boxes* and *requesters*. Dialog boxes are individual windows that pop up on the screen and display a message string.

Dialog boxes contain one or more push buttons for user input. Requesters are individual windows that pop up on the display, obtain user input using the mouse and/or keyboard, and return information to the calling function.

Dialog Boxes

All MATLAB dialog boxes are based on the **dialog** function whose help text is

```
» help dialog
DIALOG displays a dialog box.
        FIG = DIALOG(P1,V1,...) displays a dialog box.
        Valid Param/Value pairs include
        Style              error | warning | help | question
        Name               string
        Replace            on | off
        Resize             on | off
        BackgroundColor    ColorSpec
        ButtonStrings      'Button1String | Button2String | ... '
        ButtonCalls        'Button1Callback | Button2Callback | ... '
        ForegroundColor    ColorSpec
        Position           [x y width height]
                           [x y] - centers around screen point
        TextString         string
        Units              pixels | normal | cent | inches | points
        UserData           matrix

        Note: Until dialog becomes built-in, set and get
              are not valid for dialog objects.

              At most three buttons are allowed.

              The callbacks are ignored for "question" dialogs.

              If ButtonStrings / ButtonCalls are unspecified then it
              defaults to a single "OK" button which removes the figure.

                  There's still problems with making the question
                  dialog modal.

                  The entire Parameter name must be passed in.
                  (i.e. no automatic completion).

                  Nothing beeps yet.
              See also ERRORDLG, HELPDLG, WARNDLG, QUESTDLG
```

Note that dialog boxes are not Handle Graphics objects in themselves, but rather are M-files constructed from a number of Handle Graphics objects. They are *figures*, containing an *axes* along with frame, edit, and pushbutton *uicontrol* objects. In future versions of MATLAB, `dialog` may become a built-in function with more capabilities. The default dialog box is a `Help` dialog consisting of an edit text box containing the string `'Default help string'` and a push button labeled `'OK'`. Type » `dialog` for an example.

Predefined dialog boxes are created by the `helpdlg`, `errordlg`, `warndlg`, and `questdlg` functions. `helpdlg` and `warndlg` accept a text string and window-title string as arguments. `errordlg` accepts a `'Replace'` argument as well. All except `questdlg` create a similar dialog box with individual default titles and text strings. A single button labeled `'OK'` closes the dialog box when clicked.

The help text for `helpdlg` is

```
» help helpdlg

HELPDLG Displays a help dialog box.
        HANDLE = HELPDLG(HELPSTRING,DLGNAME) displays the
        message HelpString in a dialog box with title DLGNAME.
     If a Help dialog with that name is already on the screen,
        it is brought to the front. Otherwise it is created.

        See also: DIALOG
```

The help text for `warndlg` is

```
» help warndlg

WARNDLG Creates a warning dialog box.
        HANDLE = WARNDLG(WARNSTR,DLGNAME) creates a warning
        dialog box which displays WARNSTR in a window named
        DLGNAME. A pushbutton labeled OK must be pressed to
        make the warning box disappear.

        See also: DIALOG
```

The help text for `errordlg` is

```
» help errordlg

ERRORDLG Creates an error dialog box.
        HANDLE = ERRORDLG(ERRORSTR,DLGNAME,Replace) creates an
        error dialog box which displays ERRORSTR in a window
        named DLGNAME. A pushbutton labeled OK must be pressed
        to make the error box disappear. If REPLACE='on' and
        an error dialog with Name DLGNAME already exits, it is
        simply brought to the front (no new dialog is created).

        See also: DIALOG
```

The question dialog box is a little different. The three previous functions display a single push button and return the handle of the dialog box *figure* object. The **questdlg** function displays either two or three push buttons and returns the label string of the push button selected by the user. The help text for **questdlg** is

```
» help questdlg

QUESTDLG Creates a question dialog box.
        CLICK = QUESTDLG(Q,YES,NO,CANCEL,DEFAULT) creates a question
        dialog box which displays Q. Up to three pushbuttons,
        with strings given by YES, NO, and CANCEL, will appear
        along with Q in the dialog. The dialog will be destroyed
        returning the string CLICK depending on which button is
        clicked. DEFAULT is the default button number.
```

The following function fragment illustrates the use of a question dialog box from within a function:

```
question1 = 'Change color map to copper?';
response1 = questdlg(question1,'Sure','Nope','Maybe',2);
if strcmp(response1,'Sure')
    colormap(copper);
elseif strcmp(response1,'Maybe')
    warndlg('That response does not compute!');
    response2 = questdlg(['Please make up your mind.|' question1],'Yes','No');
    if strcmp(response2,'Yes')
        colormap(copper);
    end
end
```

Requesters

Requesters are used to get input from the user using a dialog box. MATLAB requesters are built-in GUI functions that use the platform's native windowing system to generate familiar-looking requesters.

The **uigetfile** and **uiputfile** built-in functions are available on all platforms. They are used to obtain filenames interactively, which are then typically used by the calling function to read data from a file or save data to a file. The help text for **uigetfile** is

```
» help uigetfile

UIGETFILE Interactively retrieve a filename by displaying a dialog box.
   [FILENAME, PATHNAME] = UIGETFILE('filterSpec', 'dialogTitle', X, Y)
   displays a dialog box for the user to fill in, and returns the
   filename and path strings. A successful return occurs only if
   the file exists. If the user selects a file that does not exist,
   an error message is displayed, and control returns to the dialog box.
   The user may then enter another filename, or press the Cancel button.

   All parameters are optional, but if one is used, all previous
   parameters must also be used.

   The filterSpec parameter determines the initial display of files in
   the dialog box. For example '*.m' lists all the MATLAB M-files.

   Parameter 'dialogTitle' is a string containing the title of the dialog
   box.

   The X and Y parameters define the initial position of the dialog
   box in units of pixels. Some systems may not support this option.

   The output variable FILENAME is a string containing the name of the
   file selected in the dialog box. If the user presses the Cancel button
   or if any error occurs, it is set to 0.

   The output parameter PATHNAME is a string containing the path of
   the file selected in the dialog box. If the user presses the Cancel
   button or if any error occurs, it is set to 0.

   See also UIPUTFILE.
```

The following example shows how **uigetfile** can be used within a function to retrieve an ASCII data file interactively, and plot the sine of the data:

```
% Ask the user for a file name.

[datafile datapath] = uigetfile('*.dat','Choose a data file');

% If the user selected an existing file, read the data.
% (The extra quotes avoid problems with spaces in file or path names
% on the Macintosh platform.)
% Then determine the variable name from the file name,
% copy the data to a variable, and plot the data.
```

```
if datafile
    eval(['load(''' datapath datafile ''')']);
    x = eval(strtok(datafile,'.'));
    plot(x,sin(x));
end
```

The requester will not accept the name of a file that does not exist. The only way this situation can occur is if the user types a file name into the text box in the requester. The Macintosh version, however, has no text box. The user must choose an existing file or press the **Cancel** button to exit the requester.

The `uiputfile` function is very similar to the `uigetfile` function. The arguments are similar, and they both return filenames and path strings. The help text for `uiputfile` is

```
» help uiputfile

UIPUTFILE Interactively retrieve a filename by displaying a dialog box.
    [FILENAME, PATHNAME] = UIPUTFILE('initFile', 'dialogTitle')
    displays a dialog box and returns the filename and path strings.

    The initFile parameter determines the initial display of files
    in the dialog box. Full file name specifications as well as wildcards
    are allowed. For example, 'newfile.m' initializes the display to
    that particular file and lists all other existing .m files. This may
    be used to provide a default file name. A wildcard specification such
    as '*.m' lists all the existing MATLAB M-files.

    Parameter 'dialogTitle' is a string containing the title of the dialog
    box.

    The output variable FILENAME is a string containing the name of the
    file selected in the dialog box. If the user presses the Cancel button
    or if any error occurs, it is set to 0.

    The output variable PATHNAME is a string containing the name of the
    path selected in the dialog box. If the user presses the Cancel button
    or if any error occurs, it is set to 0.

    [FILENAME, PATHNAME] = UIPUTFILE('initFile', 'dialogTitle',X,Y)
    places the dialog box at screen position [X,Y] in pixel units.
    Not all systems support this option.

    Example:
        [newmatfile, newpath] = uiputfile('*.mat', 'Save As');

    See also UIGETFILE.
```

If the user selects a file that already exists, a dialog box appears asking if the user wants to delete the existing file. A **No** response returns to the original requester for another try. A **Yes** response closes the requester and dialog box and returns the file name to the calling function. The file is not deleted by the requester. The calling function must delete or overwrite the file if necessary.

It is important to remember that neither of these functions actually reads or writes any files. The calling function must do that. These functions simply return a file name and path to the calling function.

On MS-Windows and Macintosh platforms, `uisetcolor` lets the user interactively choose a color and optionally apply that color to an object. As of MATLAB version 4.2c, X Window System platforms do not support `uisetcolor`.

The help text for `uisetcolor` is

```
» help uisetcolor


UISETCOLOR Interactively set a ColorSpec by displaying a dialog box.
   C = UISETCOLOR(ARG, 'dialogTitle') displays a dialog box for the
   user to fill in, and applies the selected color to the input
   graphics object.

   The parameters are optional and may be specified in any order.

   ARG may be either a handle to a graphics object or an RGB triple.
   If a handle is used, it must specify a graphics object that supports
   color. If RGB is used, it must be a valid RGB triple (e.g., [1 0 0]
   for red). In both cases, the color specified is used to initialize
   the dialog box. If no initial RGB is specified, the dialog box
   initializes the color to black.

   If parameter 'dialogTitle' is used, it is a string containing the
   title of the dialog box.

   The output value C is the selected RGB triple. If the input parameter
   is a handle, the graphics object's color is set to the RGB color
   selected.

   If the user presses Cancel from the dialog box, or if any error
   occurs, the output value is set to the input RGB triple, if provided;
   otherwise, it is set to 0.

   Example:
      C = uisetcolor(hText, 'Set Text Color')

   NOTE: This function is only available in MS-Windows and Macintosh
      versions of MATLAB.
```

The *Mastering MATLAB Toolbox* includes a function mentioned earlier called **mmsetc** that works on all platforms and provides much of the same functionality as **uisetcolor**. **mmsetc** even lets the user select an object with a mouse click and apply the chosen color to the selected object. Here is the help text for **mmsetc**:

```
» help mmsetc

 MMSETC Obtain an RGB triple interactively from a color sample.
  MMSETC displays a dialog box for the user to select
  a color interactively and displays the result.

 X = MMSETC returns the selected color in X.

 MMSETC([r g b]) uses the RGB triple as the initial
 RGB value for modification.

 MMSETC C -or-
 MMSETC('C') where C is a color spec (y,m,c,r,g,b,w,k), uses
 the specified color as the initial value.

 MMSETC(H) where the input argument H is the handle of
 a valid graphics object that supports color, uses the color
 property of the object as the initial RGB value.

 MMSETC select -or-
 MMSETC('select') waits for the user to click on a valid
 graphics object that supports color, and uses the color
 property of the object as the initial RGB value.

 If the initial RGB value was obtained from an object or
 object handle, the 'Done' pushbutton will apply the
 resulting color property to the selected object.

 If no initial color is specified, black will be used.

 Examples:
        mmsetc
        mycolor=mmsetc
        mmsetc([.25 .62 .54])
        mmsetc(H)
        mmsetc g
        mmsetc red
        mmsetc select
        mycolor=mmsetc('select')
```

The following example applies a single-color color map interactively using the *Mastering MATLAB Toolbox* functions **mmap** and **mmsetc**:

```
» mesh(peaks)

» colormap(mmap(mmsetc))
```

The final requester `uisetfont` is available on MS-Windows and Macintosh platforms. `uisetfont` lets the user interactively choose font attributes and apply them to
an object. As of MATLAB version 4.2c, X Window System platforms do not support
`uisetfont`.

    The help text for `uisetfont` is

```
» help uisetfont

  UISETFONT Interactively set a font by displaying a dialog box.
     H = UISETFONT(HIN, 'dialogTitle') displays a dialog box for the
     user to fill in, and applies the selected font to the input
     graphics object.

     The parameters are optional and may be specified in any order.

     If parameter HIN is used, it must specify a handle to a text or
     axis graphics object. The font properties currently assigned to
     this object are used to initialize the font dialog box.

     If parameter 'dialogTitle' is used, it is a string containing the
     title of the dialog box.

     The output H is a handle to a graphics object. If HIN is specified,
     H is identical to HIN. If HIN is not specified, a new text object
     is created with the selected font properties, and its handle is
     returned.

     If the user presses Cancel from the dialog box, or if any error
     occurs, the output value is set to the input handle, if provided;
     otherwise, it is set to 0.

     Example:
        uisetfont(hText, 'Update Font')

     NOTE: This function is only available in MS-Windows and Macintosh
        versions of MATLAB.
```

The *Mastering MATLAB Toolbox* includes a function called **mmsetf** that works on all platforms and provides similar functionality to `uisetfont`. **mmsetf** even lets the user select a *text* object with a mouse click, then applies the chosen font attributes to the selected
object. The help text for **mmsetf** is

```
» help mmsetf

 MMSETF Choose font characteristics interactively.
 MMSETF displays a dialog box for the user to select
 font characteristics.

 X = MMSETF returns the handle of the text object or 0
     if an error occurs or 'Cancel' is pressed.

 MMSETF(H) where the input argument H is the handle of
 a valid text or axes object, uses the font characteristics
 of the object as the initial values.

 MMSETF select -or-
 MMSETF('select') waits for the user to click on a valid
 graphics object, and uses the font characteristics
 of the object as the initial values.

 If the initial values were obtained from an object or
 object handle, the 'Done' pushbutton will apply the
 resulting text properties to the selected object.

 If no initial object handle is specified, a zero is returned in X.

 Examples:
        mmsetf
        mmsetf(H)
        mmsetf select
        Hx_obj=mmsetf('select')
```

The font selection is limited in **mmsetf**, however. Different platform types and even individual computers of the same type have different fonts installed. The MATLAB **uisetfont** function is built into MATLAB and uses the operating system of the computer platform to list the available fonts. Since **mmsetf** is a MATLAB GUI function with no ability to determine which fonts are available, a limited selection of usually available fonts has been used. Font sizes are limited for the same reason.

The sample text string in the **mmsetf** requester shows the effect of each attribute change. If after making a selection the text string does not appear as you had expected, one or more of the font attributes you have chosen (font name, size, etc.) is not available on your computer. What you see is what you get.

21.10 USER-CONTRIBUTED GUI M-FILES

A number of MATLAB users have taken advantage of the GUI tools available in MATLAB to write some interesting GUI functions. Most of them are available to the user community on the MATLAB *anonymous FTP site* in the `/pub/contrib/graphics` directory. The Internet Resources chapter contains detailed instructions for obtaining files from this repository.

Some of the M-files and M-file collections contain a large number of Handle Graphics GUI objects, and illustrate the usefulness of *uimenus* and *uicontrols*. One of the most impressive is the `matdraw` collection by Keith Rogers. It can be found in the `/pub/contrib/graphics/matdraw2.0` directory. This collection adds a tool bar and palette of drawing tools to a *Figure* window. The effect is similar to a drawing program integrated into MATLAB.

Another collection that deserves investigation is the `guimaker` collection by Patrick Marchand. This is a set of interactive tools that lets you use a GUI to create GUI functions. You can create, size, and place GUI objects using the mouse. Version 1.0 is distributed as freeware, meaning that it is available to be used free of charge. Version 2.0 and later, however, are distributed as shareware. You can use it for free for 30 days; then you should register it and pay a small fee. Both versions are available on the MATLAB *anonymous FTP site* in the `/pub/contrib/graphics` directory.

These are just a few of the many examples of GUI functions available. Check them out. Maybe you have an idea for a killer GUI application that others would like to use. Chapter 23 contains detailed instructions to help you access existing MATLAB resources and even become an M-file contributor.

21.11 SUMMARY

Graphical User Interface design is not for everyone. However, if you have a need for one, MATLAB makes it possible to create impressive GUI functions from M-files alone. There is no need to use MEX files, high-level languages, or library calls to create a pleasing interface to a useful MATLAB function.

The font-control limitations of GUI objects should be addressed in the next MATLAB release. There is also some indication that dialog boxes will become built-in functions in the future. All of these improvements will make writing GUI functions easier and make them both more platform-independent and pleasing to the eye.

The functions discussed in this chapter are summarized in the following tables.

Handle Graphics GUI Functions

`uimenu(handle,'PropertyName',Value)`	Create menus for *figures*
`uicontrol(handle,'PropertyName',Value)`	Get object properties
`dialog('PropertyName',Value)`	Display a dialog box

`helpdlg('HelpString','DlgName')`	Display a 'help' dialog box
`warndlg('WarnString','DlgName')`	Display a 'warning' dialog box
`errordlg('ErrString','DlgName',Replace)`	Display an 'error' Dialog box
`questdlg('QString',S1,S2,S3,Default)`	Display a 'question' dialog box
`uigetfile(Filter,DlgName,X,Y)`	Retrieve a filename interactively
`uiputfile(InitFile,DlgName,X,Y)`	Retrieve a filename interactively
`uisetcolor(Handle,DlgName)`	Choose a color interactively
`uisetcolor([r g b],DlgName)`	Choose a color interactively
`uisetfont(Handle,DlgName)`	Choose font properties interactively

These are the *Mastering MATLAB Toolbox* functions mentioned in this chapter:

Mastering MATLAB Handle Graphics GUI Functions

`mmenus`	Example *uimenu* function
`mmclock(X,Y)`	Example *uicontrol* function
`mmsetclr`	Limited script version of the `mmsetc` function
`mmview3d`	Add azimuth and elevation sliders to a *figure*
`mmcxy`	Show x-y coordinates using the mouse

`mmtext('String')`	Drag and place *text* using the mouse
`mmdraw('PropertyName',Value)`	Draw a *line* and set properties using the mouse
`mmsetc(Handle)`	Set color attributes using the mouse
`mmsetc(ColorSpec)`	
`mmsetc('select')`	
`mmsetf(Handle)`	Set font attributes using the mouse
`mmsetf('select')`	

22

The Symbolic Math Toolbox

The *Symbolic Math Toolbox* available with MATLAB is different from all other Toolboxes. Rather than being highly specific to a particular specialty or subspecialty, it is useful to a wide audience. Morevover, it is different because it performs symbolic analysis using character strings, rather than numerical analysis on numerical arrays. For these reasons, a tutorial for this Toolbox is included in this text.

22.1 INTRODUCTION

The *Symbolic Math Toolbox* is a collection of *tools* (functions) for manipulating and solving symbolic expressions. There are tools to combine, simplify, differentiate, integrate, and solve algebraic and differential equations. Other tools are used in linear algebra to derive exact results for inverses, determinants, and canonical forms, and to find the eigenvalues of symbolic matrices without the error introduced by numerical computations. Variable-precision arithmetic, which calculates symbolically and returns a result to any specified accuracy, is also supported.

412

The tools in the *Symbolic Math Toolbox* are built upon the powerful software program called Maple, originally developed at the University of Waterloo in Canada. When you ask MATLAB to perform some symbolic operation, it asks Maple to do it and then returns the result to the MATLAB *Command* window. As a result, performing symbolic manipulations in MATLAB is a natural extension of the way you use MATLAB to crunch numbers.

22.2 SYMBOLIC EXPRESSIONS

Symbolic expressions are MATLAB character strings, or arrays of character strings, that represent numbers, functions, operators, and variables. The variables are not required to have predefined values. *Symbolic equations* are symbolic expressions containing an equals sign. *Symbolic arithmetic* is the practice of solving these symbolic equations by applying known rules and identities to the given symbols, exactly the way you learned to solve them in algebra and calculus. *Symbolic matrices* are arrays whose elements are symbolic expressions.

MATLAB represents symbolic expressions internally as character strings to differentiate them from numeric variables or operations; otherwise, they look almost exactly like basic MATLAB commands. Here are some examples of symbolic expressions along with their MATLAB equivalents:

Symbolic Expression	**MATLAB Representation**
$\dfrac{1}{2x^n}$	`'1/(2*x^n)'`
$y = \dfrac{1}{\sqrt{2x}}$	`y = '1/sqrt(2*x)'`
$\cos(x^2) - \sin(2x)$	`'cos(x^2) - sin(2*x)'`
$M = \begin{bmatrix} a & b \\ c & d \end{bmatrix}$	`M = sym('[a,b;c,d]')`
$f = \displaystyle\int_a^b \dfrac{x^3}{\sqrt{1-x}}\,dx$	`f = int('x^3/sqrt(1-x)','a','b')`

MATLAB symbolic functions let you manipulate these expressions in many ways. For example,

```
» diff('cos(x)') % differentiate cos(x) with respect to x
ans =
    -sin(x)

» M=sym('[a,b;c,d]') % create a symbolic matrix M
M =
    [a,b]
    [c,d]

» determ(M) % find the determinant of the symbolic matrix M
ans =
    a*d-b*c
```

Notice that in the first example above, the symbolic expression was defined *implicitly* by using single quotes to tell MATLAB that `'cos(x)'` is a string and imply that `diff('cos(x)')` is a symbolic expression rather than a numeric expression, while in the second example the **sym** function was used to *explicitly* tell MATLAB that `M=sym('[a,b;c,d]')` is a symbolic expression. Often the explicit **sym** function is not required where MATLAB can determine the argument type on its own.

As described in Chapter 8, in MATLAB the form `function argument` is equivalent to `function('argument')` where `function` is a function and `argument` is a character string. For example, MATLAB can figure out that `diff  cos(x)` and `diff('cos(x)')` both mean `diff(sym('cos(x)'))`, but the first form is certainly easier to type. However, at times the **sym** function *is* necessary. In the second example above,

```
» M=[a,b;c,d] % M is a numeric matrix using values of a through d
??? Undefined function or variable a.

» M='[a,b;c,d]' % M is a character string, but not a symbolic matrix
M =
    [a,b;c,d]

» M=sym('[a,b;c,d]') % M is a symbolic matrix
M =
    [a,b]
    [c,d]
```

In the example above, **M** was defined three ways: numerically (if **a**, **b**, **c**, and **d** had been predefined), as a character string, and as a symbolic matrix. Many symbolic functions are smart enough to convert character strings to symbolic expressions automatically. But in some cases, especially when creating a symbolic array, the **sym** function must be used to specifically convert a character string to a symbolic expression. The implicit form, e.g., `diff cos(x)`, is most useful for simple tasks that do not reference earlier results. However, the simplest form (without quotes) requires an argument that is a single character string containing no imbedded spaces.

```
» diff x^2+3*x+5 % the argument is equivalent to 'x^2+3*x+5'
ans =
     2*x+3
```

```
» diff x^2 + 3*x + 5 % spaces break the argument into separate strings
??? Error using ==> diff
Too many input arguments.
```

Symbolic expressions without variables are called *symbolic constants*. When symbolic constants are displayed, they are often difficult to distinguish from integers. For example,

```
» f=symop('(3*4-2)/5+1') % reduce a symbolic constant to its simplest form
f =
     3
```

```
» isstr(f) % is f a string? (1=yes, 0=no)
ans =
     1
```

In this case, **f** represents the symbolic constant **'3'**; not the number **3**. As described in Chapter 6, MATLAB stores strings as the ASCII representation of the characters. Consequently, if you perform a numeric operation on a string, it uses the ASCII value of each character in the operation. Since the number 51 is the ASCII representation of the character '3', adding 1 to **f** numerically will not produce the expected result.

```
                          » f+1
                          ans =
                              52
```

Symbolic Variables

When working with symbolic expressions containing more than one variable, exactly one variable is the *independent* variable. If MATLAB is not told which variable is the independent variable, it selects one based on the following rule:

> **The default independent variable in a symbolic expression is the unique, lower case letter, other than i and j, that is not part of a word. If there is no such character, x is chosen. If the character is not unique, the one closest to x alphabetically is chosen. If there is a tie, the one later in the alphabet is chosen.**

The default independent variable, sometimes known as the *free* variable, in the expression **'1/(5+cos(x))'** is **'x'**; the free variable in the expression **'3*y+z'** is **'y'**; and the free variable in the expression **'a+sin(t)'** is **'t'**. The free symbolic variable in the expression **'sin(pi/4)-cos(3/5)'** is **'x'** because this expression is a symbolic constant, containing no symbolic variables. You can ask MATLAB to tell you which variable in a symbolic expression it thinks is the independent variable by using the **symvar** function.

```
» symvar('a*x+y') % find the default symbolic variable
ans =
    x


» symvar('a*t+s/(u+3)') % u is closest to 'x'
ans =
    u


» symvar('sin(omega)') % 'omega' is not a single character
ans =
    x


» symvar('3*i+4*j') % i and j are equal to sqrt(-1)
ans =
    x


» symvar('y+3*s','t') % find the variable closest to t rather than x
ans =
    s
```

If **symvar** cannot find a default symbolic variable using the rule, it assumes there is none and returns **x**. This is true for expressions containing multi-character variables, such as **alpha** or **s2**, as well as symbolic constants, which contain no variables.

Most commands give you the option to specify the independent variable if desired.

```
» diff('x^n') % differentiate with respect to the default variable 'x'
ans =
    x^n*n/x


» diff('x^n','n') % differentiate x^n with respect to 'n'
ans =
    x^n*log(x)


» diff('sin(omega)') % differentiate using the default variable (x)
ans =
    0


» diff('sin(omega)','omega') % specify the independent variable
ans =
    cos(omega)
```

22.3 OPERATIONS ON SYMBOLIC EXPRESSIONS

Once you have created a symbolic expression, you will probably want to change it in some way. You may wish to extract part of an expression, combine two expressions, or find the numeric value of a symbolic expression. There are many symbolic tools that let you accomplish these tasks.

All symbolic functions (with a few specific exceptions discussed later) act on symbolic expressions and symbolic arrays, and return symbolic expressions or arrays. The result may sometimes look like a number, but it is a symbolic expression internally represented by a character string. As we discussed earlier, you can find out if what looks like a number is an integer or a string by using the MATLAB `isstr` function.

Extracting Numerators and Denominators

If your expression is a *rational polynomial* (a ratio of two polynomials), or can be expanded to a rational polynomial (including those with a denominator of 1), you can extract the numerator and denominator using **numden**. For example, given the expressions:

$$m = x^2 \quad f = \frac{ax^2}{b - x} \quad g = \frac{3}{2}x^2 + \frac{2}{3}x - \frac{3}{5} \quad h = \frac{x^2 + 3}{2x - 1} + \frac{3x}{x - 1} \quad k = \begin{bmatrix} \dfrac{3}{2} & \dfrac{2x + 1}{3} \\ \dfrac{4}{x^2} & 3x + 4 \end{bmatrix}$$

numden combines and rationalizes the expression if necessary, and returns the resulting numerator and denominator. The MATLAB statements to do this are

```
» m='x^2' % create a simple expression
m =
    x^2

» [n,d]=numden(m) % extract the numerator and denominator
n =
    x^2
d =
    1

» f='a*x^2/(b-x)' % create a rational expression
f =
    a*x^2/(b-x)
```

```
» [n,d]=numden(f) % extract the numerator and denominator
n =
     a*x^2
d =
     b-x
```

The first two expressions produce the expected result.

```
» g='3/2*x^2+2/3*x-3/5' % rationalize and extract the parts
g =
     3/2*x^2+2/3*x-3/5

» [n,d]=numden(g)
n =
     45*x^2+20*x-18
d =
     30

» h='(x^2+3)/(2*x-1)+3*x/(x-1)' % the sum of rational polynomials
h =
     (x^2+3)/(2*x-1)+3*x/(x-1)

» [n,d]=numden(h) % rationalize and extract
n =
     x^3+5*x^2-3
d =
     (2*x-1)*(x-1)
```

These two expressions, for **g** and **h**, are *rationalized,* or turned into a single expression with a numerator and denominator, before the parts are extracted.

```
» k=sym('[3/2,(2*x+1)/3;4/x^2,3*x+4]') % try a symbolic array
k =
     [   3/2,(2*x+1)/3]
     [4/x^2,      3*x+4]

» [n,d]=numden(k)
n =
     [3, 2*x+1]
     [4, 3*x+4]
d =
     [  2, 3]
     [x^2, 1]
```

This expression **k** is a symbolic array. **numden** returns two new arrays, **n** and **d**, where **n** is the array of numerators and **d** is the array of denominators. If you use the form **s=numden(f)**, **numden** returns only the numerator into the variable **s**.

Standard Algebraic Operations

A number of standard algebraic operations can be performed on symbolic expressions. The functions **symadd**, **symsub**, **symmul**, and **symdiv** add, subtract, multiply, and divide two expressions, and **sympow** raises one expression to the power of another. For example, given two functions,

$$f = 2x^2 + 3x - 5 \qquad g = x^2 - x + 7$$

```
» f='2*x^2+3*x-5' % define the symbolic expressions
f =
    2*x^2+3*x-5

» g='x^2-x+7'
g =
    x^2-x+7

» symadd(f,g) % find an expression for f + g
ans =
    3*x^2+2*x+2

» symsub(f,g) % find an expression for f - g
ans =
    x^2+4*x-12

» symmul(f,g) % find an expression for f * g
ans =
    (2*x^2+3*x-5)*(x^2-x+7)

» symdiv(f,g) % find an expression for f / g
ans =
    (2*x^2+3*x-5)/(x^2-x+7)

» sympow(f,'3*x') % find an expression for f³
ans =
    (2*x^2+3*x-5)^3
```

Another general-purpose function lets you create new expressions from other symbolic variables, expressions, and operators. **symop** takes up to 16 comma-separated arguments, each of which can be a symbolic expression, numeric value, or operator (`'+'`, `'-'`, `'*'`, `'/'`, `'^'`, `'('`, or `')'`). **symop** then catenates the arguments and returns the resulting expression.

```
» f='cos(x)' % create an expression
f =
    cos(x)
```

```
» g='sin(2*x)' % create another expression
f =
     sin(2*x)

» symop(f,'/',g,'+',3) % combine them
ans =
     cos(x)/sin(2*x)+3
```

All of these operations work with array arguments as well.

Advanced Operations

MATLAB has the capability to perform more advanced operations on symbolic expressions. The **compose** function combines *f(x)* and *g(x)* into *f(g(x))*, the **finverse** function finds the functional inverse of an expression, and the **symsum** function finds the symbolic summation of an expression.

Given the expressions

$$f = \frac{1}{1 + x^2} \qquad g = \sin(x) \qquad h = \frac{1}{1 + u^2} \qquad k = \sin(v)$$

```
» f='1/(1+x^2)'; % create the four expressions

» g='sin(x)';

» h='1/(1+u^2)';

» k='sin(v)';

» compose(f,g) % find an expression for f(g(x))
ans =
     1/(1+sin(x)^2)

» compose(g,f) % find an expression for g(f(x))
ans =
     sin(1/(1+x^2))
```

compose can also be used on functions that have different independent variables.

```
» compose(h,k,'u','v') % given h(u), k(v), find h(k(v))
ans =
     1/(1+sin(v)^2)
```

The functional inverse of an expression, say *f(x)*, is the expression *g(x)* that satisfies the condition *g(f(x))=x*. For example, the functional inverse of e^x is *ln(x)* since *ln(e^x)=x*. The func-

tional inverse of *sin(x)* is *arcsin(x)*, and the functional inverse of $\dfrac{1}{tan(x)}$ is $arctan\left(\dfrac{1}{x}\right)$.
The **finverse** function returns the functional inverse of an expression and warns you if
the result is not unique.L

```
» finverse('1/x') % the inverse of 1/x is 1/x since '1/(1/x) = x'
ans =
    1/x

» finverse('x^2') % g(x^2)=x has more than one solution
Warning: finverse(x^2) is not unique
ans =
    x^(1/2)

» finverse('a*x+b') % find the solution to 'g(f(x))=x'
ans =
    -(b-x)/a

» finverse('a*b+c*d-a*z','a') % find solution to 'g(f(a))=a'
ans =
    -(c*d-a)/(b-z)
```

The **symsum** function finds the symbolic summation of an expression. There are four
forms of the function: **symsum(f)** returns $\sum\limits_{0}^{x-1} f(x)$, **symsum(f,'s')** returns $\sum\limits_{0}^{s-1} f(s)$),
symsum(f,a,b) returns $\sum\limits_{a}^{b} f(x)$, and the most general form, **symsum(f,'s',a,b)** returns $\sum\limits_{a}^{b} f(s)$.

Let us try $\sum\limits_{0}^{x-1} x^2$, which should return $\dfrac{x^3}{3} - \dfrac{x^2}{2} + \dfrac{x}{6}$:

```
» symsum('x^2')
ans =
    1/3*x^3-1/2*x^2+1/6*x
```

How about $\sum\limits_{1}^{n} (2n-1)^2$, which should return $\dfrac{n(2n-1)(2n+1)}{3}$:

```
» symsum('(2*n-1)^2',1,'n')
ans =
    11/3*n+8/3-4*(n+1)^2+4/3*(n+1)^3

» factor(ans) % change the form (we will revisit 'factor' later on)
ans =
    1/3*n*(2*n-1)*(2*n+1)
```

Finally, let's try $\sum\limits_{1}^{\infty} \dfrac{1}{(2n-1)^2}$, which should return $\dfrac{\pi^2}{8}$:

```
» symsum('1/(2*n-1)^2',1,inf)
ans =
    1/8*pi^2
```

Conversion Functions

This section presents tools to convert from symbolic expressions to numeric values, and back again. These are some of the very few symbolic functions that can return numeric values. Notice, however, that some symbolic functions automatically convert a number into its symbolic representation if it is one of a number of arguments to the function.

The **sym** function can take a numeric argument and convert it into symbolic representation. The **numeric** function does the opposite. It converts a symbolic constant (a symbolic expression with no variables) to a numeric value.

```
» phi='(1+sqrt(5))/2' % the 'golden' ratio
phi =
    (1+sqrt(5))/2

» numeric(phi) % convert to a numeric value
ans =
    1.6180
```

As described in Chapter 6, the **eval** function passes a character string to MATLAB to evaluate. Therefore, **eval** is another function that can be used to convert a symbolic constant into a number or to evaluate an expression.

```
» eval(phi) % execute the string '(1+sqrt(5))/2'
ans =
    1.6180
```

As expected, **numeric** and **eval** return the same numeric value.

The symbolic function **sym2poly** converts a symbolic polynomial to its MATLAB equivalent coefficient vector. The function **poly2sym** does the reverse and lets you specify the variable to use in the resulting expression.

```
» f='2*x^2+x^3-3*x+5' % f is the symbolic polynomial
f =
    2*x^2+x^3-3*x+5

» n=sym2poly(f) % extract the numeric coefficient vector
n =
    1    2   -3    5

» poly2sym(n) % recreate the polynomial in x (the default)
ans =
    2*x^2+x^3-3*x+5
```

```
» poly2sym(n,'s') % recreate the polynomial in s
ans =
    s^3+2*s^2-3*s+5
```

Variable Substitution

Suppose you have a symbolic expression in x, and you want to change the variable to y. MATLAB gives you a tool to make substitutions in symbolic expressions, called **subs**. The format is **subs(f,new,old)**, where **f** is a symbolic expression, and **new** and **old** are characters, character strings, or other symbolic expressions. The string **new** will replace each occurrence of the string **old** in the expression **f**. Here are some examples.

```
» f='a*x^2+b*x+c' % create a function f(x)
f =
    a*x^2+b*x+c

» subs(f,'s','x') % substitute 's' for 'x' in the expression f
ans =
    a*s^2+b*s+c

» subs(f,'alpha','a') % substitute 'alpha' for 'a' in f
ans =
    alpha*x^2+b*x+c

» g='3*x^2+5*x-4' % create another function
g =
    3*x^2+5*x-4

» h =subs(g,'2','x') % substitute '2' for 'x' in g
h =
    18

» isstr(h) % show that the result is a symbolic expression
ans =
    1
```

The last example shows how **subs** makes the substitution and then tries to simplify the expression. Since the result of the substitution was a symbolic constant, MATLAB could reduce it to a single symbolic value. Notice that since **subs** is a symbolic function, it returns a symbolic expression, actually a symbolic constant, even though it looks like a number. To get a number, we need to use the **numeric** or **eval** function to convert the string.

```
» numeric(h) % convert a symbolic expression to a number
ans =
    18
```

```
» isstr(ans) % show that the result is a numeric value
ans =
     0
```

22.4 DIFFERENTIATION AND INTEGRATION

Differentiation and integration are central to the study and application of calculus, and are used extensively in many engineering disciplines. MATLAB symbolic tools can help solve many of these kinds of problems.

Differentiation

Differentiation of a symbolic expression uses the `diff` function in one of four forms.

```
» f='a*x^3+x^2-b*x-c' % define a symbolic expression
f =
    a*x^3+x^2-b*x-c

» diff(f) % differentiate with respect to the default variable x
ans =
    3*a*x^2+2*x-b

» diff(f,'a') % differentiate f with respect to a
ans =
    x^3

» diff(f,2) % differentiate f twice with respect to x
ans =
    6*a*x+2

» diff(f,'a',2) % differentiate f twice with respect to a
ans =
    0
```

The `diff` function also operates on arrays. If **F** is a symbolic vector or matrix, `diff(F)` differentiates each element in the array.

```
» F=sym('[a*x,b*x^2;c*x^3,d*s]') % create a symbolic array
F =
    [    a*x,b*x^2]
    [ c*x^3,   d*s]

» diff(F) % differentiate the elements with respect to x
ans =
    [      a, 2*b*x]
    [3*c*x^2,     0]
```

Note that the **diff** function is also used in MATLAB to compute the numerical differences of a numeric vector or matrix. For a numeric vector or matrix **M**, **diff(M)** computes the numerical differences **M(2:m,:)-M(1:m-1,:)** as shown:

```
» M=[(1:8).^2] % create a vector
M =
      1     4     9    16    25    36    49    64

» diff(M) % find the differences between elements
ans =
      3     5     7     9    11    13    15
```

If the expression or variable argument to **diff** is numeric, MATLAB is smart enough to compute the numerical difference; if the argument is a symbolic string or variable, MATLAB differentiates the expression.

Integration

The integration function **int(f)**, where **f** is a symbolic expression, attempts to find another symbolic expression **F** such that **diff(F)=f**. As you probably found from your study of calculus, integration is more complicated than differentiation. The integral or antiderivative may not exist in closed form, or it may exist but the software can't find it, or the software could find it eventually, but runs out of memory or time. When MATLAB cannot find the antiderivative, it will return the command unevaluated.

```
» int('log(x)/exp(x^2)') % attempt to integrate
ans =
      int(log(x)/exp(x^2),x)
```

The integration function, like differentiation, has more than one form. The form **int(f)** attempts to find an antiderivative with respect to the default independent variable. The form **int(f,'s')** attempts to find an antiderivative with respect to the symbolic variable **s**. The forms **int(f,a,b)** and **int(f,'s',a,b)** where **a** and **b** are numeric values attempt to find symbolic expressions for the definite integral from **a** to **b**. The forms **int(f,'m','n')** and **int(f,'s','m','n')**, where **m** and **n** are symbolic variables, attempt to find symbolic expressions for the definite integral from **m** to **n**.

```
» f='sin(s+2*x)') % create a symbolic function
f =
    sin(s+2*x)

» int(f) % integrate with respect to x
ans =
    -1/2*cos(s+2*x)
```

```
» int(f,'s') % integrate with respect to s
ans =
    -cos(s+2*x)

» int(f,pi/2,pi) % integrate with respect to x from π/2 to π
ans =
    -cos(s)

» int(f,'s',pi/2,pi) % integrate with respect to s from π/2 to π
ans =
    cos(2*x)-sin(2*x)

» int(f,'m','n') % integrate with respect to x from m to n
ans =
    -1/2*cos(s+2*n)+1/2*cos(s+2*m)
```

Like the **diff** function, the integration function **int** operates on each element of a symbolic array.

```
» f=sym('[a*x,b*x^2;c*x^3,d*s]') % create a symbolic array
f =
    [   a*x,b*x^2]
    [c*x^3, d*s]

» int(f) % integrate the array elements with respect to x
ans =
    [1/2*a*x^2,  1/3*b*x^3]
    [1/4*c*x^4,      d*s*x]
```

22.5 PLOTTING SYMBOLIC EXPRESSIONS

In numerous situations it's beneficial to visualize a symbolic expression. MATLAB provides the function **ezplot** for this task.

```
» y='-16*x^2+64*x+96' % expression to plot
y =
    -16*x^2+64*x+9

» ezplot(y)
```

As you can see, **ezplot** graphs the given symbolic function over the domain $-2\pi \le \times \le 2\pi$ and scales the y-axis accordingly. It also adds grids and labels. In this case, we are only interested in times between 0 and 6. Let's try again and specify the time range.

```
» ezplot(y,[0 6])  % plot y for 0<x<6
```

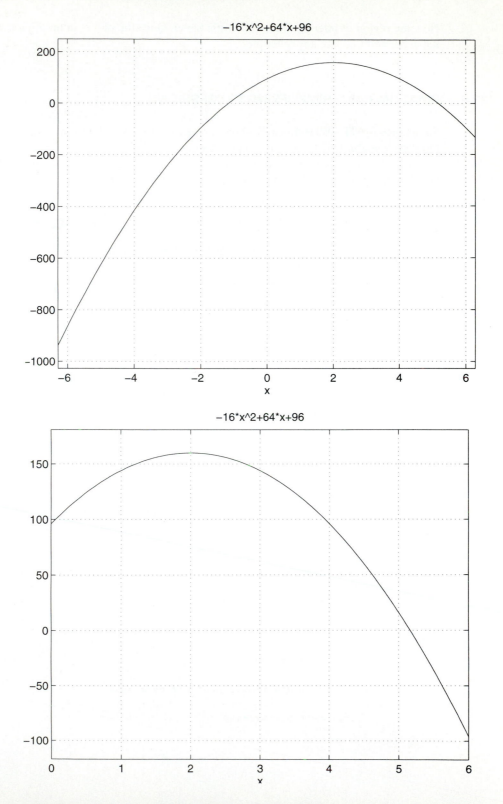

Now the region of interest shows up a little better. Once the plot is in the *Figure* window, it can be modified like any other plot.

22.6 FORMATTING AND SIMPLIFYING EXPRESSIONS

Sometimes MATLAB returns a symbolic expression that is difficult to read. A number of tools are available to help make the expression more readable. The first is the **pretty** function. This command attempts to display a symbolic expression in a form that resembles textbook mathematics. Let's look at a Taylor series expansion in 6 terms.

```
» f=taylor('log(x+1)/(x-5)') % 6 terms is the default
f =
-1/5*x+3/50*x^2-41/750*x^3+293/7500*x^4-1207/37500*x^5+0*(x^6)

» pretty(f)

                      2    41   3    293   4    1207   5        6
      - 1/5 x + 3/50 x  - --- x  + ---- x  - ----- x  + 0(x )
                          750      7500      37500
```

Symbolic expressions can be presented in many equivalent forms. Some forms may be preferable to others in different situations. MATLAB uses a number of commands to simplify or change the form of symbolic expressions.

```
» f=sym('(x^2-1)*(x-2)*(x-3)') % create a function
f =
    (x^2-1)*(x-2)*(x-3)

» collect(f) % collect all like terms
ans =
    x^4-5*x^3+5*x^2+5*x-6

» horner(ans) % change to Horner or nested representation
ans =
    -6+(5+(5+(-5+x)*x)*x)*x

» factor(ans) % express as a product of polynomials
ans =
    (x-1)*(x-2)*(x-3)*(x+1)

» expand(f) % distribute products over sums
ans =
    x^4-5*x^3+5*x^2+5*x-6
```

`simplify` is a powerful, general-purpose tool that attempts to simplify an expression by the application of many different kinds of algebraic identities involving sums, integral and fractional powers, trig, exponential and log functions, and Bessel, hypergeometric, and gamma functions. A few examples should illustrate the power of this function.

```
» simplify('log(2*x/y)')
ans =
      log(2)+log(x)-log(y)

» simplify('sin(x)^2+3*x+cos(x)^2-5')
ans =
      3*x-4

» simplify('(-a^2+1)/(1-a)')
ans =
      a+1
```

The last function to be discussed here is one of the most powerful, but least orthodox, of all the simplification tools. The function **simple** tries several different simplification tools and then selects the form that has the fewest number of characters in the resulting expression. Let's try a cube root.

$$ f = \sqrt[3]{\frac{1}{x^3} + \frac{6}{x^2} + \frac{12}{x} + 8} $$

```
» f='(1/x^3+6/x^2+12/x+8)^(1/3)' % create the expression
f =
      1/x^3+6/x^2+12/x+8)^(1/3)

» simple(f) % simplify it
simplify:
      (2*x+1)/x
ans =
      (2*x+1)/x

» simple(ans) % try it once again - another method may help
combine(trig):
      2+1/x
ans =
      2+1/x
```

As you can see, **simple** tries a number of simplifications that may help reduce the expression, and lets you see the result of each try. Sometimes it helps to apply **simple** more than once to try a different simplification operation on the result of the first, as was done above. **simple** is especially useful for expressions containing trig functions. Let's try one.

$$ \cos(x) + \sqrt{-\sin(x)^2} $$

```
» simple('cos(x)+sqrt(-sin(x)^2)') % simplify a trig expression
simplify:
     cos(x)+(cos(x)^2-1)^(1/2)
radsimp:
     cos(x)+i*sin(x)
combine(trig):
     cos(x)+(-1/2+1/2*cos(2*x))^(1/2)
factor:
     cos(x)+(-sin(x)^2)^(1/2)
expand:
     cos(x)+(-sin(x)^2)^(1/2)
convert(exp):
     1/2*exp(i*x)+1/2/exp(i*x)+1/4*4^(1/2)*exp(exp(i*x)-1/exp(i*x))
convert(tan):
(1-tan(1/2*x)^2)/(1+tan(1/2*x)^2+(-4*tan(1/2*x)^2/(1+tan(1/2*x)^2)^2)^(1/2)
ans =
     cos(x)+i*sin(x)
» simple(ans) % one more time!
convert(exp):
     exp(i*x)
convert(tan):
     (1-tan(1/2*x)^2)/(1+tan(1/2*x)^2+2*i*(tan(1/2*x)/(1+tan(1/2*x)^2)
ans =
     exp(i*x)
```

22.7 VARIABLE PRECISION ARITHMETIC

Round-off error can be introduced in any operation on numeric values since numeric precision is limited by the number of digits preserved by each numeric operation. Repeated or multiple numeric operations can therefore accumulate error. Operations on symbolic expressions, however, are highly accurate since they do not perform numeric computations and there is no round-off error. Using **eval** or **numeric** on the result of a symbolic operation can introduce round-off error only in the converted result.

MATLAB relies exclusively on the computer's floating-point arithmetic for number crunching. Although fast and easy on the computer's memory, floating-point operations are limited by the number of digits supported and can introduce round-off error in each operation; they cannot produce exact results. The relative accuracy of individual arithmetic operations in MATLAB is about 16 digits. The symbolic capabilities of Maple, on the other hand, can carry out operations to any arbitrary number of digits. As the default number of digits is increased, however, additional time and computer storage are required for each computation.

Maple defaults to 16 digits of accuracy unless told differently. The function **digits** returns the current value of the global **Digits** parameter. The default number of digits of accuracy for Maple functions can be changed using **digits(n)**, where **n** is the number of digits of accuracy desired. The downside of increasing accuracy this way is that every Maple function will subsequently carry out computations to the new accuracy, increasing computation time. The display of results will not change; only the default accuracy of the underlying Maple functions will be affected.

Another function *is* available, however, that will let you perform a single computation to any arbitrary accuracy while leaving the global **Digits** parameter unchanged. The variable precision arithmetic function, or **vpa**, evaluates a single symbolic expression to the default or to any specified accuracy and displays a numeric result to the same accuracy.

```
» format long % let's see all the usual digits

» pi % how about π to numeric accuracy
ans =
    3.14159265358979

» digits % display the default 'Digits' value
Digits = 16

» vpa('pi') % how about π to 'Digits' accuracy
ans =
    3.141592653589793

» digits(18) % change the default to 18 digits
» vpa('pi') % evaluate π to 'Digits' digits
ans =
    3.14159265358979324

» vpa('pi',20) % how about π to 20 digits
ans =
    3.1415926535897932385

» vpa('pi',50) % how about π to 50 digits
ans =
    3.1415926535897932384626433832795028841971693993751

» vpa('2^(1/3)',200) % the cube root of 2 to 200 digits
ans =

1.2599210498948731647672106072782283505702514647015079800819751121552996765
1395948372939656243625509415431025603561566525939902404061373722845911030426
9355246960642616625000977474526565480306867185405
```

The **vpa** function applied to a symbolic matrix evaluates each element to the number of digits specified as well.

```
» A=sym('[1/4,log(sqrt(2));exp(1),3/7]')
A =
    [   1/4,log(sqrt(2))]
    [exp(1),        3/7]

» vpa(A,20) % evaluate to 20 digits
ans =
    [.25000000000000000000, .34657359027997265471]
    [2.7182818284590452354, .42857142857142857143]
```

22.8 SOLVING EQUATIONS

Symbolic equations can be solved using symbolic tools available in MATLAB. Some of them have been introduced earlier, and more will be examined in this section.

Solving a Single Algebraic Equation

We have seen earlier that MATLAB contains tools for solving symbolic expressions. If the expression is not an equation (it does not contain an equals sign), the **solve** function sets the symbolic expression equal to zero before solving it:

```
» solve('a*x^2+b*x+c') % solve for the roots of the quadratic equation
ans =
    [1/2/a*(-b+(b^2-4*a*c)^(1/2)]
    [1/2/a*(-b-(b^2-4*a*c)^(1/2)]
```

The result is a symbolic vector whose elements are the two solutions. If you wish to solve for something other than the default variable x, **solve** will let you specify it.

```
» solve('a*x^2+b*x+c','b') % solve for b
ans =
    -(a*x^2+c)/x
```

Symbolic equations containing equals signs can also be solved.

```
» f=solve('cos(x)=sin(x)') % solve for x
f =
    1/4*pi

» t=solve('tan(2*x)=sin(x)')
t =
    [                      0]
    [acos(1/2+1/2*3^(1/2))]
    [acos(1/2-1/2*3^(1/2))]
```

and numeric solutions found.

```
» numeric(f)
ans =
     0.7854

» numeric(t)
ans =
              0
              0 + 0.8314i
        1.9455
```

Notice that when solving equations of periodic functions, there are an infinite number of solutions. **solve** restricts its search for solutions in these cases to a limited range near zero, and returns a non-unique subset of solutions.

If a symbolic solution cannot be found, a variable precision solution will be computed.

```
» x=solve('exp(x)=tan(x)')
x =
     1.306326940423079
```

Several Algebraic Equations

Several algebraic equations can be solved at the same time. A statement of the form `solve(S1,S2,...,Sn)` solves n equations for the default variables. A statement of the form `solve(S1,S2,...,Sn,'v1,v2,...,vn')` solves n symbolic equations for the n unknowns listed as `'v1,v2,...,vn'`.

How about tackling a typical school algebra problem?

Diane wants to go to the movies, so she dumps out her piggy bank and counts her coins. She finds that:

- *The number of dimes plus half the total number of nickels and pennies is equal to the number of quarters.*
- *The number of pennies is 10 more than the number of nickels, dimes, and quarters.*
- *The number of quarters and dimes is equal to the number of pennies plus 1/4 of the number of nickels.*
- *The number of quarters and pennies is one more than the number of nickels plus 8 times the number of dimes.*

If the movie ticket costs $3.00, popcorn is $1.00, and a candy bar is 50 cents, does she have enough to get all three?

First, create a set of linear equations from the information given above. Let p, n, d, and q be the number of pennies, nickels, dimes, and quarters, respectively.

$$d + \frac{n + p}{2} = q \qquad p = n + d + q - 10 \qquad q + d = p + \frac{n}{4}$$

$$q + p = n + 8d - 1$$

Next, create MATLAB symbolic equations, and solve for the variables.

```
» eq1='d+(n+p)/2=q';

» eq2='p=n+d+q-10';

» eq3='q+d=p+n/4';

» eq4='q+p=n+8*d-1';

» [pennies,nickels,dimes,quarters]=solve(eq1,eq2,eq3,eq4,'p,n,d,q')
pennies =
    16
nickels =
    8
dimes =
    3
quarters =
    15
```

So Diane has 16 pennies, 8 nickels, 3 dimes, and 15 quarters. That means she has

```
» money=.01*16+.05*8+.10*3+.25*15
money =
    4.6100
```

which is enough for a ticket, popcorn, and a candy bar with 11 cents left over.

Single Differential Equation

Ordinary differential equations are sometimes difficult to solve. MATLAB gives you a powerful tool to help you find solutions to differential equations.

The function **dsolve** computes symbolic solutions to ordinary differential equations. Since we are working with differential equations, we need a way to include differentials in an expression. Therefore, the syntax of **dsolve** is a little different from most other functions. The equations are specified by using the letter **D** to denote differentiation, and **D2**, **D3**, etc., to denote repeated differentiation. Any letters following **D**'s are dependent variables. The equation $d^2y/dx^2 = 0$ is represented by the symbolic expression **D2y=0**. The independent variable can be specified or will default to the one chosen by the **symvar** rule. For example, the general solution to the first-order equation $dy/dx = 1 + y^2$ can be found by

```
» dsolve('Dy=1+y^2') % find the general solution
ans =
    -tan(-x+C1)
```

where **C1** is a constant of integration. Solving the same equation with the initial condition *y(0)*=1 will produce

```
» y=dsolve('Dy=1+y^2','y(0)=1') % add an initial condition
y =
    tan(x+1/4*pi)
```

The independent variable can be specified using this form:

```
» dsolve('Dy=1+y^2','y(0)=1','v') % find solution to dy/dv
ans =
    tan(v+1/4*pi)
```

Let's try a second-order differential equation with two initial conditions.

$$\frac{d^2y}{dx^2} = \cos(2x) - y \qquad \frac{dy}{dx}(0) = 0 \qquad y(0) = 1$$

```
» y=dsolve('D2y=cos(2*x)-y','Dy(0)=0','y(0)=1')
y =
    -2/3*cos(x)^2+1/3+4/3*cos(x)
```

```
» y=simple(y) % y looks like it can be simplified
y =
    -1/3*cos(2*x)+4/3*cos(x)
```

Often, a differential equation to be solved contains terms of more than one order and is presented in the following form:

$$\frac{d^2y}{dx^2} - 2\frac{dy}{dx} - 3y = 0$$

The general solution is

```
» y=dsolve('D2y-2*Dy-3*y=0')
y =
    C1*exp(-x)+C2*exp(3*x)
```

Applying the initial conditions *y(0) = 0* and *y(1) = 1* gives

```
» y=dsolve('D2y-2*Dy-3*y=0','y(0)=0,y(1)=1')
y =
    1/(exp(-1)-exp(3))*exp(-x)-1/(exp(-1)-exp(3))*exp(3*x)
```

```
» y=simple(y) % this looks like a candidate for simplification
y =
     -(exp(-x)-exp(3*x))/(exp(3)-exp(-1))

» pretty(y) % pretty it up
```

$$- \frac{\exp(-x) - \exp(3 x)}{\exp(3) - \exp(-1)}$$

Now plot the result in an interesting region.

```
» ezplot(y,[-6 2])
```

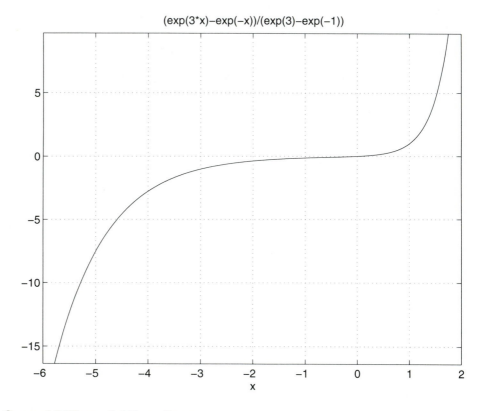

$(\exp(3{*}x){-}\exp({-}x))/(\exp(3){-}\exp({-}1))$

Several Differential Equations

The function **dsolve** can also handle several differential equations at once. Here is a pair of linear, first-order equations.

$$\frac{df}{dx} = 3f + 4g \qquad \frac{dg}{dx} = -4f + 3g$$

The general solutions are

```
» [f,g]=dsolve('Df=3*f+4*g','Dg=-4*f+3*g')
f =
     C1*exp(3*x)*sin(4*x)+C2*exp(3*x)*cos(4*x)
g =
     -C2*exp(3*x)*sin(4*x)+C1*exp(3*x)*cos(4*x)
```

Adding initial conditions $f(0) = 0$ and $g(0) = 1$, we get

```
» [f,g]=dsolve('Df=3*f+4*g','Dg=-4*f+3*g','f(0)=0,g(0)=1')
f =
   exp(3*x)*sin(4*x)
g =
   exp(3*x)*cos(4*x)
```

22.9 LINEAR ALGEBRA AND MATRICES

In this section, we present an introduction to symbolic matrices and the tools MATLAB supplies for solving problems using linear algebra.

Symbolic Matrices

Symbolic matrices and vectors are arrays whose elements are symbolic expressions. They can be generated with the **sym** function

```
» A=sym('[a,b,c;b,c,a;c,a,b]')
A =
     [ a,  b,  c]
     [ b,  c,  a]
     [ c,  a,  b]

» G=sym('[cos(t),sin(t);-sin(t),cos(t)]')
G =
     [  cos(t),  sin(t)]
     [ -sin(t),  cos(t)]
```

The **sym** function can also expand a formula that specifies individual elements. **Note that in this case only, the _i_ and _j_ are row and column positions, respectively, and do not affect the default values of _i_ and _j_ (which represent $\sqrt{-1}$).** The following examples create 3-by-3 matrices whose elements depend on their row and column positions:

```
» S=sym(3,3,'(i+j)/(i-j+s)') % create a matrix using a formula
S =
    [     2/s, 3/(-1+s), 4/(-2+s)]
    [3/(1+s),      4/s, 5/(-1+s)]
    [4/(2+s),  5/(1+s),      6/s]

» S=sym(3,3,'m','n','(m-n)/(m-n-t)') % use m and n in another formula
S =
    [       0, -1/(-1-t), -2/(-2-t)]
    [1/(1-t),         0, -1/(-1-t)]
    [2/(2-t),   1/(1-t),         0]
```

The **sym** function can also convert a numeric matrix to its symbolic form.

```
» M=[1.1,1.2,1.3;2.1,2.2,2.3;3.1,3.2,3.3] % a numeric matrix
M =
    1.1000   1.2000   1.3000
    2.1000   2.2000   2.3000
    3.1000   3.2000   3.3000

» S=sym(M) % convert to symbolic form
S =
    [11/10,  6/5, 13/10]
    [21/10, 11/5, 23/10]
    [31/10, 16/5, 33/10]
```

If the elements of the numeric matrix can be specified as the ratio of small integers, the **sym** function will use the rational (fractional) representation. If the elements are irrational, **sym** will represent the elements as floating-point numbers in symbolic form.

```
» E=[exp(1) sqrt(2)]
E =
    2.7183      1.4142

» sym(E)
ans =
    [3060513257434036*2^(-50),  3184525836262886*2^(-51)]
```

The size (number of rows and columns) of a symbolic matrix can be found using **symsize**. This function returns a numeric value or vector, not a symbolic expression. The four forms of **symsize** are illustrated below.

```
» S=sym('[a,b,c;d,e,f]') % create a symbolic matrix
S =
    [a,b,c]
    [d,e,f]
```

```
» d=symsize(S) % returns the size of S as the 2-element vector d
d =
    2    3

» [m,n]=symsize(S) % return the number of rows in m, and columns in n
m =
    2
n =
    3

» m=symsize(S,1) % return the number of rows
m =
    2

» n=symsize(S,2) % return the number of columns
n =
    3
```

Numeric arrays use the form **N(m,n)** to access a single element, but symbolic array elements must be referenced using symbolic functions, such as **sym(S,m,n)**. It would be nice to be able to use the same syntax, but MATLAB representation of symbolic expressions gets in the way. A symbolic array is represented internally as an array of strings, which in turn are arrays of characters; **S(m,n)** returns a single character. Individual elements of symbolic arrays, therefore, must be referenced by a symbolic function such as **sym**, rather than directly.

```
» G=sym('[ab,cd;ef,gh]') % create a 2-by-2 symbolic matrix
G =
    [ab,cd]
    [ef,gh]

» G(1,2) % this is the second character of the first row of G
ans =
    a

» r=sym(G,1,2) % this is the second expression in the first row of G
r =
    cd
```

Remember that the symbolic matrix **G** in the example above is actually stored in the computer as a 2-by-7 array of characters. The first row is **'[ab,cd]'**, so the second character is **'a'**.

Finally, **sym** can be used to change an element of a symbolic array.

```
» sym(G,2,2,'pq') % change the (2,2) element in G from 'gh' to 'pq'
ans =
    [ab,cd]
    [ef,pq]
```

Algebraic Operations

A number of common algebraic operations can be performed on symbolic matrices using the **symadd**, **symsub**, **symmul**, and **symdiv** functions. Powers are computed using **sympow**, and the transpose of a symbolic matrix is computed using **transpose**.

```
» G=sym('[cos(t),sin(t);-sin(t),cos(t)]') % create a symbolic matrix
G =
    [  cos(t),  sin(t)]
    [ -sin(t),  cos(t)]

» symadd(G,'t') % add 't' to each element
ans =
    [  cos(t)+t,  sin(t)+t]
    [ -sin(t)+t,  cos(t)+t]

» symmul(G,G) % multiply G by G; sympow(G,2) does the same thing
ans =
    [cos(t)^2-sin(t)^2,    2*cos(t)*sin(t)]
    [ -2*cos(t)*sin(t),  cos(t)^2-sin(t)^2]

» simple(ans) % try to simplify
ans =
    [ cos(2*t), sin(2*t)]
    [-sin(2*t), cos(2*t)]
```

Next we'll show that **G** is an orthogonal matrix by showing that the transpose of **G** is its inverse.

```
» I=symmul(G,transpose(G)) % multiply G by its transpose
I =
    [cos(t)^2+sin(t)^2,                      0]
    [                0, cos(t)^2+sin(t)^2]

» simplify(I) % there appears to be a trig identity here
ans =
    [1, 0]
    [0, 1]
```

which is the identity matrix, as expected.

Linear Algebra Operations

The inverse and determinant of symbolic matrices are computed by the functions **inverse** and **determ**.

```
» H=sym(hilb(3)) % the symbolic form of the numeric 3-by-3 Hilbert matrix
H =
    [  1, 1/2, 1/3]
    [1/2, 1/3, 1/4]
    [1/3, 1/4, 1/5]

» determ(H) % find the determinant of H
ans =
    1/2160

» J=inverse(H) % find the inverse of H
J =
    [  9,  -36,   30]
    [-36,  192, -180]
    [ 30, -180,  180]

» determ(J) % find the determinant of the inverse
ans =
    2160
```

The function **linsolve** is used for solving simultaneous linear equations; it is the symbolic equivalent of the *basic* MATLAB backslash operator. **linsolve(A,B)** solves the matrix equation $A*X=B$ for a square matrix X. Returning to the earlier coin problem,

Diane wants to go to the movies, so she dumps out her piggy bank and counts her coins. She finds that

- *The number of dimes plus half the total number of nickels and pennies is equal to the number of quarters.*
- *The number of pennies is 10 more than the number of nickels, dimes, and quarters.*
- *The number of quarters and dimes is equal to the number of pennies plus 1/4 of the number of nickels.*
- *The number of quarters and pennies is one more than the number of nickels plus 8 times the number of dimes.*

Create a set of linear equations as we did last time. Let p, n, d, and q be the number of pennies, nickels, dimes, and quarters, respectively.

$$d + \frac{n+p}{2} = q \qquad p = n + d + q - 10 \qquad q + d = p + \frac{n}{4} \qquad q + p = n + 8d - 1$$

Rearrange the expressions into the order p, n, d, q.

$$\frac{p}{2} + \frac{n}{2} + d - q = 0 \qquad p - n - d - q = -10 \qquad -p - \frac{n}{4} + d + q = 0$$

$$p - n - 8d + q = -1$$

Next, create a symbolic array of the coefficients of the equations.

```
» A=sym('[1/2,1/2,1,-1;1,-1,-1,-1;-1,-1/4,1,1;1,-1,-8,1]')
A =
    [1/2, 1/2, 1,-1]
    [  1,  -1,-1,-1]
    [ -1,-1/4, 1, 1]
    [  1,  -1,-8, 1]

» B=sym('[0;-10;0;-1]') % Create the symbolic column vector B
B =
    [  0]
    [-10]
    [  0]
    [ -1]

» X=linsolve(A,B) % solve the symbolic system A*X=B for X
X =
    [16]
    [ 8]
    [ 3]
    [15]
```

The result is the same; Diane has 16 pennies, 8 nickels, 3 dimes, and 15 quarters.

Other Features

The **symop** function catenates its arguments and evaluates the resulting expression.

```
» f='cos(x)' % create an expression
f=
    cos(x)

» symop('atan(',f,'+',a,')','^2')
ans =
    atan(cos(x)+a)^2
```

Be careful if you mix arrays and scalars when using **symop**. For example,

```
» M = sym('[a b; c d]')
M =
    [ a,b]
    [ c,d]

» symop(M,'+','t')
ans =
    [a+t,  b]
    [ c, d+t]
```

adds **t** to the diagonal of **M**.

The function **charpoly** finds the characteristic polynomial of a matrix.

```
» G=sym('[1,1/2;1/3,1/4]') % create a symbolic matrix
G =
    [  1,1/2]
    [1/3,1/4]

» charpoly(G) % find the characteristic polynomial of G
ans =
    x*2-5/4*x+1/12
```

Eigenvalues and eigenvectors of symbolic matrices can be found using the **eigensys** function.

```
» F=sym('[1/2,1/4;1/4,1/2]') % create a symbolic matrix
F =
    [1/2,1/4]
    [1/4,1/2]

» eigensys(F) % find the eigenvalues of F
ans =
    [3/4]
    [1/4]

» [V,E]=eigensys(F) % find eigenvalues E and eigenvectors V
V =
    [-1, 1]
    [ 1, 1]
E =
    [1/4]
    [3/4]
```

The Jordan canonical form of a matrix is the diagonal matrix of eigenvalues; the columns of the transformation matrix are eigenvectors. For a given matrix **A**, **jordan(A)** attempts to find a non-singular matrix **V**, so that **inv(V)*A*V** is the Jordan canonical form. The **jordan** function has two forms.

```
» jordan(F) % find the Jordan form of F, above
ans =
    [1/4,   0]
    [  0, 3/4]

» [V,J]=jordan(F) % find the Jordan form and eigenvectors
V =
    [ 1/2, 1/2]
    [-1/2, 1/2]
J =
    [1/4,   0]
    [  0, 3/4]
```

The columns of **V** on the previous page are some of the possible eigenvectors of **F**.

Since **F**, on the previous page is non-singular, the null-space basis of **F** is the empty matrix, and the column-space basis is the identity matrix.

```
» F=sym('[1/2,1/4;1/4,1/2]') % recreate F
F =
     [1/2,1/4]
     [1/4,1/2]

» nullspace(F) % the nullspace of F is the empty matrix
ans =
     []

» colspace(F) % find the column space of F
ans =
     [1,  0]
     [0,  1]
```

Singular values of a matrix can be found using the **singvals** function.

```
» A=sym(magic(3)) % Generate a 3-by-3 matrix
A =
     [8,  1,  6]
     [3,  5,  7]
     [4,  9,  2]

» singvals(A) % find the singular value expressions
ans =
     [15.00000000000000]
     [6.928203230275511]
     [3.464101615137752]
```

The function **jacobian(w,v)** computes the Jacobian of **w** with respect to **v**. The (i,j)-th entry of the result is $\dfrac{df(i)}{dv(j)}$. Note that when **f** is a scalar, the Jacobian of **f** is the gradient of **f**.

```
» jacobian('u*exp(v)',sym('u,v'))
ans =
     [exp(v,  u*exp(v)]
```

22.10 SUMMARY

The following tables summarize features in the *Symbolic Math Toolbox*.

Operations on Symbolic Expressions

numeric	Symbolic to numeric conversion
pretty	Show pretty symbolic output
subs	Substitute for subexpression
sym	Create symbolic matrix or expression
symadd	Symbolic addition
symdiv	Symbolic division
symmul	Symbolic multiplication
symop	Symbolic operations
sympow	Power of symbolic expression
symrat	Rational approximation
symsub	Symbolic subtraction
symvar	Find symbolic variables

Symbolic Simplification

collect	Collect terms
expand	Expand
factor	Factor
simple	Find shortest form
simplify	Simplify
symsum	Sum series

Symbolic Polynomials

charploy	Characteristic polynomial
horner	Nested polynomial representation
numden	Extract numerator or denominator
poly2sym	Polynomial vector to symbolic
sym2poly	Symbolic to polynomial vector

Symbolic Calculus

`diff`	Differentiation
`int`	Integration
`jacobian`	Jacobian matrix
`taylor`	Taylor series expansion

Symbolic Variable Precision Arithmetic

`digits`	Set variable precision accuracy
`vpa`	Variable precision arithmetic

Symbolic Equation Solutions

`compose`	Functional composition
`dsolve`	Solution of differential equations
`finverse`	Functional inverse
`linsolve`	Solution of simultaneous linear equations
`solve`	Solution of algebraic equations

Symbolic Linear Algebra

`charploy`	Characteristic polynomial
`determ`	Matrix determinant
`eigensys`	Eigenvalues and eigenvectors
`inverse`	Matrix inverse
`jordan`	Jordan canonical form
`linsolve`	Solution of simultaneous linear equations
`transpose`	Matrix transpose

Internet Resources

There are a wealth of MATLAB resources available for those who have access to the Internet, be it full access or simply electronic mail forwarding. These resources are accessible through direct Internet connections from educational institutions, government offices, and commercial firms, as well as through commercial Internet-access providers and on-line services such as America On-Line, Delphi, CompuServe, and others.

23.1 USENET NEWSGROUP

One of the most popular information sources and forums on the Internet is a bulletin-board system called a NetNews Feed, USENET, or simply News. This is a huge collection of messages or *postings* that is circulated from computer to computer around the globe. A News feed consists of thousands of individual postings collected in thousands of *newsgroups* or topics. Many sites make all newsgroups available, while others are more selective due to disk-space limitations or company policy concerns.

If you have access to NetNews, you can read postings from other MATLAB users and post comments and questions yourself. The MATLAB newsgroup is called `comp.soft-sys.matlab.` By subscribing to this newsgroup you can browse through questions, comments, and answers from other users and from the staff at *The MathWorks*. In fact, MATLAB developers frequently post to this forum. If you get a limited selection of newsgroups at your site and cannot find this one, contact your local News administrator and ask that `comp.soft-sys.matlab` be included.

Be aware that this newsgroup is *unmoderated*. While this means that you or anyone else can post your comments and questions, there is no method to filter out inappropriate postings. Thousands of people around the world read this newsgroup, so please post carefully and appropriately.

If you do not have access to NetNews but do have FTP capability, as will be discussed next, a digest of postings to this newsgroup can be found on the MATLAB anonymous FTP site (`ftp.mathworks.com`) in the `/pub/doc/cssm-digest/` directory.

If a News feed is not available and FTP is not an option, but you do have an electronic mail (E-mail) connection to the Internet, *The MathWorks* maintains a mail-to-news gateway that allows posting to the newsgroup via E-mail. Messages sent by E-mail to `comp.soft-sys.matlab@mathworks.com` are posted to the newsgroup for you. If you cannot read the newsgroup, be sure to include in your posting a request that replies be sent directly to you by E-mail, and supply your E-mail address.

23.2 ANONYMOUS FTP

One of the most useful resources for MATLAB users is the *Anonymous FTP* site maintained by *The MathWorks, Inc.* FTP is the *file transfer protocol* used to connect to a host computer and then transfer files to and from your computer. Anonymous FTP sites allow users to connect to them and transfer files without requiring a user account and password on the host computer.

There are a number of computer programs that use graphical user interfaces to make FTP connections and transfer files, including `WinFTP` on PCs, `Fetch` on Macintoshes, and `ftptool` on Unix machines. As will be discussed later, Web browsers like `Mosaic` and `NetScape` are also used to access this site. If none of these programs are available at your site, you must use the original `ftp` program which is command-line oriented. Because the original `ftp` program is generally available on Unix workstations and is available for PCs and Macintoshes, its use is described here.

The MATLAB FTP site is located at `ftp.mathworks.com`. It contains user-contributed M-files, product information, documentation including *Frequently Asked Questions* (FAQs), patches, and bug fixes. To connect to the site, `ftp` to the site, log in using the name `'anonymous'` or `'ftp'`, and enter your E-mail address when asked for a password.

The following is a short sample `ftp` session. Items you type in are shown in **bold.** This example connects to the site, lists the files and directories available, changes to ASCII mode to transfer a text file, retrieves the file **README**, and closes the connection.

```
ftp ftp.mathworks.com
Connected to ftp.mathworks.com.
220 ftp FTP server (Version wu-2.4(1) Thu April 14 16:25:23 EDT) ready.
Name (ftp.mathworks.com:user): anonymous
331 Guest login ok, send your complete e-mail address as password.
Password: user@host.maine.edu
230
230 Welcome to the MathWorks Library!
...
230 Guest login ok, access restrictions apply.
Remote system type is UNIX.
Using binary mode to transfer files.
ftp> dir
200 PORT command successful.
150 Opening ASCII mode data connection for /bin/ls.
226 Transfer complete.
total 27
-rw-r--r--  1 admin   daemon       488 Jun 24  1994 .welcome.msg
-rw-r--r--  1 admin   ftpusers    2172 Sep 28  1994 README
-rw-r--r--  1 admin   ftpusers    2626 Sep 28  1994 README.incoming
drwxr-xr-x  2 root    daemon       512 Jun 16  1994 bin
drwxr-xr-x  2 root    daemon       512 Jun 28  1993 dev
drwxr-xr-x  2 root    daemon       512 Jun  7  1993 etc
drwxrwx-wx 38 admin   102        14336 Apr 21 17:23 incoming
lrwxrwxrwx  1 root    daemon         3 Nov  6  1993 matlab -> pub
drwxr-xr-x 12 admin   ftpusers     512 Mar 27 15:17 pub
drwxr-xr-x  3 root    daemon       512 Apr 23  1993 usr
ftp> ascii
200 Type set to A.
ftp> get README
200 PORT command successful.
150 Opening ASCII mode data connection for README (2172 bytes).
226 Transfer complete.
2172 bytes received.
ftp> bye
221 Goodbye.
```

Names listed with a `'d'` in the first column are directories. Those with `'-'` in the first column are files. An `'l'` in the first column signifies a link, which can point to another directory or file somewhere on the site. Information presented in the directory listing includes owner, group, file or directory size, modification date, and file or directory name.

If you have trouble connecting, you may not have access to an Internet nameserver, which translates host names into IP addresses. In that case try using the IP address for `ftp.mathworks.com`, instead. It is `144.212.100.10`.

```
ftp 144.212.100.10
Connected to ftp.mathworks.com
```

The `ftp` program supports a large number of commands and options. The most useful commands are listed below:

Basic FTP Commands

`help`	List all available `ftp` commands
`help command`	Return a short description of *command*
`cd /pub/contrib`	Change to the */pub/contrib* directory on the remote site
`lcd localdir`	Change to the *localdir* directory on your local machine
`ascii`	Transfer files in ASCII text mode; convert line-ending carriage-return/line-feeds to the default for text files on your machine
`binary`	Transfer in binary mode; no text conversions are performed
`get remotefile`	Retrieve the file named *remotefile* from the remote site
`put localfile`	Send the file named *localfile* to the remote site
`dir`	List the current directory on the remote site; full details
`ls`	List the current directory on the remote site; limited details
`bye, quit`	Close the connection and exit `ftp`

One nice feature of the MATLAB site that is not available at many other anonymous FTP sites is the ability to perform file archiving and file compression on-the-fly. Compression formats supported are `gzip` (`file.gz`) from the GNU project at MIT, standard Unix `compress` (`file.Z`), and `zip` (`file.zip`) for PCs and Unix. Archiving is available in standard Unix `tar` format (`dir.tar`), Unix `shar` format (`dir.sh`), and `zip` format (`dir.zip`) for PCs and Unix workstations. Versions of each of these formats are available for Unix workstations and PCs, and most are available for Macintosh computers.

Archiving and compression can be performed simultaneously. For example, the entire contents of the `/pub/contrib/math` directory can be retrieved in Unix **compressed tar** format by issuing the commands

```
cd /pub/contrib
get math.tar.Z
```

Other combinations are available as well. The commands

```
get math.zip
get math.sh
get math.tar.gz
```

retrieve the directory as a `zip` archive, a `shar` archive, and as a `gzipped tar` archive, respectively. Unfortunately for Macintosh users, the MATLAB site does not support `Stuffit` (`dir.sit`) archives or `BinHex` (`file.hqx`) encoding.

The following table describes features of the MATLAB FTP site.

Important Files and Directories on the MATLAB Anonymous FTP Site

Name	Type	Contents
`/README`	text	Information for new users
`/README.incoming`	text	Information for user submissions
`/incoming`	dir	Drop box directory for user M-file submissions
`/pub`	dir	Main directory of files
`/matlab`	link	Link to `/pub`
`/pub/INDEX`	text	Information about directory contents
`/pub/NEWFILES`	text	New files on the site
`/pub/ls-lR`	text	Recursive list of all files on the site
`/pub/ftphelp`	text	Beginner's guide for new `ftp` users
`/pub/books`	dir	M-files associated with MATLAB-base books
`/pub/conference`	dir	Upcoming MATLAB conference information
`/pub/contrib`	dir	User-contributed M-files and MEX files
`/pub/doc`	dir	Documentation, tutorials, help files, FAQs, etc.

`/pub/mathworks`	dir	Bug fixes and M-files from MATLAB developers
`/pub/mosaic`	dir	Mosaic WWW browser programs for Unix
`/pub/pentium`	dir	Collection of papers relating to the Pentium bug
`/pub/proceedings`	dir	Proceedings of past MATLAB conferences
`/pub/product-info`	dir	MATLAB, SIMULINK, and Toolbox information
`/pub/tech-support`	dir	Technical support bulletins and other information

If you have trouble connecting or need help with the FTP site, send E-mail to the site maintainer at `ftpadmin@mathworks.com`.

23.3 WORLD WIDE WEB

The newest, easiest, and most flexible way to get information, images, and files over the Internet is called the *World Wide Web, WWW,* or simply the *Web.* The Web is a collection of server sites that deliver *hypertext* documents to your local computer upon request. Hypertext documents can incorporate imbedded graphics and links to other documents. These links can be activated simply by clicking on the highlighted text or graphic with the mouse. The linked document is then retrieved for viewing and can be saved to your local computer, if desired. Some of the links can connect to *Gopher* sites (text-based information servers) and to FTP sites around the world.

If you have access to a full-service Internet connection, you have the option of this kind of graphical interface to MATLAB information. *The MathWorks* has a World Wide Web server that can be accessed with a Web client program such as *Mosaic, NetScape,* or the text-based *Lynx.* Use the menu item **Open Location . . .** or **Open URL . . .** and enter the string `http://www.mathworks.com` to connect to the MathWorks *home page.* The home page is the top-level hypertext document that contains links to many other documents and files on the site. Once you have connected, you can save the access information as a *bookmark* to save time and effort the next time you want to connect.

From the MathWorks home page, hypertext links can retrieve information about *The MathWorks, Inc.* and its products, an on-line copy of the quarterly newsletter, a listing of books based on MATLAB (over 100 at this writing) along with their associated M-files, information about trade shows that the staff will be attending, and on-line copies of the latest MATLAB conference proceedings. There is an archive of *Frequently Asked Questions* (FAQs) and responses about MATLAB and SIMULINK. You can also read the Technical Notes section contributed by *The MathWorks* Technical Support Staff that contains tips and techniques relating to a number of topics including memory management, graphics and printing, MEX files, Toolboxes, and integrating MATLAB with other software.

The MathWorks home page also has a direct connection to the anonymous FTP site discussed earlier. You can browse the directories, read the index files, and actually retrieve files using your Web browser. If you do not have a Web browser program, versions of *Mosaic* for a number of workstations are available on the FTP site in the directory `/pub/mosaic`. Versions of *Mosaic* for PC and Macintosh users are available via anonymous FTP from `ftp.ncsa.uiuc.edu`. Versions of *NetScape* for workstations and microcomputers are available from `ftp.netscape.com`. Be aware that while *Mosaic* is free software, there are some restrictions on the free use of *NetScape* by commercial firms.

23.4 MATLIB AUTOMATED E-MAIL RESPONSE SYSTEM

Those without FTP capability but who have an E-mail connection to the Internet are not left out. *The MathWorks* maintains an automated file-by-mail service called MATLIB that will send files from the FTP site by E-mail. Send E-mail to `matlib@mathworks.com` with the single word `help` in the *body* of the message to get more detailed instructions about accessing the server. The MATLIB file-by-mail service is a computer program that gets its instructions from the body of your message; the subject line is ignored. Therefore, it is very important that the syntax of your message be correct.

Many of the instructions that you can use in your message are a subset of standard FTP commands including `cd`, `ls`, `dir`, `get`, and `quit`. Others are used to specify the encoding, archiving, and compression methods to use. Still others direct the `matlib` program to reply to a different E-mail address or to limit response message sizes.

One example is the `reply-to` command. This is optional, and if used will cause the MATLIB service to send files to the E-mail address specified by this command. If not present, it will reply to the sender's E-mail address. For example,

```
reply-to user@host.maine.edu
```

will send the response to `user@host.maine.edu` rather than the originating E-mail address.

Binary files cannot be sent by E-mail and are encoded (converted to an ASCII text representation) before mailing. Binary files are sent using *uuencode* by default, but can be sent using *mime* encoding if requested. Check with your computer system administrator to see which methods are supported at your site. Compression and archiving are also available using *compress, gzip, zip, shar,* and *tar.* As an example, consider a message sent to `matlib@mathworks.com` containing the following lines:

```
dir
cd /pub/contrib
get INDEX
```

```
get integration.tar.Z
get math.zip
cd games
get fifteens.m.gz
cd ..
get diffeq.sh
quit
```

This script instructs MATLIB to first retrieve a directory listing of the top-level directory. Then change to the **/pub/contrib/** directory and retrieve the **INDEX** file. Then retrieve the entire **/pub/contrib/integration/** directory as a **compressed tar** file. Note that this is a binary file, so it is **uuencoded** before it is sent. Next retrieve a **uuencoded zip** archive of the entire contents of the **/pub/contrib/math** directory. Now change to the **games** directory and retrieve the M-file **fifteens.m** using **gzip** compression. Finally, change to the next higher directory (**/pub/contrib**), retrieve the entire **diffeq** directory as a **shar** archive, and quit.

Responses are limited to 100 Kbytes each, so larger files or archives will be sent as multiple E-mail messages. You can limit the size of messages even further using the **size** command if your E-mail message size is limited. Refer to the response from the **help** message for more detailed information.

23.5 THE MATHWORKS MATLAB DIGEST

The *MATLAB Digest* is a monthly electronic newsletter sent to subscribers via E-mail. This newsletter contains MATLAB news, articles contributed by MATLAB developers and users, hints and tips, and responses to user questions. To subscribe to the Digest, send E-mail to **subscribe@mathworks.com** with a request to be added to the mailing list or type » **subscribe** at the MATLAB prompt. The **subscribe** command asks questions and uses the answers to generate a printed request form that can be mailed or FAXed to *The MathWorks*. Unix implementations offer the option to automatically E-mail your request rather than printing it out. Back issues of the newsletter are available on the MATLAB FTP site in the directory **/pub/doc/tmw-digest**.

23.6 THE MATLAB NEWSLETTER

The quarterly written publication *MATLAB News & Notes* is available to anyone who uses MATLAB. If you become a subscribing user, you will receive the Digest and the quarterly newsletter, as well as free technical support. You do not have to be a registered user to become a subscribing user, and it is a free service. Use the form on the World Wide Web site, type » **subscribe** at the MATLAB prompt, or send E-mail to **subscribe@mathworks.com** with your request. The newsletter usually has news, tips, articles by Cleve Moler and others, and a calendar of events.

23.7 THE MATHWORKS E-MAIL AND NETWORK ADDRESSES

The following is a list of useful E-mail addresses and network resources.

E-mail Addresses at *The MathWorks, Inc.*

`support@mathworks.com`	Technical support
`bugs@mathworks.com`	Bug reports
`doc@mathworks.com`	Documentation error reports
`suggest@mathworks.com`	Product enhancement suggestions
`service@mathworks.com`	Order status, license renewals, passcodes
`subscribe@mathworks.com`	Subscribing user information
`info@mathworks.com`	Sales, pricing, and general information
`micro-updates@mathworks.com`	PC and Macintosh updates
`matlib@mathworks.com`	File-by-mail server
`digest@mathworks.com`	MATLAB Digest submissions
`ftpadmin@mathworks.com`	*The Mathworks* FTP site maintainer
`webmaster@mathworks.com`	*The Mathworks* World Wide Web maintainer

MATLAB Network Resources

`www.mathworks.com`	World Wide Web site
`ftp.mathworks.com`	Anonymous FTP site
`144.212.100.10`	WWW and FTP Internet address
`novell.felk.cvut.cz`	Mirrors `ftp.mathworks.com`
`192.108.154.33`	Mirror Internet address

Other Network Resources

`mm@eece.maine.edu`	E-mail address for questions and comments to the authors about *Mastering MATLAB* and the *Mastering MATLAB Toolbox*
`ftp.ncsa.uiuc.edu`	PC and Macintosh versions of *Telnet* and *FTP,* and *Mosaic* for PC, Macintosh, and Unix
`ftp.netscape.com`	*NetScape* Web browser for PC, Mac, and Unix
`sumex-aim.stanford.edu`	Mac versions of *Mosaic, Telnet(FTP), Fetch, gzip, zip, shar,* and *compress*
`sics.se`	Mirrors the files at `sumex-aim.stanford.edu.`
`gatekeeper.dec.com`	Large anonymous FTP site
`sunsite.unc.edu`	Large anonymous FTP site
`ftp.wustl.edu`	Large anonymous FTP site
`nic.funet.fi`	Large anonymous FTP site
`ftp.luth.se`	Large anonymous FTP site
`ftp.cdrom.com`	Large anonymous FTP site
`ftp.uws.edu.au`	Large anonymous FTP site

Appendices

APPENDIX A MATLAB QUICK REFERENCE TABLES

The following tables are duplicates of those found elsewhere in this text. The page number where each original table appears is given in the **upper-right corner of the tables**.

7

MATLAB

Command	average_cost	Comments
`format long`	`35.83333333333334`	16 digits
`format short e`	`3.5833e+01`	5 digits plus exponent
`format long e`	`3.58333333333334e+01`	16 digits plus exponent
`format hex`	`4041eaaaaaaaaaab`	hexadecimal
`format bank`	`35.83`	2 decimal digits
`format +`	`+`	positive, negative, or zero
`format rat`	`215/6`	rational approximation
`format short`	`35.8333`	default display

8

Variable Naming Rules	Comments/Examples
Variable names are case sensitive	`Items`, `items`, `itEms`, and `ITEMS` are all different MATLAB variables
Variable names can contain up to 19 characters; characters beyond the 19th are ignored	`howaboutthisvariablename`

Variable names must start with a letter, followed by any number of letters, digits, or underscores. Punctuation characters are not allowed since many have special meaning to MATLAB	`how_about_this_variable_name`
	`X51_483`
	`a_b_c_d_e`

9

Special Variables	Value
ans	The default variable name used for results
pi	The ratio of the circumference of a circle to its diameter
eps	The smallest number such that when added to one creates a number greater than one on the computer
flops	Count of floating point operations
inf	Which stands for infinity, e.g., 1/0
NaN	which stands for Not-a-Number, e.g., 0/0
i (and) **j**	**i=j=R(-1)**
nargin	Number of function input arguments used
nargout	Number of function output arguments used
realmin	The smallest usable positive real number
realmax	The largest usable positive real number

14

Common Functions

abs(x)	Absolute value or magnitude of complex number
acos(x)	Inverse cosine
acosh(x)	Inverse hyperbolic cosine
angle(x)	Four quadrant angle of complex
asin(x)	Inverse sine
asinh(x)	Inverse hyperbolic sine
atan(x)	Inverse tangent
atan2(x,y)	Four quadrant inverse tangent
atanh(x)	Inverse hyperbolic tangent
ceil(x)	Round toward plus infinity
conj(x)	Complex conjugate
cos(x)	Cosine
cosh(x)	Hyperbolic cosine
exp(x)	Exponential: e^x
fix(x)	Round toward zero

`floor(x)`	Round toward minus infinity
`gcd(x,y)`	Greatest common divisor of integers x and y
`imag(x)`	Complex imaginary part
`lcm(x,y)`	Least common multiple of integers x and y
`log(x)`	Natural logarithm
`log10(x)`	Common logarithm
`real(x)`	Complex real part
`rem(x,y)`	Remainder after division: `rem(x,y)` gives the remainder of `x/y`
`round(x)`	Round toward nearest integer
`sign(x)`	Signum function: return sign of argument, e.g., `sign(1.2)=1`, `sign(-23.4)=-1`, `sign(0)=0`
sin(x)	Sine
`sinh(x)`	Hyperbolic sine
`sqrt(x)`	Square root
`tan(x)`	Tangent
`tanh(x)`	Hyperbolic tangent

17

M-file Functions

`disp(ans)`	Display results without identifying variable names
`echo`	Control the *Command* window echoing of script file commands
`input`	Prompt user for input
`keyboard`	Give control to keyboard temporarily (Type `return` to quit.)
`pause`	Pause until user presses any keyboard key
`pause(n)`	Pause for **n** seconds
`waitforbuttonpress`	Pause until user presses mouse button or keyboard key

20

File Management Functions

`cd`	Show present working directory or folder
`p=cd`	Return present working directory in `p`
`cd path`	Change to directory or folder given by path
`chdir`	Same as `cd`
`chdir path`	Same as `cd path`
`delete test`	Delete the m-file `test.m`
`dir`	List all files in the current directory or folder
`ls`	Same as `dir`
`matlabroot`	Return directory path to matlab executable program
`path`	Display or modify matlab's search path
`pwd`	Show present directory
`type test`	Display the m-file `test.m` in the *Command* window
`what`	Return a listing of all m-files and mat-files in the current directory or folder
`which test`	Display the directory path to `test.m`.

21

MATLAB Search Path

In general, when you enter » `cow`, MATLAB does the following:

(1) It checks to see if `cow` is a **variable** in the MATLAB workspace; if not,

(2) It checks to see if `cow` is a **built-in function;** if not,

(3) It checks to see if a MEX-file `cow.mex` exists in the **current directory;** if not,

(4) It checks to see if an M-file named `cow.m` exists in the **current directory;** if not,

(5) It checks to see if `cow.mex` or `cow.m` exists anywhere on the **MATLAB search path,** by searching the path in the order in which it is specified

22

Command Window Control Commands

clc	Clear the *Command* window
diary	Save *Command* window text to a file
home	Move cursor to upper-left corner
more	Page the *Command* window

31

Simple Array Construction

x=[2 2*pi sqrt(2) 2-3j]	Create row vector **x** containing elements specified
x=first:last	Create row vector **x** starting with **first**, counting by one, ending at or before **last**
x=first:increment:last	Create row vector **x** starting with **first**, counting by **increment**, ending at or before **last**
x=linspace(first,last,n)	Create row vector **x** starting with **first**, ending at **last**, having **n** elements
x=logspace(first,last,n)	Create logarithmically-spaced row vector **x** starting with 10^{first}, ending at 10^{last}, having **n** elements

39

Element-by-Element Array Mathematics

Illustrative data:	**a = [a$_1$ a$_2$ ... a$_n$], b = [b$_1$ b$_2$ ... b$_n$], c= <a scalar>**
Scalar addition	**a+c = [a$_1$+c a$_2$+c ... a$_n$+c]**
Scalar multiplication	**a*c = [a$_1$*c a$_2$*c ... a$_n$*c]**
Array addition	**a+b = [a$_1$+b$_1$ a$_2$+b$_2$ ... a$_n$+b$_n$]**

Array multiplication	$a.*b = [a_1*b_1\ \ a_2*b_2\ \ \ldots\ \ a_n*b_n]$
Array right division	$a./b = [a_1/b_1\ \ a_2/b_2\ \ \ldots\ \ a_n/b_n]$
Array left division	$a.\backslash b = [a_1\backslash b_1\ \ a_2\backslash b_2\ \ \ldots\ \ a_n\backslash b_n]$
Array powers	$a.\hat{}c = [a_1\hat{}c\ \ a_2\hat{}c\ \ \ldots\ \ a_n\hat{}c]$
	$c.\hat{}a = [c\hat{}a_1\ \ c\hat{}a_2\ \ \ldots\ \ c\hat{}a_n]$
	$a.\hat{}b = [a_1\hat{}b_1\ \ a_2\hat{}b_2\ \ \ldots\ \ a_n\hat{}b_n]$

47

Array Addressing

A(r,c)	Addresses a subarray within **A** defined by the index vector of desired rows in **r** and index vector of desired columns in **c**
A(r,:)	Addresses a subarrray within **A** defined by the index vector of desired rows in **r** and all columns
A(:,r)	Addresses a subarray within **A** defined by all rows and the index vector of desired columns in **c**
A(:)	Addresses all elements of **A** as a column vector taken column by column
A(i)	Addresses a subarray within **A** defined by the single index vector of desired elements in **i**, as if **A** was the column vector, **A(:)**
A(x)	Addresses a subarray within **A** defined by the logical array **x. x** must contain only the values **0** and **1**, and must be the same size as **A**

49

Array Searching

i=find(x)	Return indices of the array **x** where its elements are nonzero
[r,c]=find(X)	Return row and column indices of the array **x** where its elements are nonzero

51

Array Size

whos	Display variables that exist in the workspace and their sizes
s=size(A)	Return a two-element vector **s**, whose first element is the number of rows in **A** and whose second element is the number of columns in **A**
[r,c]=size(A)	Return two scalars **r** and **c** containing the number of rows and columns in **A**, respectively
r=size(A,1)	Return the number of rows in **A** in the variable **r**
c=size(A,2)	Return the number of columns in **A** in the variable **c**
n=length(A)	Return **max(size(A))** in the variable **n**

51

Array Manipulation Functions

flipud(A)	Flip a matrix upside-down
fliplr(A)	Flip a matrix left to right
rot90(A)	Rotate a matrix counterclockwise 90 degrees
reshape(A,m,n)	Return an **m**-by-**n** matrix whose elements are taken columnwise from **A**. **A** must contain **m*n** elements
diag(A)	Extract the diagonal of the matrix **A** as a column vector
diag(v)	Create a diagonal matrix with the vector **v** on its diagonal
tril(A)	Extract lower triangular part of the matrix **A**
triu(A)	Extract upper triangular part of the matrix **A**

60

Matrix Functions

balance(A)	Scale to improve eigenvalue accuracy
cdf2rdf(A)	Complex diagonal form to real block diagonal form
chol(A)	Cholesky factorization

`cond(A)`	Matrix condition number
`condest(A)`	1-norm matrix condition number estimate
`d=eig(A),`	Eigenvalues and eigenvectors
`[V,D]=eig(A)`	
`det(A)`	Determinant
`expm(A)`	Matrix exponential
`expm1(A)`	M-file implementation of **expm**
`expm2(A)`	Matrix exponential using Taylor series
`expm3(A)`	Matrix exponential using eigenvalues and eigenvectors
`funm(A,'fun')`	Compute general matrix function
`hess(A)`	Hessenberg form
`inv(A)`	Matrix inverse
`logm(A)`	Matrix logarithm
`lscov(A,b,V)`	Least squares with known covariance
`lu(A)`	Factors from Gaussian elimination
`nnls(A,b)`	Nonnegative least squares
`norm(A)`	Matrix and vector norms:
`norm(A,1)`	1-norm,
`norm(A,2)`	2-norm (Euclidean),
`norm(A,inf)`	Infinity,
`norm(A,p)`	P-norm (vectors only),
`norm(A,'fro')`	F-norm
`null(A)`	Null space
`orth(A)`	Orthogonalization
`pinv(A)`	Pseudoinverse
`poly(A)`	Characteristic polynomial
`polyvalm(A)`	Evaluate matrix polynomial
`qr(A)`	Orthogonal-triangular decomposition
`qrdelete(Q,R,j)`	Delete column from qr factorization
`qrinsert(Q,R,j,x)`	Insert column in qr factorization
`qz(A,B)`	Generalized eigenvalues
`rank(A)`	Number of linearly independent rows or columns
`rcond(A)`	Reciprocal condition estimator
`rref(A)`	Reduced row echelon form
`rsf2csf(U,T)`	Real schur form to complex schur form
`schur(A)`	Schur decomposition

`sqrtm(A)`	Matrix square root
`svd(A)`	Singular value decomposition
`trace(A)`	Sum of diagonal elements

65

Special Matrices

`[]`	The empty matrix
`compan`	Companion matrix
`eye`	Identity matrix
`gallery`	Several small test matrices
`hadamard`	Hadamard matrix
`hankel`	Hankel matrix
`hilb`	Hilbert matrix
`invhilb`	Inverse Hilbert matrix
`magic`	Magic square
`ones`	Matrix containing all ones
`pascal`	Pascal triangle matrix
`rand`	Uniformly distributed random matrix with elements between 0 and 1
`randn`	Normally distributed random matrix with elements having zero mean and unit variance
`rosser`	Symmetric eigenvalue test matrix
`toeplitz`	Toeplitz matrix
`vander`	Vandermonde matrix
`wilkinson`	Wilkinson eignenvalue test matrix
`zeros`	Matrix containing all zero elements

65

Sparse Matrix Functions

`colmmd`	Reorder by column minimum degree
`colperm`	Reorder by ordering columns based on nonzero count
`condest`	Estimate 1-norm matrix condition

dmperm	Reorder by Dulmage-Mendelsohn decomposition
find	Find indices of nonzero entries
gplot	Graph theory plot of sparse matrix
nnz	Number of nonzero entries
nonzeros	Nonzero entries
normest	Estimate 2-norm
nzmax	Storage allocated for nonzeros
randperm	Random permutation
spalloc	Allocate memory for nonzeros
spaugment	Form least-squares augmented system
spconvert	Convert sparse to external format
spdiags	Sparse matrix formed from diagonals
speye	Sparse identity matrix
spones	Replace nonzeros with ones
spparms	Set sparse matrix routine parameters
sprandn	Sparse random matrix
sprandsym	Sparse symmetric random matrix
sprank	Structural rank
spy	Visualize sparse structure
symbfact	Symbolic factorization analysis
symmd	Reorder by symmetric minimum degree
symrcm	Reorder by reverse Cuthill-McKee algorithm

68

Relational Operator	Description
<	less than
<=	less than or equal to
>	greater than
>=	greater than or equal to
==	equal to
~=	not equal to

70

Logical Operator	Description
&	AND
|	OR
~	NOT

71

Other Relational and Logical Functions

`xor(x,y)`	Exclusive OR operation. Return ones where either **x** or **y** is nonzero (True). Return zeros where both **x** and **y** are zero (False) or both are nonzero (True)
`any(x)`	Return one if *any* element in a vector **x** is nonzero. Return one for each column in a matrix **x** that has nonzero elements
`all(x)`	Return one if *all* elements in a vector **x** are nonzero. Return one for each column in a matrix **x** that has all nonzero elements

72

Test Functions

`finite`	Return True where elements are finite
`isempty`	Return True if argument is empty
`isglobal`	Return True if argument is a global variable
`ishold`	Return True if current plot hold state is ON
`isieee`	Return True if computer performs IEEE arithmetic
`isinf`	Return True where elements are infinite
`isletter`	Return True where elements are letters of the alphabet
`isnan`	Return True where elements are NaNs
`isreal`	Return True if argument has no imaginary part
`isspace`	Return True where elements are whitespace characters
`issparse`	Return True if argument is a sparse matrix

`isstr`	Return True if argument is a character string
`isstudent`	Return True if Student Edition of MATLAB
`isunix`	Return True if computer is UNIX
`isvms`	Return True if computer is VMS

73

NaNs and Empty Matrices

Data	a = [1	2	nan	inf	nan]
Expression	**Result**				
`2*a`	[2	4	NaN	∞	NaN]
`(a==nan)`	[0	0	0	0	0]
`(a~=nan)`	[1	1	1	1	1]
`isnan(a)`	[0	0	1	0	1]
`y=find(a==0)`	y=[]				
`isempty(y)`	1				
`(y==0)`	0				
`find(y==0)`	[]				
`(y==0)`	1				
`j=find(y~=0)`	j=1				
`y(j)`	Error!y(j) does not exist				

80

String Conversions

`abs`	String to ASCII
`dec2hex`	Decimal number to hexadecimal string
`fprintf`	Write formatted text to file or screen
`hex2dec`	Hex string to decimal number
`hex2num`	Hexadecimal string to IEEE floating-point number
`int2str`	Integer to string
`lower`	String to lower case
`num2str`	Number to string

setstr	ASCII to string
sprintf	Convert number to string with format control
sscanf	Convert string to number under format control
str2mat	Strings to a text matrix
str2num	String to number
upper	String to upper case

83

String Functions

eval(string)	Evaluate string as a MATLAB command
eval (try,catch)	Evaluate **try**, if error occurs evaluate **catch**
blanks(n)	Return a string of n blanks or spaces
deblank	Remove trailing blanks from a string
feval	Evaluate function given by string
findstr	Find one string within another
isletter	Return True where alphabet characters exist
isspace	Return True where whitespace characters exist
isstr	Return True if input is a string
lasterr	Return string of last MATLAB error issued
strcmp	Return True if strings are identical
strrep	Replace one string with another
strtok	Find first token in a string

82

Number Format Conversion Examples

Command	Result
fprintf('%.0e\n',pi)	3e+00
fprintf('%.1e\n',pi)	3.1e+00
fprintf('%.3e\n',pi)	3.142e+00
fprintf('%.5e\n',pi)	3.14159e+00
fprintf('%.10e\n',pi)	3.1415926536e+00

```
fprintf('%.0f\n',pi)       3
fprintf('%.1f\n',pi)       3.1
fprintf('%.3f\n',pi)       3.142
fprintf('%.5f\n',pi)       3.14159
fprintf('%.10f\n',pi)      3.1415926536

fprintf('%.0g\n',pi)       3
fprintf('%.1g\n',pi)       3
fprintf('%.3g\n',pi)       3.14
fprintf('%.5g\n',pi)       3.1416
fprintf('%.10g\n',pi)      3.141592654

fprintf('%8.0g\n',pi)             3
fprintf('%8.1g\n',pi)             3
fprintf('%8.3g\n',pi)          3.14
fprintf('%8.5g\n',pi)         3.1416
fprintf('%8.10g\n',pi)     3.141592654
```

93

Control Flow Structures

`for x = array`
`    commands`
`end`
	A For Loop that on each iteration assigns **x** to the ith column of **array** and executes **commands**

`while expression`
`    commands`
`end`
	A While Loop that executes **commands** as long as **all** elements of **expression** are True or nonzero

`if expression`
`    commands`
`end`
	A simple If-Else-End structure where **commands** is executed if **all** elements in **expression** are True or nonzero

`if expression`
`    commands evaluated`
`        if True`
	An If-Else-End structure with two paths. One group of commands is executed if **expression** is True

```
else
    commands evaluated
        if False
end
```
or nonzero. The other set is executed if **expression** is False or zero

```
if expression1
    commands evaluated if
        expression1 is True
elseif expression2
    commands evaluated if
        expression2 is True
elseif expression3
    commands evaluated if
        expression3 is True
elseif expression4
    commands evaluated if
        expression4 is True
elseif...
        .
        .
        .
else
    commands evaluated if
        no other expression
            is True
end
```
The most general If-Else-End structure. Only the commands associated with the first True expression are evaluated

```
break
```
Terminates execution of For Loops and While Loops

112

Data Analysis Functions

`corrcoef(x)`	Correlation coefficients
`cov(x)`	Covariance matrix
`cplxpair(x)`	Sort vector into complex conjugate pairs
`cross(x,y)`	Vector cross product
`cumprod(x)`	Cumulative product of columns
`cumsum(x)`	Cumulative sum of columns

`del2(A)`	Five-point discrete Laplacian
`diff(x)`	Compute differences between elements
`dot(x,y)`	Vector dot product
`gradient(Z,dx,dy)`	Approximate gradient
`histogram(x)`	Histogram or bar chart
`max(x), max(x,y)`	Maximum component
`mean(x)`	Mean or average value of columns
`median(x)`	Median value of columns
`min(x), min(x,y)`	Minimum component
`prod(x)`	Product of elements in columns
`rand(x)`	Uniformly distributed random numbers
`randn(x)`	Normally distributed random numbers
`sort(x)`	Sort columns in ascending order
`std(x)`	Standard deviation of columns
`subspace(A,B)`	Angle between two subspaces
`sum(x)`	Sum of elements in each column

124

Polynomial Functions

`conv(a,b)`	Multiplication
`[q,r]=deconv(a,b)`	Division
`poly(r)`	Construct polynomial from roots
`polyder(a)`	Differentiate polynomial or rational polynomials
`polyfit(x,y,n)`	Fit polynomial to data
`polyval(p,x)`	Evaluate polynomial at `x`
`[r,p,k]=residue(a,b)`	Partial fraction expansion
`[a,b]=residue(r,p,k)`	
`roots(a)`	Find polynomial roots

140

Curve Fitting and Interpolation Functions

`polyfit(x,y,n)`	Least-squares curve-fitting of data describing y=f(x) to an nth order polynomial
`interp1(x,y,xo)`	1-D linear interpolation
`interp1(x,y,xo,'spline')`	1-D cubic spline interpolation
`interp1(x,y,xo,'cubic')`	1-D cubic interpolation
`interp2(x,y,Z,xi,yi)`	2-D linear interpolation
`interp2(x,y,Z,xi,yi,'cubic')`	2-D cubic interpolation
`interp2(x,y,Z,xi,yi,'nearest')`	2-D nearest neighbor interpolation

152

Cubic Spline Functions

`yi=spline(x,y,xi)`	Cubic spline interpolation of y=f(x) at points in **xi**
`pp=spline(x,y)`	Return piecewise polynomial representation of y=f(x)
`yi=ppval(pp,xi)`	Evaluate piecewise polynomial at points in **xi**
`[breaks,coefs,npolys,ncoefs]` `=unmkpp(pp)`	Unmake piecewise polynomial representation
`pp=mkpp(breaks,coefs)`	Make piecewise polynomial representation

173

Numerical Analysis Functions

`fplot('fname',[lb ub])`	Plot function between lower and upper bounds
`fmin('fname',[lb ub])`	Find scalar minimum between lower and upper bounds

`fmins('fname',Xo)`	Find vector minimum near **Xo**
`fzero('fname',xo)`	Find zero in scalar function near **xo**
`trapz(x,y)`	Trapezoidal integration of area under $y = f(x)$ given data points in **x** and **y**
`diff(x)`	Difference between array elements
`[t,y]=` `ode23('fname',to,tf,yo)`	Solution of a set of differential equations using a 2nd/3rd order Runge-Kutta algorithm
`[t,y]=` `ode45('fname',to,tf,yo)`	Solution of a set of differential equations using a 4th/5th order Runge-Kutta algorithm

177

Signal Processing Functions

conv	Convolution
conv2	2-D convolution
fft	Fast Fourier Transform
fft2	2-D Fast Fourier Transform
ifft	Inverse Fast Fourier Transform
ifft2	2-D Inverse Fast Fourier Transform
filter	Discrete time filter
filter2	2-D discrete time filter
abs	Magnitude
angle	Four quadrant phase angle
unwrap	Remove phase angle jumps at 360 degree boundaries
fftshift	Shift FFT results so negative frequencies appear first
nextpow2	Next higher power of two

189

MATLAB Low Level File I/O Functions

fclose	Close file
feof	Test for end-of-file

ferror	Inquire file I/O error status
fgetl	Read line from file, ignoring new line character
fgets	Read line from file, including new line character
fopen	Open file
fprintf	Write formatted data to file or screen
fread	Read binary data from file
frewind	Rewind to start of file
fscanf	Read formatted data from file
fseek	Set file position indicator
ftell	Get file position indicator
fwrite	Write binary data to file

192

MATLAB Debugging Functions

dbclear	Remove breakpoint
dbcont	Resume execution after breakpoint
dbdown	Drop down on workspace level
dbquit	Quit debug mode
dbstack	List who called whom
dbstatus	List all breakpoints
dbstep	Execute one or more lines
dbstop	Set breakpoint
dbtype	List M-file with line numbers
dbup	Move up one workspace level

196

Plot Line Types and Colors

Symbol	Color	Symbol	Linestyle
y	yellow	.	point
m	magenta	o	circle

Symbol	Color	Symbol	Linestyle
c	cyan	x	x-mark
r	red	+	plus
g	green	*	star
b	blue	-	solid line
w	white	:	dotted line
k	black	-.	dash-dot line
		--	dashed line

201

axis Commands

`axis([xmin xmax ymin ymax])`	Set the maximum and minimum values of the axes using values given in the row vector. If `xmin` or `ymin` is set to `-inf`, the minimum is autoscaled. If `xmax` or `ymax` is set to `inf`, the maximum is autoscaled
`axis auto` `axis('auto')`	Return the axis scaling to its automatic defaults: `xmin=min(x)`, `xmax=max(x)`, etc.
`axis(axis)`	Freeze scaling at the current limits, so that if `hold` is turned on, subsequent plots use the same axis limits
`axis xy` `axis('xy')`	Use the (default) *Cartesian* coordinate form, where the *system origin* (the smallest coordinate pair) is at the lower-left corner. The horizontal axis increases left to right, and the vertical axis increases bottom to top
`axis ij` `axis('ij')`	Use the *matrix* coordinate form, where the *system origin* is at the top left corner. The horizontal axis

	increases left to right, but the vertical axis increases top to bottom
`axis square` `axis('square')`	Set the current plot to be a square rather than the default rectangle
`axis equal` `axis('equal')`	Set the scaling factors for both axes to be equal
`axis image` `axis('image')`	Set the aspect ratio and the axis limits so the image in the current axes has square pixels
`axis normal` `axis('normal')`	Turn off **axis equal** and **axis square**
`axis off` `axis('off')`	Turn off all axis labeling, grid, and tic marks. Leave the title and any labels placed by the **text** and **gtext** commands
`axis on` `axis('on')`	Turn on axis labeling, tic marks, and grid
`v=axis`	Return the current axis limits in the vector **v**

228

2-D Plotting Functions

`bar`	Bar graph
`compass`	Plot of vectors emanating from the origin
`errorbar`	Linear plot with errorbars
`feather`	Plot of vectors emanating from a line
`fill`	Draw filled 2-D polygons (basic)
`hist`	Histogram plot
`image`	Display image
`loglog`	Plot with both axes scaled logarithmically (basic)
`plot`	Simple linear plot (basic)
`polar`	Plot in polar coordinates
`rose`	Angle histogram plot
`semilogx`	Plot with x-axis scaled logarithmically (basic)
`semilogy`	Plot with y-axis scaled logarithmically (basic)
`stairs`	Staircase or zero-order-hold plot
`stem`	Discrete sequence plot

229

2-D Plotting Tools

axis	Modify axis properties
clf	Clear *Figure* windows
close	Close *Figure* windows
figure	Create or select *Figure* windows
ginput	Get data coordinates with mouse
grid	Place grid
gtext	Place text with mouse
hold	Hold current plot
subplot	Create subplots within a *Figure* window
text	Place text at given location
title	Place title
xlabel	Place x-axis label
ylabel	Place y-axis label
zoom	Magnify or shrink plot axes

237

view Function

view(az,el) **view([az,el])**	Set view to azimuth **az** and elevation **el**
view([x,y,z])	Set view looking toward the origin along the vector **[x,y,z]** in Cartesian coordinates, e.g., **view([0 0 1])=view(0,90)**
view(2)	Sets the default 2-D view, **az=0**, **el=90**
view(3)	Sets the default 3-D view, **az=-37.5**, **el=30**
[az,el]=view	Returns the current azimuth **az** and elevation **el**
view(T)	Uses the 4-by-4 transformation matrix **T** to set the view
T=view	Returns the current 4-by-4 transformation matrix

264

3-D Plotting Functions

`contour`	2-D contour plot, i.e., `contour3` viewed from above
`contour3`	Contour plot
`fill3`	Filled polygons
`mesh`	Mesh plot
`meshc`	Mesh plot with underlying contour plot
`meshz`	Mesh plot with zero plane
`pcolor`	2-D pseudocolor plot, i.e., `surf` plot viewed from above
`plot3`	Line plot
`quiver`	2-D arrow velocity plot
`surf`	Surface plot
`surfc`	Surface plot with underlying contour plot
`surfl`	Surface plot with lighting
`waterfall`	Mesh plot lacking cross lines

265

3-D Plotting Tools

`axis`	Modify axis properties
`clf`	Clear *Figure* windows
`clabel`	Place contour labels
`close`	Close *Figure* windows
`figure`	Create or select *Figure* windows
`getframe`	Capture movie frame
`grid`	Place grid
`griddata`	Interpolate (potentially scattered) data for plots
`hidden`	Hidden line removal in `mesh` plots
`hold`	Hold current plot
`meshgrid`	Data generation for 3-D plots
`movie`	Play movie
`moviein`	Create frame matrix for movie storage
`shading`	`faceted`, `flat`, `interp` shading in `surf` and `pcolor` plots
`subplot`	Create subplots within a *Figure* window

text	Place text at given location
title	Place title
view	Change viewpoint of plot
xlabel	Place x-axis label
ylabel	Place y-axis label
zlabel	Place z-axis label

268

Simple Colors

Red	Green	Blue	Color
0	0	0	black
1	1	1	white
1	0	0	red
0	1	0	green
0	0	1	blue
1	1	0	yellow
1	0	1	magenta
0	1	1	cyan
2/3	0	1	violet
1	1/2	0	orange
.5	0	0	dark red
.5	.5	.5	medium gray

268

Standard Color Maps

hsv	Hue-saturation-value (begins and ends with red)
hot	Black to red to yellow to white
cool	Shades of cyan and magenta
pink	Pastel shades of pink
gray	Linear gray-scale
bone	Gray-scale with a tinge of blue
jet	A variant of **hsv** (begins with blue and ends with red)

`copper`	Linear copper-tone
`prism`	Prism, alternating red, orange, yellow, green, and blue-violet
`flag`	Alternating red, white, blue, and black

285

Color as a Fourth Dimension in `surf`, `mesh`, and `pcolor` Plots

`surf(X,Y,Z,fun(X,Y,Z))`	Color is applied according to function `fun(X,Y,Z)`
`surf(X,Y,Z)=surf(X,Y,Z,Z)`	Default action, color is applied to the z-axis
`surf(X,Y,Z,X)`	Color is applied to x-axis
`surf(X,Y,Z,Y)`	Color is applied to y-axis
`surf(X,Y,Z,X.^2+Y.^2)`	Color is applied according to the radial distance from the `x=0`, `y=0` origin in the `z=0` plane
`surf(X,Y,Z,del2(Z))`	Color is applied according to Laplacian of surface
`[dZdx,dZdy]=gradient(Z);` `surf(X,Y,Z,dZdx)`	Color is applied according to surface slope in x-direction
`dZ=sqrt(dZdx.^2 + dZdy.^2);` `surf(X,Y,Z,dZ)`	Color is applied according to magnitude of surface slope

---

286

Color and Lighting Functions

`colormap(map)`	Install a color map in the current *Figure* window
`colorbar`	Display a vertical or horizontal color scale on the current figure
`rgbplot(map)`	Line plot of the red, green, and blue components of a color map
`brighten(a)`	Brighten current color map if `0<a<1`, darken if `-1<a<0`
`m=brighten(map,a)`	Return brightened color map **m**
`[cmin,cmax]=caxis`	Return color axis limits
`caxis([cmin,cmax])`	Set color axis limits

342

Handle Graphics Functions

`set(handle,'PropertyName',Value)`	Set object properties
`get(handle,'PropertyName')`	Get object properties
`reset(handle)`	Reset object properties to defaults
`delete(handle)`	Delete object and all its children
`gcf`	Get handle to the current *figure*
`gca`	Get handle to the current *axes*
`gco`	Get handle to the current object
`findobj('PropertyName',Value)`	Get handles to objects with specified property values
`waitforbuttonpress`	Wait for key or button press over a figure
`figure('PropertyName',Value)`	Create *figure* object
`axes('PropertyName',Value)`	Create *axes* object
`line(X,Y,'PropertyName',Value)`	Create *line* object
`text(X,Y,S,'PropertyName',Value)`	Create *text* object
`patch(X,Y,C,'PropertyName',Value)`	Create *patch* object
`surface(X,Y,Z,'PropertyName',Value)`	Create *surface* object
`image(C,'PropertyName',Value)`	Create *image* object

409

Handle Graphics GUI Functions

`uimenu(handle,'PropertyName',Value)`	Create menus for figures
`uicontrol(handle,'PropertyName',Value)`	Get object properties
`dialog('PropertyName',Value)`	Display a dialog box

`helpdlg('HelpString','DlgName')`	Display a 'help' dialog box
`warndlg('WarnString','DlgName')`	Display a 'warning' dialog box
`errordlg('ErrString','DlgName',` `Replace)`	Display an 'error' dialog box
`questdlg('QString',S1,S2,S3,` `  Default)`	Display a 'question' dialog box
`uigetfile(Filter,DlgName,X,Y)`	Retrieve a filename interactively
`uiputfile(InitFile,DlgName,X,Y)`	Retrieve a filename interactively
`uisetcolor(Handle,DlgName)`	Choose a color interactively
`uisetcolor([r g b],DlgName)`	Choose a color interactively
`uisetfont(Handle,DlgName)`	Choose font properties interactively

APPENDIX B HANDLE GRAPHICS PROPERTY TABLES

*indicates property is undocumented.
{ }indicates default values.

316

Root Object Properties

`BlackAndWhite`		Automatic hardware checking flag
	`on:`	Assume display is monochrome; do not check
	`{off}:`	Check the display type
* `BlackOutUnusedSlots`		Values are `[{no}	yes]`
* `CaptureMap`		
`CaptureMatrix`		Read-only matrix of image data of the region enclosed by the `CaptureRect` rectangle. Use `image` to display

	CaptureRect	Size and position of rectangle to capture. A four-element vector **[left,bottom,width, height]** in units specified by the **Units** property
*	**CaseSen**	Values are **[{on} | off]**
	CurrentFigure	The handle of the current *Figure* window
	Diary	Session logging
	on:	Copy all keyboard input and most output to a file
	{off}:	Do not save input and output to a file
	DiaryFile	A string containing the file name of the **Diary** file The default file name is **diary**
	Echo	Script echoing mode
	on:	Display each line of a script file as it executes
	{off}:	Do not echo unless **echo on** is specified
	Format	Number display format
	{short}:	Fixed-point format with 5 digits
	shortE:	Floating-point format with 5 digits
	long:	Scaled fixed-point format with 15 digits
	longE:	Floating-point format with 15 digits
	hex:	Hexadecimal format
	bank:	Fixed format of dollars and cents
	+:	Displays + and − symbols
	rat:	Approximation by ratio of small integers
	FormatSpacing	Output spacing
	{loose}:	Display extra line feeds
	compact:	Suppress extra line feeds
*	**HideUndocumented**	Control display of undocumented properties
	no:	Display undocumented properties
	yes:	Do not display undocumented properties

	PointerLocation	Read-only vector **[left, bottom]** or **[x, y]** of pointer location relative to the lower-left corner of the screen in units specified by the **Units** property
	PointerWindow	Handle of the *Figure* window containing the mouse pointer. If not in a *Figure* window, contains 0
	ScreenDepth	Integer specifying depth of the screen in bits, e.g., 1 for monochrome, 8 for 256 colors or grayscales
	ScreenSize	Position vector **[left, bottom, width, height]** where **[left, bottom]** is always **[0 0]** and **[width, height]** are the screen dimensions in units specified by the **Units** property
*	**StatusTable**	Vector
*	**TerminalHideGraphCommand**	Text string
	TerminalOneWindow	Used by the terminal graphics driver
	no:	Terminal has multiple windows
	yes:	Terminal has only one window
*	**TerminalDimensions**	Vector **[width, height]** of terminal dimensions
	TerminalProtocol	Terminal type set at startup, then read-only
	none:	Not in terminal mode, not connected to X server
	x:	X display server found; X Windows mode
	tek401x:	Tektronix 4010/4014 emulation mode
	tek410x:	Tektronix 4100/4105 emulation mode
*	**TerminalShowGraphCommand**	Text string
	Units	Unit of measurement for **position** property values
	inches:	Inches
	centimeters:	Centimeters

	normalized:	Normalized coordinates, where the lower-left corner of the screen maps to **[0 0]** and the upper-right corner maps to **[1 1]**
	points:	Typesetting points; equal to 1/72 of an inch
	{pixels}:	Screen pixels; the smallest unit of resolution of the computer screen
* **UsageTable**		Vector
ButtonDownFcn		MATLAB callback string, passed to the **eval** function whenever the object is selected; initially an empty matrix
Children		Read-only vector of handles to all *figure* objects
Clipping		Data clipping
	{on}:	No effect for *root* objects
	off:	No effect for *root* objects
Interruptible		**ButtonDownFcn** callback string interruptibility
	{no}:	Not interruptible by other callbacks
	yes:	Interruptible by other callbacks
Parent		Handle of parent object, always the empty matrix
* **Selected**		Values are **[on | off]**
* **Tag**		Text string
Type		Read-only object identification string; always **'root'**
UserData		User-specified data. Can be matrix, string, etc.
Visible		Object visibility
	{on}:	No effect for *root* objects
	off:	No effect for *root* objects

320

Figure Object Properties

BackingStore		Store a copy of the *Figure* window for fast redraw
	{on}:	Copy from backing store when previously covered portions of a *figure* are exposed. Faster window refresh, but uses more memory
	off:	Redraw previously covered portions of the *figure*. Slower window refresh, but saves memory
* **CaptureMap**		Matrix
* **Clint**		Matrix
Color		*Figure* background color. A 3-element RGB vector or one of MATLAB's predefined color names. The default color is **black**
Colormap		An m-by-3 matrix of RGB vectors. See the **colormap** function
* **Colortable**		Matrix; probably contains a copy of the system color table
CurrentAxes		The handle of the *figure's* current *axes*
CurrentCharacter		The most recent key pressed on the keyboard when the mouse pointer was in the *Figure* window
CurrentMenu		The handle of the most recently selected menu item
CurrentObject		The handle of the most recently selected object within the *Figure* window. This is the handle returned by the function **gco**
CurrentPoint		A position vector **[left, bottom]** or **[x, y]** of the point within the *Figure* window where the mouse pointer was located when the mouse button was last pressed or released

FixedColors		An n-by-3 matrix of RGB values that define the colors using slots in the system color table. The initial fixed colors are **black** and **white**
* Flint		Undocumented
InvertHardcopy		Change *figure* element colors for printing
	{on}:	Change black *figure* background to white, and lines, text, and axes to black for printing
	off:	Color of printed output exactly matches display
KeyPressFcn		MATLAB callback string passed to the **eval** function whenever a key is pressed while the mouse pointer is within the *Figure* window
MenuBar		Display MATLAB menus at the top of the *Figure* window or at the top of the screen on some systems
	{figure}:	Display default MATLAB menus
	none:	Do not display default MATLAB menus
MinColormap		Minimum number of color table entries to use. This affects the system color table, and if set too low, can cause unselected *figures* to display in false colors
Name		*Figure* window title (**not** axis title). By default, the empty string. If set to ***string***, the window title becomes: **Figure No. n:** ***string***
NextPlot		Determines drawing action for new plots
	new:	Create a new *figure* before drawing
	{add}:	Add new objects to the current *figure*
	replace:	Reset all *figure* object properties except **Position** to defaults and delete all children before drawing

NumberTitle		Prepend the *figure* number to the *figure* title
	{on}:	The window title is **Figure No. N**, with : *string* appended if the **Name** property is set to *string*
	off:	The window title is the **Name** property string only
PaperUnits		Unit of measurement for **Paper** . . . properties
	{inches}:	Inches
	centimeters:	Centimeters
	normalized:	Normalized coordinates
	points:	Points. One point is 1/72 of an inch
PaperOrientation		Paper orientation for printing
	{portrait}:	Portrait orientation; longest page dimension is vertical
	landscape:	Landscape orientation; longest page dimension is horizontal
PaperPosition		Position vector **[left, bottom, width, height]** representing the location of the figure on the printed page. **[left, bottom]** represents the location of the lower-left corner of the figure with respect to the lower-left corner of the printed page. **[width, height]** are the dimensions of the printed figure. Units are specified by the **PaperUnits** property
PaperSize		Vector **[width, height]** representing the dimensions of the paper to be used for printing. Units are specified by the **PaperUnits** property. The default **PaperSize is [8.5, 11]**
PaperType		Paper type for printed figures. When **PaperUnits** is set to normalized, MATLAB uses **PaperType** to scale figures for printing

	{usletter}:	Standard U.S. letter paper
	uslegal:	Standard U.S. legal paper
	a3:	European A3 paper
	a4letter:	European A4 letter paper
	a5:	European A5 paper
	b4:	European B4 paper
	tabloid:	Standard U.S. tabloid paper
Pointer		Mouse pointer shape
	crosshair:	Crosshair pointer
	{arrow}:	Arrow pointer
	watch:	Watch pointer
	topl:	Arrow pointing top-left
	topr:	Arrow pointing top-right
	botl:	Arrow pointing bottom-left
	botr:	Arrow pointing bottom-right
	circle:	Circle
	cross:	Double-line cross
	fleur:	Four-headed arrow or compass
Position		Position vector **[left, bottom, width, height]** where **[left, bottom]** represents the location of the lower-left corner of the figure with respect to the lower-left corner of the computer screen and **[width, height]** are the screen dimensions. Units are specified by the **Units** property
Resize		Allow/disallow interactive *Figure* window resizing
	{on}:	The window may be resized using the mouse
	off:	The window cannot be resized using the mouse
ResizeFcn		MATLAB callback string passed to the **eval** function whenever the window is resized using the mouse
* **Scrolled**		Values are **[{on}	off]**
SelectionType		A read-only string providing information about the method used for the last mouse button selection.

	The actual key and/or button press is platform-dependent
{normal}:	Click (press and release) the left, or only, mouse button
extended:	Hold down the **shift** key and make multiple normal selections; click both buttons of a two-button mouse; or click the middle button of a three-button mouse
alt:	Hold down the **control** key and make a normal selection; or click the right button of a two or three-button mouse
open:	Double-click any mouse button
ShareColors	Share color table slots
no:	Do not share color table slots with other windows
{yes}:	Reuse existing color table slots whenever possible
* **StatusTable**	Vector
Units	Unit of measurement for **position** property values
inches:	Inches
centimeters:	Centimeters
normalized:	Normalized coordinates, where the lower-left corner of the screen maps to **[0 0]** and the upper right corner maps to **[1 1]**
points:	Typesetting points; equal to 1/72 of an inch
{pixels}:	Screen pixels; the smallest unit of resolution of the computer screen
* **UsageTable**	Vector
WindowButtonDownFcn	MATLAB callback string passed to the **eval** function whenever a mouse button is pressed while the mouse pointer is within a *figure*
WindowButtonMotionFcn	MATLAB callback string passed to the **eval** function whenever a mouse pointer is moved while the mouse pointer is within a *figure*

WindowButtonUpFcn		MATLAB callback string passed to the **eval** function whenever a mouse button is released while the mouse pointer is within a *figure*
* **WindowID**		Large integer
ButtonDownFcn		MATLAB callback string, passed to the **eval** function whenever the figure is selected; initially an empty matrix
Children		Read-only vector of handles to all children of the *figure; axes, uicontrol,* and *uimenu* objects
Clipping		Data clipping
	{on}:	No effect for *figure* objects
	off:	No effect for *figure* objects
Interruptible		Specifies if *figure* callback strings are interruptible
	{no}:	Not interruptible by other callbacks
	yes:	Callback strings are interruptible by other callbacks
Parent		Handle of the *figure's* parent object, always **0**
* **Selected**		Values are **[on | off]**
* **Tag**		Text string
Type		Read-only object identification string; always **'figure'**
UserData		User-specified data. Can be matrix, string, etc.
Visible		Visibility of *Figure* window
	{on}:	Window is visible on the screen
	off:	Window is not visible

Axes Object Properties   325

AspectRatio	Aspect ratio vector **[axis_ratio, data_ratio]** where **axis_ratio** is the aspect

		ratio of the *axes* object (width/height), and **data_ratio** is the ratio of the lengths of data units along the horizontal and vertical axes. If set, MATLAB creates the largest axes that will fit within the rectangle defined by **Position** while preserving these ratios. The default value for **AspectRatio** is **[NaN, NaN]**.
Box		*Axes* bounding box
	on:	Enclose axes in a box or cube
	{off}:	Do not enclose axes
CLim		Color limit vector **[cmin cmax]** that determines the mapping of data to the colormap. **cmin** is the data value that maps to the first colormap entry, and **cmax** maps to the last. See the **caxis** function
CLimMode		Color-limits mode
	{auto}:	Color limits span the full range of data of the axes children
	manual:	Color limits do not automatically change. Setting **Clim** sets **ClimMode** to manual
Color		Axes background color. A 3-element RGB vector or one of MATLAB's predefined color names. The default color is **none** which uses the *figure* background color
ColorOrder		An m-by-3 matrix of RGB values. If a line color is not specified with **plot** and **plot3**, these colors will be used. The default **ColorOrder** is yellow, magenta, cyan, red, green, blue
CurrentPoint		Coordinate matrix containing a pair of points in the *axes* data space that define a line in 3-D that extends from the *front* to the *back* of the *axes* volume. The form is [xb yb

		zb; xf yf zf]. Units are specified by the **Units** property. The points [xf yf zf] are the coordinates of the last mouse click on the *axes* object
	DrawMode	Object-rendering order
	{normal}:	Sort objects and draw them from back to front based on the current **view**
	fast:	Draw objects in the order created; do not sort first
*	**ExpFontAngle**	Values are [{normal} | italic | oblique]
*	**ExpFontName**	Default is **Helvetica**
*	**ExpFontSize**	Default is **8** points
*	**ExpFontStrikeThrough**	Values are [on | {off}]
*	**ExpFontUnderline**	Values are [on | {off}]
*	**ExpFontWeight**	Values are [light | {normal} | demi | bold]
	FontAngle	Italics for *axes* text
	{normal}:	Regular font angle
	italic:	Italics
	oblique:	Italics on some systems
	FontName	Name of the font used for *axes* tick labels. Axes labels do not change fonts until they are redisplayed by setting the **XLabel**, **YLabel**, and **ZLabel** properties. The default font is **Helvetica**
	FontSize	Size in points used for *axes* labels and titles. Default is **12** points
*	**FontStrikeThrough**	Values are [on | {off}]
*	**FontUnderline**	Values are [on | {off}]
	FontWeight	Bolding for *axes* text
	light:	Light font weight
	{normal}:	Normal font weight
	demi:	Medium or bold font weight
	bold:	Bold font weight
	GridLineStyle	Line style of grid lines
	-:	Solid lines
	--:	Dashed lines

`{:}:`	Dotted lines
`-.:`	Dash-dot lines
* `Layer`	Values are `[top	{bottom}]`.
`LineStyleOrder`	A text string specifying the order of line styles to use to plot multiple lines on the axes. For example, `'-.|:|--|-'` will cycle through dash-dot, dotted, dashed, and solid lines. The default `LineStyleOrder` is `'-'`; solid lines only
`LineWidth`	Width of X, Y, and Z-axis lines. Default is `0.5` points
* `MinorGridLineStyle`	Values are `[-|--| {:}	-.]`
`NextPlot`	Action to be taken for new plots
`new:`	Create new *axes* before drawing
`add:`	Add new objects to the current *axes*. See **hold**
`{replace}:`	Delete the current *axes* and its children, and replace it with a new *axes* object before drawing
`Position`	Position vector `[left, bottom, width, height]` where `[left, bottom]` represents the location of the lower-left corner of the axes with respect to the lower-left corner of the figure object and `[width, height]` are the axes dimensions. Units are specified by the **Units** property
`TickLength`	Vector `[2Dlength 3Dlength]` representing the length of axes tick marks in 2-D and 3-D views. The length is relative to the axis length. Default is `[0.01 0.025]` representing 1/100 of the axis length in 2-D views and 25/1000 of the axis length in 3-D views
`TickDir`	`[{in}	out]`
`in:`	Tick marks point inward from the axis line. Default for 2-D view

	out:	Tick marks point outward from the axis line. Default for 3-D view
Title		The handle of the *axes* title text object
Units		Unit of measurement for position property values
	inches:	Inches
	centimeters:	Centimeters
	{normalized}:	Normalized coordinates, where the lower-left corner of the object maps to **[0 0]** and the upper right corner maps to **[1 1]**
	points:	Typesetters points; equal to 1/72 of an inch
	pixels:	Screen pixels; the smallest unit of resolution of the computer screen
View		A vector **[az el]** representing the viewpoint of the observer. **az** is the azimuth or rotation of the viewpoint in degrees to the right of the negative **Y** axis. **el** is the elevation in degrees above the plane of the **X-Y** axis. See the 3-D graphics chapter for details
XColor		An RGB vector or predefined MATLAB color string that specifies the color of the **X** axis line, labels, tick marks, and grid lines. Default is **white**
XDir		Direction of increasing **X** values
	{normal}:	**X** values increase from left to right
	reverse:	**X** values increase from right to left
XForm		A 4-by-4 view transformation matrix. Setting **view** affects **XForm**
XGrid		**X**-axis grid lines
	on:	Grid lines are drawn at each tick mark on the **X** axis
	{off}:	No grid lines are drawn
XLabel		The handle of the **X**-axis label text object
XLim		Vector **[xmin xmax]** specifying the minimum and maximum **X**-axis values

XLimMode		X-axis limits mode
	{auto}:	**XLim** is automatically calculated to include all **XData** of all *axes* children
	manual:	X-axis limits are taken from **XLim**
* **XMinorGrid**		Values are [**on** | **{off}**]
* **XMinorTicks**		Values are [**on** | **{off}**]
XScale		X-axis scaling
	{linear}:	Linear scaling
	log:	Logarithmic scaling
XTick		Vector of data values at which tick marks will be drawn on the **X** axis. Setting **XTick** to the empty matrix will suppress tick marks
XTickLabels		A matrix of text strings to be used to label tick marks on the **X** axis. If empty, MATLAB will use the data values at the tick marks
XTickLabelMode		X-axis tick-mark labeling mode
	{auto}:	**X**-axis tick-labels span the **XData**
	manual:	Take **X**-axis tick labels from **XTickLabels**
XTickMode		X-axis tick-mark-spacing mode
	{auto}:	**X**-axis tick-mark spacing to span **XData**
	manual:	**X**-axis tick-mark spacing from **XTick**
YColor		An RGB vector or predefined MATLAB color string that specifies the color of the **Y** axis line, labels, tick marks, and grid lines. Default is **white**
YDir		Direction of increasing Y values
	{normal}:	**Y** values increase from bottom to top
	reverse:	**Y** values increase from top to bottom
YGrid		Y-axis grid lines
	on:	Grid lines are drawn at each tick mark on the **Y** axis
	{off}:	No grid lines are drawn

YLabel		The handle of the **Y**-axis label text object
YLim		Vector **[ymin ymax]** specifying the minimum and maximum **Y**-axis values
YLimMode		**Y**-axis limits mode
	{auto}:	**YLim** is automatically calculated to include all **YData** of all axes children
	manual:	**Y**-axis limits are taken from **YLim**
* **YMinorGrid**		Values are **[on | {off}]**
* **YMinorTicks**		Values are **[on | {off}]**
YScale		**Y**-axis scaling
	{linear}:	Linear scaling
	log:	Logarithmic scaling
YTick		Vector of data values at which tick marks will be drawn on the **Y** axis. Setting **YTick** to the empty matrix will suppress tick marks
YTickLabels		A matrix of text strings to be used to label tick marks on the **Y** axis. If empty, MATLAB will use the data values at the tick marks
YTickLabelMode		**Y**-axis tick-mark labeling mode
	{auto}:	**Y**-axis tick labels span the **YData**
	manual:	Take **Y**-axis tick labels from **YTickLabels**
XTickMode		**Y**-axis tick-mark spacing mode
	{auto}:	**Y**-axis tick-mark spacing to span **YData**
	manual:	**Y**-axis tick-mark spacing from **YTick**
ZColor		An RGB vector or predefined MATLAB color string that specifies the color of the **Z** axis line, labels, tick marks, and grid lines. Default is **white**
ZDir		Direction of increasing **Z** values
	{normal}:	**Z** values increase from bottom to top (3-D) or pointing out of the screen (2-D)

	reverse:	**Z** values increase from top to bottom (3-D) or pointing into the screen (2-D)
ZGrid		**Z**-axis grid lines
	on:	Grid lines are drawn at each tick mark on the **Z** axis
	{off}:	No grid lines are drawn
ZLabel		The handle of the **Z**-axis label text object
ZLim		Vector **[zmin zmax]** specifying the minimum and maximum **Z**-axis values
ZLimMod		**Z**-axis limits mode
	{auto}:	**ZLim** is automatically calculated to include all **ZData** of all axes children
	manual:	**Z**-axis limits are taken from **ZLim**
* **ZMinorGrid**		Values are **[on | {off}]**
* **ZMinorTicks**		Values are **[on | {off}]**
ZScale		**Z**-axis scaling
	{linear}:	Linear scaling
	log:	Logarithmic scaling
ZTick		Vector of data values at which tick marks will be drawn on the **Z** axis. Setting **ZTick** to the empty matrix will suppress tick marks
ZTickLabels		A matrix of text strings to be used to label tick marks on the **Z** axis. If empty, MATLAB will use the data values at the tick marks
ZTickLabelMode		**Z**-axis tick-mark labeling mode
	{auto}:	**Z**-axis tick labels span the **ZData**
	manual:	Take **Z**-axis tick labels from **ZTickLabels**
XTickMode		**Z**-axis tick-mark spacing mode
	{auto}:	**Z**-axis tick-mark spacing to span **ZData**
	manual:	**Z**-axis tick-mark spacing from **ZTick**
ButtonDownFcn		MATLAB callback string, passed to the **eval** function whenever the

		axes is selected; initially an empty matrix
Children		Read-only vector of handles to all children of the *axes* except axes labels and titles: *line, surface, image, patch,* and *text* objects
Clipping		Data clipping
	{on}:	No effect for *axes* objects
	off:	No effect for *axes* objects
Interruptible		Specifies if **ButtonDownFcn** callback string is interruptible
	{no}:	Not interruptible by other callbacks
	yes:	**ButtonDownFcn** callback string is interruptible
Parent		Handle of the figure containing the *axes* object
* **Selected**		Values are **[on | off]**
* **Tag**		Text string
Type		Read-only object identification string; always **'axes'**
UserData		User-specified data. Can be matrix, string, etc.
Visible		Visibility of axes lines, tick marks, and labels
	{on}:	Axes are visible on the screen
	off:	Axes are not visible

332

Line Object Properties

Color		*Line* color. A 3-element RGB vector or one of MATLAB's predefined color names. The default color is **white**
EraseMode		Erase and redraw mode
	{normal}:	Redraws the affected region of the display ensuring that all objects are rendered correctly. This is the

		most accurate mode but is also the slowest
	background:	The line is erased by drawing it in the *figure's* background color. This can damage objects behind the erased line
	xor:	The line is drawn and erased by performing an exclusive OR (XOR) with the color of the screen beneath it. Incorrect color can be used when drawing over other objects
	none:	The line is not erased when moved or deleted
LineStyle		Line style control
	{-}:	Solid line is drawn through all data points
	--:	Dashed line is drawn through all data points
	: :	Dotted line is drawn through all data points
	-.:	Dash-dot line is drawn through all data points
	+:	Plus symbol is used as a marker at all data points
	o:	Circle is used as a marker at all data points
	*:	Star symbol is used as a marker at all data points
	. :	A solid dot is used as a marker at all data points
	x:	An X-mark is used as a marker at all data points
LineWidth		Line width in points; defaults to **0.5** points
MarkerSize		Marker size in points; defaults to **6** points
Xdata		Vector of **x**-coordinates for the line
Ydata		Vector of **y**-coordinates for the line
Zdata		Vector of **z**-coordinates for the line
ButtonDownFcn		MATLAB callback string, passed to the **eval** function whenever

		the line is selected; initially the empty matrix
Children		The empty matrix. Line objects have no children
Clipping		Data clipping mode
	{on}:	Any portion of the line outside the axes limits is not displayed
	off:	Line data is not clipped
Interruptible		Specifies if **ButtonDownFcn** callback string is interruptible
	{no}:	Not interruptible by other callbacks
	yes:	**ButtonDownFcn** callback string is interruptible
Parent		Handle of the *axes* containing the *line* object
* **Selected**		Values are **[on | off]**
* **Tag**		Text string
Type		Read-only object identification string; always **'line'**
UserData		User-specified data. Can be matrix, string, etc.
Visible		Visibility of line
	{on}:	Line is visible on the screen
	off:	Line is not visible

334

Text Object Properties

Color		Line color. A 3-element RGB vector or one of MATLAB's predefined color names. The default color is **white**
EraseMode		Erase and redraw mode
	{normal}:	Redraws the affected region of the display ensuring that all objects are rendered correctly. This is the

		most accurate mode but is also the slowest
	background:	The text is erased by drawing it in the figure's background color. This can damage objects behind the erased text
	xor:	The text is drawn and erased by performing an exclusive OR (XOR) with the color of the screen beneath it. Incorrect color can be used when drawing over other objects
	none:	The text is not erased when moved or deleted
Extent		Text position vector **[left, bottom, width, height]** where **[left, bottom]** represents the location of the lower-left corner of the text with respect to the lower-left corner of the axes object and **[width, height]** are the dimensions of a rectangle enclosing the text string. Units are specified by the **Units** property
FontAngle		Italics for text object
	{normal}:	Regular font angle
	italic:	Italics
	oblique:	Italics on some systems
FontName		Name of the font used for *axes* objects. The default font is **Helvetica**
FontSize		Size in points of text objects. Default is **12** points
* **FontStrikeThrough**		Values are **[on | {off}]**
* **FontUnderline**		Values are **[on | {off}]**
FontWeight		Bolding for *text* object
	light:	Light font weight
	{normal}:	Normal font weight
	demi:	Medium or bold font weight
	bold:	Bold font weight
HorizontalAlignment		Horizontal text alignment

	{left}:	Text is left-justified with respect to its **Position**
	center:	Text is centered with respect to its **Position**
	right:	Text is right-justified with respect to its **Position**
Position		Two- or three-element vector [x y (z)] specifying the location of the *text* object in three dimensions. Units are specified by the **Units** property
Rotation		Text orientation in degrees of rotation
	{0}:	Horizontal orientation
	±90:	Rotate text $\pm$ 90 degrees
	±180:	Rotate text $\pm$ 180 degrees
	±270:	Rotate text $\pm$ 270 degrees
String		The text string that is displayed
Units		Unit of measurement for position property values
	inches:	Inches
	centimeters:	Centimeters
	normalized:	Normalized coordinates, where the lower-left corner of the axes maps to **[0 0]** and the upper-right corner maps to **[1 1]**
	points:	Typesetters points; equal to 1/72 of an inch
	pixels:	Screen pixels; the smallest unit of resolution of the computer screen
	{data}:	Data units of the parent axes
VerticalAlignment		Vertical text alignment
	top:	String is placed at the top of the specified **Y**-position
	cap:	Font's capital-letter height is placed at the specified **Y**-position
	{middle}:	String is placed at the middle of the specified **Y**-position
	baseline:	Font's baseline is placed at the specified **Y**-position

	bottom:	String is placed at the bottom of the specified **Y**-position
ButtonDownFcn		MATLAB callback string, passed to the **eval** function whenever the text is selected; initially the empty matrix
Children		The empty matrix. *Text* objects have no children
Clipping		Data clipping mode
	{on}:	Any portion of the text outside the axes is not displayed
	off:	Text data is not clipped
Interruptible		Specifies if **ButtonDownFcn** callback string is interruptible
	{no}:	Not interruptible by other callbacks
	yes:	**ButtonDownFcn** callback string is interruptible
Parent		Handle of the *axes* object containing the *text* object
* **Selected**		Values are [on | off]
* **Tag**		Text string
Type		Read-only object identification string; always **'text'**
UserData		User-specified data. Can be matrix, string, etc.
Visible		Visibility of text
	{on}:	Text is visible on the screen
	off:	Text is not visible

336

Surface Object Properties

CData	Matrix of values that specify the color at every point in **ZData**. If **CData** is not the same size as **ZData**, the image contained in

		CData is mapped to the surface defined by **ZData**
EdgeColor		Surface edge color control
	none:	Edges are not drawn
	{flat}:	Edges are a single color determined by the first **CData** entry for that face. The default is **black**
	interp:	Each edge color is determined by linear interpolation through the values at the vertices
	A ColorSpec:	A 3-element RGB vector or one of MATLAB's predefined color names specifying a single color for edges. The default color is **black**
EraseMode		Erase-and-redraw mode
	{normal}:	Redraws the affected region of the display ensuring that all objects are rendered correctly. This is the most accurate mode but is also the slowest
	background:	The surface is erased by drawing it in the figure's background color. This can damage objects behind the erased surface
	xor:	The surface is drawn and erased by performing an exclusive OR (XOR) with the color of the screen beneath it. Incorrect color can be used when drawing over other objects
	none:	The surface is not erased when moved or deleted
FaceColor		Surface face color control
	none:	Faces are not drawn, but edges may be drawn
	{flat}:	The first **CData** entry determines face color
	interp:	Each face color is determined by linear interpolation through the mesh points on the surface
	A ColorSpec:	An RGB vector or one of MATLAB's predefined color

		names specifying a single color for faces
LineStyle		Edge line style control
	{-}:	Solid line is drawn through all mesh points
	--:	Dashed line is drawn through all mesh points
	: :	Dotted line is drawn through all mesh points
	-.:	Dash-dot line is drawn through all mesh points
	+:	Plus symbol is used as a marker at all mesh points
	o:	Circle is used as a marker at all mesh points
	***:**	Star symbol is used as a marker at all mesh points
	. :	A solid dot is used as a marker at all mesh points
	x:	An X-mark is used as a marker at all mesh points
LineWidth		Edge line width in points; defaults to **0.5** points
MarkerSize		Edge line marker size in points; defaults to **6** points
MeshStyle		Draw row and/or column lines
	{both}:	Draw all edges
	row:	Draw row edges only
	column:	Draw column edges only
* **PaletteModel**		Values are **[{scaled} | direct | bypass]**
XData		**X**-coordinates of surface points
YData		**Y**-coordinates of surface points
ZData		**Z**-coordinates of surface points
ButtonDownFcn		MATLAB callback string, passed to the **eval** function whenever the surface is selected; initially the empty matrix
Children		The empty matrix. *Surface* objects have no children

Clipping		Data clipping mode
	{on}:	Any portion of the surface outside the axes is not displayed
	off:	Surface data is not clipped
Interruptible		Specifies if **ButtonDownFcn** callback string is interruptible
	{no}:	Not interruptible by other callbacks
	yes:	**ButtonDownFcn** callback string is interruptible
Parent		Handle of the *axes* object containing the *surface* object
* Selected		Values are **[on | off]**
* Tag		Text string
Type		Read-only object identification string; always **'surface'**
UserData		User-specified data. Can be matrix, string, etc.
Visible		Visibility of surface
	{on}:	Surface is visible on the screen
	off:	Surface is not visible

339

Patch Object Properties

CData		Matrix of values that specify the color at every point along the edge of the patch. Only used if **EdgeColor** or **FaceColor** is set to **interp** or **flat**
EdgeColor		Patch edge color control
	none:	Edges are not drawn
	{flat}:	Edges are a single color determined by the average of the color data for the patch. The default color is **black**
	interp:	Edge color is determined by linear interpolation through the values at the patch vertices
	A *ColorSpec*:	A 3-element RGB vector or one of MATLAB's predefined color names specifying a single color for edges. The default color is **black**

EraseMode	Erase-and-redraw mode
{normal}:	Redraws the affected region of the display ensuring that all objects are rendered correctly. This is the most accurate mode but is also the slowest
background:	The patch is erased by drawing it in the figure's background color. This can damage objects behind the erased patch
xor:	The patch is drawn and erased by performing an exclusive OR (XOR) with the color of the screen beneath it. Incorrect color can be used when drawing over other objects
none:	The patch is not erased when moved or deleted
FaceColor	Patch face color control
none:	Faces are not drawn, but edges may be drawn
{flat}:	The values in the color argument **c** determine face color for each patch
interp:	Each face color is determined by linear interpolation through the values specified in the **CData** property
A ColorSpec:	An RGB vector or one of MATLAB's predefined color names specifying a single color for faces
LineWidth	Edge line width in points; defaults to **0.5** points
* **PaletteModel**	Values are **[{scaled} | direct | bypass]**
XData	**X**-coordinates of points along the edge of the patch
YData	**Y**-coordinates of points along the edge of the patch
ZData	**Z**-coordinates of points along the edge of the patch
ButtonDownFcn	MATLAB callback string, passed to the **eval** function whenever the patch is selected; initially the empty matrix
Children	The empty matrix. *Patch* objects have no children
Clipping	Data clipping mode
{on}:	Any portion of the patch outside the axes is not displayed
off:	Patch data is not clipped

Interruptible	Specifies if **ButtonDownFcn** callback string is interruptible
{no}:	Not interruptible by other callbacks
yes:	**ButtonDownFcn** callback string is interruptible
Parent	Handle of the *axes* object containing the *patch* object
* Selected	Values are [**on** | **off**]
* Tag	Text string
Type	Read-only object identification string; always **'patch'**
UserData	User-specified data. Can be matrix, string, etc.
Visible	Visibility of patch
{on}:	Patch is visible on the screen
off:	Patch is not visible

341

Image Object Properties

CData	A matrix of values that specifies the color of each element of the image. **image(C)** assigns **C** to **CData**. The elements of **CData** are indices into the current color map
XData	Image **X**-data; specifies the position of the image rows. If omitted, the row indices of **CData** are used
YData	Image **Y**-data; specifies the position of the image columns. If omitted, the column indices of **CData** are used
ButtonDownFcn	MATLAB callback string, passed to the **eval** function whenever the image is selected; initially an empty matrix
Children	The empty matrix. *Image* objects have no children
Clipping	Data clipping mode

	{on}:	Any portion of the image outside the axes is not displayed
	off:	Image data is not clipped
Interruptible		Specifies if **ButtonDownFcn** callback string is interruptible
	{no}:	Not interruptible by other callbacks
	yes:	**ButtonDownFcn** callback string is interruptible
Parent		Handle of *axes* object containing the *image* object
* Selected		Values are [on | off]
* Tag		Text string
Type		Read-only object identification string; always **image**
UserData		User-specified data. Can be matrix, string, etc.
Visible		Visibility of image
	{on}:	Image is visible on the screen
	off:	Image is not visible

349

Uimenu Object Properties

Accelerator	A character specifying the keyboard equivalent or *shortcut* key for the menu item. For X-Windows, the key sequence is **Control-Char**; for Macintosh systems the sequence is **Command-Char** or ⌘ **-Char**
BackgroundColor	Uimenu background color. A 3-element RGB vector or one of MATLAB's predefined color names. The default background color is **light gray**
CallBack	MATLAB callback string, passed to the **eval** function whenever you

		select the menu item; initially the empty matrix
Checked		Checkmark for selected items
	{on}:	Checkmark appears next to selected items
	off:	Checkmarks are not displayed
Enable		Menu-enable mode
	{on}:	The menu item is enabled. Selecting the menu item will send the **CallBack** string to **eval**
	off:	The menu item is disabled and the label string is dimmed. Selecting the menu item has no effect
ForegroundColor		Uimenu foreground (text) color. A 3-element RGB vector or one of MATLAB's predefined color names. The default text color is **black**
Label		A text string containing the label of the menu item. On PC systems, a **&** preceding a character in the label defines a shortcut key that is activated with the key sequence **Alt-Char**
Position		Relative position of the *uimenu* object. Top-level menus are numbered from left to right, and submenus are numbered top to bottom
Separator		Separator-line mode
	on:	A dividing line is drawn above the menu item
	{off}:	No dividing line is drawn
* **Visible**		Visibility of menu object
	{on}:	Uimenu is visible on the screen
	off:	Uimenu is not visible
ButtonDownFcn		MATLAB callback string, passed to the **eval** function whenever the object is selected; initially the empty matrix
Children		Object handles of other *uimenu* objects (submenus)

Clipping		Clipping mode
	{on}:	No effect for *uimenu* objects
	off:	No effect for *uimenu* objects
DestroyFcn		Macintosh version 4.2c only. No documentation available
Interruptible		Specifies if **ButtonDownFcn** and **CallBack** strings are interruptible
	{no}:	Callbacks are not interruptible by other callbacks
	yes:	Callback strings are interruptible
Parent		Handle of the parent object; either a *figure* handle if the *uimenu* object is a top-level menu, or the parent *uimenu* object handle for submenus
* **Selected**		Values are [on | off]
* **Tag**		Text string
Type		Read-only object identification string; always **uimenu**
UserData		User-specified data. Can be matrix, string, etc.
Visible		Visibility of *uimenu* object
	{on}:	Uimenu is visible on the screen
	off:	Uimenu is not visible

369

Uicontrol Object Properties

BackgroundColor	Uicontrol background color. A 3-element RGB vector or one of MATLAB's predefined color names. The default background color is **light gray**
CallBack	MATLAB callback string, passed to the **eval** function whenever the uicontrol object is *activated;* initially an empty matrix
ForegroundColor	Uicontrol foreground (text) color. A 3-element RGB vector or one of

	MATLAB's predefined color names. The default text color is **black**
HorizontalAlignment	Horizontal alignment of label string
left:	Text is left-justified with respect to the uicontrol
{center}:	Text is centered with respect to the uicontrol
right:	Text is right-justified with respect to the uicontrol
Max	The largest value allowed for the **Value** property. The value depends on the uicontrol **Type**. Radio buttons and check boxes set **Value** to **Max** while the uicontrol is in the **on** state. This value defines the maximum index value of a popup menu or the maximum value of a slider. Editable text boxes are multi-line text when **Max-Min>1**. The default value is **1**
Min	The smallest value allowed for the **Value** property. The value depends on the uicontrol **Type**. Radio buttons and check boxes set **Value** to **Min** while the uicontrol is in the **off** state. This value defines the minimum index value of a popup menu or minimum value of a slider. Editable text boxes are multi-line text when **Max-Min>1**. The default is **0**
Position	Position vector **[left, bottom, width, height]** where **[left, bottom]** represents the location of the lower-left corner of the uicontrol with respect to the lower-left corner of the *figure* object and **[width, height]** are the uicontrol dimensions. Units are specified by the **Units** property

* Enable		Control-enable mode
	{on}:	The uicontrol is enabled. Activating the uicontrol will send the **CallBack** string to **eval**
	off:	The uicontrol is disabled and the label string is dimmed. Activating the uicontrol has no effect
String		Text string specifying the uicontrol label on push buttons, radio buttons, check boxes, and popup menus. For editable text boxes, this property is set to the string typed by the user. For multiple items in a popup menu or an editable text box, separate each item with a vertical bar character (l), and quote the entire strings. Not used for frames or sliders
Style		Defines the type of *uicontrol* object
	{pushbutton}:	Push button: performs an action when selected
	radiobutton:	Radio button: when used alone, toggles between two states; when used in a group, lets the user choose one option
	checkbox:	Check box: when used alone, toggles between two states; when used in a group, lets the user choose one option
	edit:	Editable text box: displays a string and lets the user change it
	text:	Static text box: displays a string
	slider:	Slider: lets the user choose a value within a range of values
	frame:	Frame: displays a border around one or more controls, forming a logical group
	popupmenu:	Popup menu containing a number of mutually-exclusive choices
Units		Unit of measurement for position property values

inches:	Inches	
centimeters:	Centimeters	
{normalized}:	Normalized coordinates, where the lower-left corner of the figure maps to **[0 0]** and the upper-right corner maps to **[1 1]**	
points:	Typesetters points; equal to 1/72 of an inch	
pixels:	Screen pixels; the smallest unit of resolution of the computer screen	

Value
Current value of the uicontrol. Radio buttons and check boxes set **Value** to **Max** when **on** and **Min** when **off**. Sliders set **Value** to the number set by the slider **(Min≤Value≤Max)**. Popup menus set **Value** to the index of the item selected **(1≤Value≤Max)**. Text objects and push buttons do not set this property

ButtonDownFcn
MATLAB callback string, passed to the **eval** function whenever the uicontrol is *selected;* initially the empty matrix

Children
Uicontrol objects have no children; always returns the empty matrix

Clipping
Clipping mode

{on}: No effect for *uicontrol* objects

off: No effect for *uicontrol* objects

DestroyFcn
Macintosh version 4.2c only. No documentation available

Interruptible
Specifies if **ButtonDownFcn** and **CallBack** strings are interruptible

{no}: Callbacks are not interruptible by other callbacks

yes: Callback strings are interruptible

Parent
Handle of the *figure* containing the *uicontrol* object

* **Selected**
Values are **[on | off]**

* **Tag**
Text string

Type	Read-only object identification string; always **uicontrol**
UserData	User-specified data. Can be matrix, string, etc.
Visible	Visibility of *uicontrol* object
{on}:	Uicontrol is visible on the screen
off:	Uicontrol is not visible, but still exists

APPENDIX C *SYMBOLIC MATH TOOLBOX* QUICK REFERENCE TABLES

The following tables summarize features in the *Symbolic Math Toolbox*.

445

Operations on Symbolic Expressions

numeric	Symbolic to numeric conversion
pretty	Show pretty symbolic output
subs	Substitute for subexpression
sym	Create symbolic matrix or expression
symadd	Symbolic addition
symdiv	Symbolic division
symmul	Symbolic multiplication
symop	Symbolic operations
sympow	Power of symbolic expression
symrat	Rational approximation
symsub	Symbolic subtraction
symvar	Find symbolic variables

445

Symbolic Simplification

collect	Collect terms
expand	Expand

`factor`	Factor
`simple`	Find shortest form
`simplify`	Simplify
`symsum`	Sum series

445

Symbolic Polynomials

`charploy`	Characteristic polynomial
`horner`	Nested polynomial representation
`numden`	Extract numerator or denominator
`poly2sym`	Polynomial vector to symbolic
`sym2poly`	Symbolic to polynomial vector

446

Symbolic Calculus

`diff`	Differentiation
`int`	Integration
`jacobian`	Jacobian matrix
`taylor`	Taylor series expansion

446

Symbolic Variable Precision Arithmetic

`digits`	Set variable precision accuracy
`vpa`	Variable precision arithmetic

446

Symbolic Equation Solutions

compose	Functional composition
dsolve	Solution of differential equations
finverse	Functional inverse
linsolve	Solution of simultaneous linear equations
solve	Solution of algebraic equations

446

Symbolic Linear Algebra

charploy	Characteristic polynomial
determ	Matrix determinant
eigensys	Eigenvalues and eigenvectors
inverse	Matrix inverse
jordan	Jordan canonical form
linsolve	Solution of simultaneous linear equations
transpose	Matrix transpose

APPENDIX D *MASTERING MATLAB TOOLBOX* QUICK REFERENCE TABLES

The following tables group *Mastering MATLAB Toolbox* functions into categories.

Mastering MATLAB Array Manipulation

shiftlr(A,n)	Shift matrix columns right $n>0$ columns
shiftlr(A,n)	Shift matrix columns left $n<0$ columns
shiftlr(A,n,1)	Circularly shift matrix columns right n columns
shiftud(A,n)	Shift matrix columns down $n>0$ rows
shiftud(A,n)	Shift matrix columns up $n<0$ rows
shiftud(A,n,1)	Circularly shift matrix columns down n columns

Mastering MATLAB Data and Numerical Analysis Functions

`mmax(A)`	Matrix maximum value
`mmin(A)`	Matrix minimum value
`mmono(v)`	Test for monotonic vector
`mmderiv(x,y)`	Derivative using weighted central differences
`mmintgrl(x,y)`	Integral using trapezoidal rule
`mminterp(tab,col,val)`	1-D table search by linear interpolation
`mmtool`	User-interactive visualization of 2-D data
`mmtable(tab,col,vals)`	1-D monotonic table search,
`mmtable(x,y,xi)`	10× faster than interp1.

Mastering MATLAB Polynomial Manipulation

`mmp2str(a)`	Polynomial vector to string conversion, `a(s)`
`mmp2str(a,'x')`	Polynomial vector to string conversion, `a(x)`
`mmp2str(a,'x',1)`	Constant and monic polynomial conversion
`mmpadd(a,b)`	Polynomial addition
`mmpsim(a)`	Polynomial simplification

Mastering MATLAB Spline Functions

`yi=spintgrl(x,y,xi)`	Cubic spline interpolation of integral of y=f(x) at points in `xi`
`ppi=spintgrl(pp)`	Return piecewise polynomial representation of integral of y=f(x), given piecewise polynomial representation of y=f(x)
`yi=spintgrl(pp,xi)`	Find piecewise polynomial representation of integral of y=f(x), given piecewise polynomial representation of y=f(x), and evaluate at points in `xi`
`yi=spderiv(x,y,xi)`	Cubic spline interpolation of derivative of y=f(x) at points in `xi`

`ppi=spderiv(pp)`	Return piecewise polynomial representation of derivative of y=f(x), given piecewise polynomial representation of y=f(x)
`yi=spderiv(pp,xi)`	Find piecewise polynomial representation of derivative of y=f(x), given piecewise polynomial representation of y=f(x) and evaluate at points in **xi**

Mastering MATLAB Fourier Series Functions

`fsderiv(Kn,Wo)`	Derivative of a Fourier series
`fseval(Kn,t,Wo)`	Evaluate Fourier series
`fsfind('fname',T,N)`	Find Fourier series coefficient vector for a time function
`[An,Bn,Ao]=fsform(Kn)` `Kn=fsform(An,Bn,Ao)`	Fourier series form conversion
`fsharm(Kn,i)`	Extract a particular Fourier series harmonic
`fsmsv(Kn)`	Compute mean square value of signal
`fsresize(Kn,N)`	Resize a Fourier series coefficient vector
`fsresp(Num,Den,Un,Wo)`	Fourier series response of a linear system to an input Fourier series, **Un**
`fsround(Kn)`	Set insignificant Fourier series coefficients to zero
`fswindow(N,'type')` `fswindow(Kn,'type')`	Generate a window function to minimize Gibb's phenomenon

Mastering MATLAB High Level Graphics

`mmcont2(X,Y,Z,N,C)`	2-D contour plot with color map
`mmcont3(X,Y,Z,N,C)`	3-D contour plot with color map
`mmfill(x,y,z,c,lb,ub)`	Fill area between two curves
`mmap(colorspec)`	Single color color map creation
`mmshow(map)`	Show color map

`mmspin3d(N)`	Make movie by 3-D azimuth spin of current figure
`mmview3d`	Viewpoint adjustment using sliders
`rainbow`	Rainbow color map

Mastering MATLAB Plotting Utilities

`mmaxes prop value...`	Modify properties of plot axes
`mmcxy (or) xy=mmcxy`	Show x-y coordinates of mouse over a plot
`mmdraw prop value...`	Draw straight lines on a plot
`mmgetxy(N)`	Get x-y coordinates using the mouse
`mmline prop value...`	Modify properties of plotted lines
`mmpage`	GUI for figure placement on printed page
`mmpaper prop value...`	Set paper attributes
`mmtile`	Tile multiple *Figure* windows
`mmtext('optional text')`	Place and/or drag text on a plot
`mmzoom`	Zoom axis in using a rubberband box
`mmzap object`	Delete text, lines, or axes using mouse
`mmfont prop value...`	Modify font properties of text

Mastering MATLAB Handle Graphics and GUI Functions

`mmgcf`	Get handle to the current figure if it exists
`mmgca`	Get handle to the current axes if it exists
`mmpx2n(X)`	Pixel to normalized coordinate transformation
`mmn2px(X)`	Normalized to pixel coordinate transformation
`mmsetc(Handle),` `mmsetc(ColorSpec),` `mmsetc('select')`	Set color attributes using mouse
`mmsetf(Handle),` `mmsetf('select')`	Set font attributes using mouse

Mastering MATLAB Example Functions

`mmenus`	Example uimenu function
`mmenu1`	Example uimenu commands
`mmenu2`	Example uimenu commands
`mmenu3`	Example uimenu commands
`mmenu4`	Example uimenu commands
`mmenu5`	Example uimenu commands
`mmcaxisd`	Color axis demo
`mmcmapd`	Color map merge demo
`mmclock(X,Y)`	Example uicontrol function
`mmctl1`	Example uicontrol commands
`mmctl2`	Example uicontrol commands
`mmctl3`	Example uicontrol commands
`mmctl4`	Example uicontrol commands
`mmctl5`	Example uicontrol commands
`mmctl6`	Example uicontrol commands
`mmctl7`	Example uicontrol commands
`mmctl8`	Example uicontrol commands
`mmsetclr`	Limited script version of the `mmsetc` function
`mmtemp`	Script file of `temps` data matrix
`mm4d`	Using color for a 4th dimension demo
`iforgot`	Recursive function call example
`expsin`	Decaying sinewave function
`sawtooth`	Sawtooth waveform function
`vdpol`	Van der Pol oscillator differential equations

APPENDIX E *MASTERING MATLAB TOOLBOX* REFERENCE

This section documents the help text of all general utility functions in the *Mastering MAT-LAB Toolbox* in alphabetical order.

```
FSDERIV Fourier Series Derivative.
  FSDERIV(Kn,Wo) returns the FS coefficients of the derivative
  of f(t) whose FS coefficients are given by Kn and whose
```

fundamental frequency is Wo rad/s.
If Wo is not given, Wo=1 is assumed.

FSEVAL Fourier Series Function Evaluation.

FSEVAL(Kn,t,Wo) computes values of a real valued function given
its complex exponential Fourier series coefficients Kn, at the
points given in t where the fundamental frequency is Wo rad/s.
K contains the Fourier coefficients in ascending order:
Kn = [k   k   ... k ... k   k]
      -N -N+1   0   N-1 N
if Wo is not given, Wo=1 is assumed.
Note: this function creates a matrix of size:
rows = length(t) and columns = (length(K)-1)/2

FSFIND Find Fourier Series Approximation.

Fn=FSFIND(FUN,T,N) computes the Complex Exponential
Fourier Series of a signal described by the function 'FUN'.
FUN is the character string name of a user created M-file function.
The function is called as f=FUN(t) where t is a vector over
the range 0<=t<=T.

The FFT is used. Choose sufficient harmonics to minimize aliasing.

T is the period of the function. N is the number of harmonics.
Fn is the vector of FS coefficients.

[Fn,nWo,f,t]=FSFIND(FUN,T,N) returns the frequencies associated
with Fn in nWo, and returns values of the function FUN
in f evaluated at the points in t over the range 0<=t<=T.

FSFIND(FUN,T,N,P) passes the data in P to the function FUN as
f=FUN(t,P). This allows parameters to be passed to FUN.

FSFORM Fourier Series Format Conversion.

Kn=FSFORM(An,Bn,Ao) converts the trigonometric FS with
An being the COSINE and Bn being the SINE coefficients to
the complex exponential FS with coefficients Kn.
Ao is the DC component and An, Bn and Ao are assumed to be real.

[Kn,i]=FSFORM(An,Bn,Ao) returns the index vector i that
identifies the harmonic number of each element of Kn.
[An,Bn,Ao]=FSFORM(Kn) does the reverse format conversion.

FSHARM Fourier Series Harmonic Component Selection.

FSHARM(Kn,N) returns the (N)th harmonic component of the
complex exponential FS given by Kn.
FSHARM(Kn) returns the DC component.

[H,i]=FSHARM(Kn,N) returns the index of the selected
harmonic H in i.

FSMSV Fourier Series Mean Square Value.
  FSMSV(Kn) uses Parseval's theorem to compute the mean
  square value of a function given its FS coefficients Kn.

FSRESIZE Resize a Fourier Series.
  FSRESIZE(Kn,N) resizes the complex exponential FS Kn to
  have N harmonics. If N is greater than the number of
  harmonics in Kn, zeros are added to the result.
  If N is less than the number of harmonics in Kn, the
  result is a truncated version of the input.

  FSRESIZE(Kn,Un) resizes the complex exponential FS Kn to
  have the same number of harmonics as the FS Un.

  [Yn,iy]=FSRESIZE(Kn,N) additionally returns the harmonic
  index of the result.

FSRESP Fourier Series Linear System Response.
  FSRESP(N,D,Un,Wo) returns the complex exponential FS of the
  output of a linear system when the input is given by a FS.
  N and D are the numerator and denominator coefficients
  respectively of the system transfer function.
  Un is the complex exponential Fourier Series of the system input.
  Wo is the fundamental frequency associated with the input.

FSROUND Round Fourier Series Coefficients.
  FSROUND(Kn) rounds the Fourier Series coefficients Kn
  to eliminate residual terms. Terms satisfying
  abs(Kn)<TOL*max(abs(Kn)) %terms of small magnitude, or
  abs(real(Kn))<TOL*abs(imag(Kn)) %terms with small real part, or
  abs(imag(Kn))<TOL*abs(real(Kn)) %terms with small imag part
  are set to zero. TOL is set equal to sqrt(eps) or can be
  specified by FSROUND(Kn,TOL).

FSWINDOW Generate Window Functions.
  FSWINDOW(N,TYPE) creates a window vector of type TYPE having
  a length equal to the scalar N.
  FSWINDOW(X,TYPE) creates a window vector of type TYPE having
  a length and orientation the same as the vector X.
  FSWINDOW(.,TYPE,alpha) provides a parameter alpha as required
  for some window types.
  FSWINDOW with no input arguments returns a string matrix whose
  i-th row is the i-th TYPE given below.

TYPE is a string designating the window type desired:
'rec' = Rectangular or Boxcar
'tri' = Triangular or Bartlet
'han' = Hann or Hanning
'ham' = Hamming
'bla' = Blackman common coefs.
'blx' = Blackman exact coefs.
'rie' = Riemann {sin(x)/x}
'tuk' = Tukey, 0< alpha < 1; alpha = 0.5 is default
'poi' = Poisson, 0< alpha < inf; alpha = 1 is default
'cau' = Cauchy, 1< alpha < inf; alpha = 1 is default
'gau' = Gaussian, 1< alpha < inf; alpha = 1 is default

Reference: F.J. Harris,"On the Use of Windows for Harmonic Analysis
   with the Discrete Fourier Transform," IEEE Proc., vol 66, no 1, Jan
   1978, pp 51-83.

MMAP Single Color Colormap.

MMAP(C,M) makes a colormap of length M starting with the
basic colorspec C. The map changes from dark to light.
MMAP(C) is the same length as the current colormap.
Example: mmap('y') is a yellow colormap
         mmap([.49 1 .83]) is an aquamarine colormap
         mmap('c',20) is a cyan colormap having length 20.
         mmap('c') is the default if C can not be interpreted.

Apply using: colormap(mmap(c,m))

MMAX Matrix Maximum Value.

MMAX(A) returns the maximum value in the matrix A.
[M,I] = MMAX(A) in addition returns the indices of
the maximum value in I = [row col].

MMAXES Set Axes Properties Using Mouse.

MMAXES waits for a mouse click on an axes then
applies the desired properties to the selected axes.
Properties are given in pairs, e.g. MMAXES name value ...
Properties:

NAME	VALUE	{default}
box	[{on} off]	for axes bounding box
color	[y m c r g b {w} k]	or an rgb in quotes: '[r g b]'
width	[points]	for axes linewidth {0.5}
tdir	[{in} out]	for tick direction
xtick	[off {on}]	to hide X-axis labels, nonreversible
xdir	[{norm} rev]	for X-axis direction (norm)
xgrid	[on {off}]	for X-axis grid

```
xscale     [{lin} log]   for X-axis scaling
zap        (n.a.)        to delete axes and plot contents
```

```
xtick, xdir, xgrid, xscale have y and z axis counterparts:
ytick, ydir, ygrid, yscale
ztick, zdir, zgrid, zscale
Examples:
MMAXES box off ygrid on      turns box off and y-axis grid on
MMAXES tdir out zscale log   sets tick direction out and z-axis to log
MMAXES color '[1 .5 0]'      sets color to orange
```

```
Clicking on an object other than an axes, or striking
a key on the keyboard aborts the command.
```

MMCONT2 2-D Contour Plot Using a Colormap.

```
MMCONT2(X,Y,Z,N,C) plots N contours of Z in 2-D using the color
specified in C. C can be a linestyle and color as used in plot,
e.g., 'r-', or C can be the string name of a colormap. X and Y
define the axis limits.
If not given, default argument values are: N = 10, C = 'hot',
X and Y = row and column indices of Z. Examples:
MMCONT2(Z)                  10 lines with hot colormap
MMCONT2(Z,20)               20 lines with hot colormap
MMCONT2(Z,'copper')         10 lines with copper colormap
MMCONT2(Z,20,'gray')        20 lines with gray colormap
MMCONT2(X,Y,Z,'jet')        10 lines with jet colormap
MMCONT2(Z,'c—')             10 dashed lines in cyan
MMCONT2(X,Y,Z,25,'pink')    25 lines in pink colormap
```

```
CS=MMCONT2(...) returns the contour matrix CS as described in
   CONTOURC.
[CS,H]=MMCONT2(...) returns a column vector H of handles to line
   objects.
```

MMCONT3 3-D Contour Plot Using a Colormap.

```
MMCONT3(X,Y,Z,N,C) plots N contours of Z in 3-D using the color
specified in C. C can be a linestyle and color as used in plot,
e.g., 'r-', or C can be the string name of a colormap. X and Y
define the axis limits.
If not given default argument values are: N = 10, C = 'hot',
X and Y = row and column indices of Z. Examples:
MMCONT3(Z)            10 lines with hot colormap
MMCONT3(Z,20)         20 lines with hot colormap
MMCONT3(Z,'copper')   10 lines with copper colormap
MMCONT3(Z,20,'gray')  20 lines with gray colormap
MMCONT3(X,Y,Z,'jet')  10 lines with jet colormap
MMCONT3(Z,'c—')       10 dashed lines in cyan
```

MMCONT3(X,Y,Z,25,'pink') 25 lines in pink colormap

CS=MMCONT3(...) returns the contour matrix CS as described in
   CONTOURC.
[CS,H]=MMCONT3(...) returns a column vector H of handles to line
   objects.

MMCXY Show x-y Coordinates Using Mouse.

MMCXY places the x-y coordinates of the mouse in the
lower left hand corner of the current 2-D figure window.
When the mouse is clicked, the coordinates are erased.
XY=MMCXY returns XY=[x,y] coordinates where mouse was clicked.
XY=MMCXY returns XY=[] if a key press was used.

MMDERIV Derivative Using Weighted Central Differences.

MMDERIV(X,Y) computes the derivative of the function y=f(x) given the
data in X and Y. X must be a vector, but Y may be a column-oriented
data matrix. The length of X must equal the length of Y if Y is a
vector, or it must equal the number of rows in Y if Y is a matrix.

X need not be equally spaced. Weighted central differences are used.
Quadratic approximation is used at the endpoints.

MMDRAW Draw a Line and Set Properties Using Mouse.

MMDRAW draws a line in the current axes using the mouse,
Click at the starting point and drag to the end point.
In addition, properties can be given to the line.
Properties are given in pairs, e.g., MMDRAW name value ...
Properties:

NAME	VALUE	{default}
color	[y m c r g b {w} k] or an rgb '[r g b]'
style	[- -- {:} -.]
mark	[o + . * x)]
width	points for linewidth {0.5}
size	points for marker size {6}

Examples:
MMDRAW color r width 2   sets color to red and width to 2 points
MMDRAW mark + size 8   sets marker type to + and size to 8 points
MMDRAW color '[1 .5 0]'   sets color to orange

MMFILL Fill Plot of Area Between Two Curves.

MMFILL(X,Y,Z,C,LB,UB) plots y=f(x) and z=g(x) and fills the
area between the two curves from LB<= X <=UB with colorspec C.
X,Y and Z are data vectors of the same length.
Missing arguments take on default values. Examples:
MMFILL(X,Y) fills area under y=f(x) with red.
MMFILL(X,Y,C) fills area under y=f(x) with colorspec C.

MMFILL(X,Y,Z) fills area between y=f(x) and z=g(x) with red.
MMFILL(X,Y,Z,C) fills area between y=f(x) and z=g(x) with C.
MMFILL(X,Y,LB,UB) fills area under y=f(x) in red between bounds.
MMFILL(X,Y,C,LB,UB) fills area under y=f(x) with C between bounds.
MMFILL(X,Y,Z,LB,UB) fills area between curves with red between bounds.

A=MMFILL(...) returns the approximate area filled by calling TRAPZ.

MMGCA Get Current Axes if it Exists.
    MMGCA returns the handle of the current axes if it exists.
If no current axes exists, MMGCA returns an empty handle.

Note that the function GCA is different. It creates a figure
and an axes and returns the axes handle if it does not exist.

MMGCF Get Current Figure if it Exists.
    MMGCF returns the handle of the current figure if it exists.
If no current figure exists, MMGCF returns an empty handle.

Note that the function GCF is different. It creates a figure
and returns its handle if it does not exist.

MMGETXY Graphical Input Using Mouse.
    XY=MMGETXY(N) gets N points from the current axes at
points selected with a mouse button press.
XY=[x,y] matrix having 2 columns and N rows.
Striking ANY key on the keyboard aborts the process.
XY=MMGETXY gathers any number of points until a
key on the keyboard is pressed.

MMIN Matrix Minimum Value.
    MMIN(A) returns the minimum value in the matrix A.
    [M,I] = MMIN(A) in addition returns the indices of
the minimum value in I = [row col].

MMINTERP 1-D Table Search by Linear Interpolation.
    Y=MMINTERP(TAB,COL,VAL) linearly interpolates the table
TAB searching for the scalar value VAL in the column COL.
All crossings are found and TAB(:,COL) need not be monotonic.
Each crossing is returned as a separate row in Y and Y has as
many columns as TAB. Naturally, the column COL of Y contains
the value VAL. If VAL is not found in the table, Y=[].

MMINTGRL Compute Integral Using Trapezoidal Rule.
    MMINTGRL(X,Y) computes the integral of the function y=f(x) given the
data in X and Y. X must be a vector, but Y may be a column-oriented
    data matrix. The length of X must equal the length of Y if Y is a

vector, or it must equal the number of rows in Y if Y is a matrix.

X need not be equally spaced. The trapezoidal algorithm is used.

MMLINE Set Line Properties Using Mouse.

MMLINE waits for a mouse click on a line then
applies the desired properties to the selected line.
Properties are given in pairs, e.g., MMLINE name value ...
Properties:

NAME	VALUE	{default}
color	[y m c r g b w k] or an rgb in quotes: '[r g b]'	
style	[- -- : -.]	
mark	[o + . * x)]	
width	points for linewidth {0.5}	
size	points for marker size (6)	
zap	(n.a.) delete selected line	

Examples:
MMLINE color r width 2   sets color to red and width to 2 points
MMLINE mark + size 8   sets marker type to + and size to 8 points
MMLINE color '[1 .5 0]' sets color to orange

Clicking on an object other than a line, or striking
a key on the keyboard aborts the command.

MMN2PX Normalized to Pixel Coordinate Transformation.

MMN2PX(X) converts the position vector X from
normalized coordinates to pixel coordinates w.r.t.
the computer screen.

MMN2PX(X,H) converts the position vector X from
normalized coordinates to pixel coordinates w.r.t.
the figure window having handle H.
X=[left bottom width height] or X=[width height]

MMONO Test for Monotonic Vector.

MMONO(X) where X is a vector returns:
2 if X is strictly increasing,
1 if X is non decreasing,
-1 if X is non increasing,
-2 if X is strictly decreasing,
0 otherwise.

MMP2STR Polynomial Vector to String Conversion.

MMP2STR(P) convert polynomial vector P into string representation.
For example: P = [2 3 4] becomes the string '2s^2 + 3s + 4'

MMP2STR(P,V) generate string using the variable V as the parameter instead of s. MMP2STR([2 3 4],'z') becomes '2z^2 + 3z + 4'

MMP2STR(P,V,1) factors the polynomial into the product of a constant
and a monic polynomial.
MMP2STR([2 3 4],[],1) becomes '2(s^2 + 1.5s + 2)'

MMPADD Polynomial Addition.
    MMPADD(A,B) adds the polynomials A and B.

MMPAGE GUI to Set Figure Paper Position.
    MMPAGE allows the user to set the current figure
    position on the printed page using a gui.
    MMPAGE(Hf) places the figure having handle Hf.

    Editable text boxes: Left Bottom Width Height
    sets the respective figure position on the page.
    Pushbuttons:
     Set Default makes the current position default for future figures.
     Get Factory changes the figure position to factory values.
     L/R Center centers the figure Left to Right.
     U/D Center centers the figure Up to Down.
     WYSIWYG makes the Width and Height identical to the Figure Window.
     Get Default sets the figure position to the Default values.
     Revert restores the figure position to that when MMPAGE was called.
     Done closes the MMPAGE GUI.

    Usage: create figure, call MMPAGE, then print.

MMPAPER Set Default Paper Properties.
    MMPAPER name value ...
    sets default paper properties for the current figure and
    succeeding figures based on name value pairs.
    Properties:
    NAME   VALUE
    units   [{inch},centimeters,points,normal]
    orient   [{portrait},landscape]
    type   [{usletter},uslegal,a3,a4letter,a5,b4,tabloid]

    Examples:
    MMPAPER units inch orient landscape
    MMPAPER type tabloid

    MMPAPER with no arguments returns the current paper defaults.

MMPSIM Polynomial Simplification, Strip Leading Zero Terms.
    MMPSIM(A) Deletes leading zeros and small coefficients in the
    polynomial A(s). Coefficients are considered small if their

magnitude is less than both one and norm(A)*1000*eps.
MMPSIM(A,TOL) uses TOL for its smallness tolerance.

<u>MMPX2N Pixel to Normalized Coordinate Transformation.</u>
MMPX2N(X) converts the position vector X from
pixel coordinates to normalized coordinates w.r.t.
the computer screen.

MMPX2N(X,H) converts the position vector X from
pixel coordinates to normalized coordinates w.r.t.
the figure window having handle H.
X=[left bottom width height] or X=[width height]

<u>MMSETC Obtain RGB Triple Interactively from a Color Sample.</u>
MMSETC displays a dialog box for the user to select
a color interactively and displays the result.

X = MMSETC returns the selected color in X.

MMSETC([r g b]) uses the RGB triple as the initial
RGB value for modification.

MMSETC C -or-
MMSETC('C') where C is a color spec (y,m,c,r,g,b,w,k), uses
the specified color as the initial value.

MMSETC(H) where the input argument H is the handle of
a valid graphics object that supports color, uses the color
property of the object as the initial RGB value.

MMSETC select -or-
MMSETC('select') waits for the user to click on a valid
graphics object that supports color, and uses the color
property of the object as the initial RGB value.

If the initial RGB value was obtained from an object or
object handle, the 'Done' pushbutton will apply the
resulting color property to the selected object.

If no initial color is specified, black will be used.

Examples:
    mmsetc
    mycolor=mmsetc
    mmsetc([.25 .62 .54])
    mmsetc(H)
    mmsetc g

```
mmsetc red
mmsetc select
mycolor=mmsetc('select')
```

MMSETF Choose Font Characteristics Interactively.

MMSETF displays a dialog box for the user to select
font characteristics.

X = MMSETF returns the handle of the text object or 0
    if an error occurs or 'Cancel' is pressed.

MMSETF(H) where the input argument H is the handle of
a valid text or axes object, uses the font characteristics
of the object as the initial values.

MMSETF select -or-
MMSETF('select') waits for the user to click on a valid
graphics object, and uses the font characteristics
of the object as the initial values.

If the initial values were obtained from an object or
object handle, the 'Done' pushbutton will apply the
resulting text properties to the selected object.

If no initial object handle is specified, a zero is returned in X.

Examples:
```
mmsetf
mmsetf(H)
mmsetf select
Hx_obj=mmsetf('select')
```

MMSHOW PCOLOR Colormap Display.

MMSHOW uses pcolor to display the current colormap.
MMSHOW(MAP) displays the colormap MAP.
MMSHOW(MAP(N)) displays the colormap MAP having N elements.

Examples: MMSHOW(hot)
           MMSHOW(pink(30))

MMSPIN3D Movie of 3-D Azimuth Rotation of Current Figure.

MMSPIN3D(N) captures and plays N frames of the current figure
through one rotation about the Z-axis at the current elevation.
M=MMSPIN3D(N) returns the movie in M for later playing with movie.
If not given, N=18 is used.
MMSPIN3D fixes the axis limits and issues axis off.

MMTEXT Place and Drag Text with Mouse.
MMTEXT waits for a mouse click on a text object,
in the current figure then allows it to be dragged
while the mouse button remains down.
MMTEXT('whatever') places the string 'whatever' on
the current axes and allows it to be dragged.

MMTEXT becomes inactive after the move is complete or
no text object is selected.

MMTILE Tile Figure Windows.
MMTILE with no arguments, tiles the current figure windows
and brings them to the foreground.
Figure size is adjusted so that 4 figure windows fit on the screen.
Figures are arranged in a clockwise fashion starting in the
upper-left corner of the display.

MMTILE(N) makes tile N the current figure if it exists.

Otherwise, the next tile is created for subsequent plotting.

Tiled figure windows are titled TILE #1, TILE #2, TILE #3, TILE #4.

MMVIEW3D GUI-Controlled Azimuth and Elevation Adjustment.
MMVIEW3D adds sliders and text boxes to the current figure window
for adjusting azimuth and elevation using the mouse.

The 'Revert' pushbutton reverts to the original view.
The 'Done' pushbutton removes all GUIs.

MMZAP Delete Graphics Object Using Mouse.
MMZAP waits for a mouse click on an object in
a figure window and deletes the object.
MMZAP or MMZAP text erases text objects.
MMZAP axes erases axes objects.
MMZAP line erases line objects.
MMZAP surf erases surface objects.
MMZAP patch erases patch objects.

Clicking on an object other than the selected type, or striking
a key on the keyboard aborts the command.

MMZOOM Simple 2-D Zoom-In Function Using RBBOX.
MMZOOM zooms in on a plot based on the size of a
rubberband box drawn by the user with the mouse.
MMZOOM x   zooms the x-axis only.

```
MMZOOM y        zooms the y-axis only.
MMZOOM reset    or
MMZOOM out      restores original axis limits.


Striking a key on the keyboard aborts the command.
MMZOOM becomes inactive after zoom is complete or aborted.
```

RAINBOW Colormap Variant to HSV.
```
  RAINBOW(M) Rainbow Colormap with M entries.


  Red - Orange - Yellow - Green - Blue - Violet


  RAINBOW by itself is the same length as the current colormap.


  Apply using: colormap(rainbow)
```

SHIFTLR Shift or Circularly Shift Matrix Columns.
```
  SHIFTLR(A,N) with N>0 shifts the columns of A to the RIGHT N columns.
  The first N columns are replaced by zeros and the last N
  columns of A are deleted.

  SHIFTLR(A,N) with N<0 shifts the columns of A to the LEFT N columns.
  The last N columns are replaced by zeros and the first N
  columns of A are deleted.

  SHIFTLR(A,N,C) where C is nonzero performs a circular
  shift of N columns, where columns circle back to the
  other side of the matrix. No columns are replaced by zeros.
```

SHIFTUD Shift or Circularly Shift Matrix Rows.
```
  SHIFTUD(A,N) with N>0 shifts the rows of A DOWN N rows.
  The first N rows are replaced by zeros and the last N
  rows of A are deleted.

  SHIFTUD(A,N) with N<0 shifts the rows of A UP N rows.
  The last N rows are replaced by zeros and the first N
  rows of A are deleted.

  SHIFTUD(A,N,C) where C is nonzero performs a circular
  shift of N rows, where rows circle back to the other
  side of the matrix. No rows are replaced by zeros.
```

SPDERIV Cubic Spline Derivative Interpolation.
```
  YI=SPDERIV(X,Y,XI) uses cubic spline interpolation to fit the
  data in X and Y, differentiates the spline, and returns values
  of the spline derivatives evaluated at the points in XI.
  PPD=SPDERIV(PP) returns the piecewise polynomial vector PPD
```

describing the cubic spline derivative of the curve described by
the piecewise polynomial in PP. PP is returned by the function
SPLINE and is a data vector containing all information to
evaluate and manipulate a spline.

YI=SPDERIV(PP,XI) differentiates the cubic spline given by
the piecewise polynomial PP, and returns the values of the
spline derivatives evaluated at the points in XI.

See also SPLINE, PPVAL, MKPP, UNMKPP, SPINTGRL

SPINTGRL Cubic Spline Integral Interpolation.
  YI=SPINTRGL(X,Y,XI) uses cubic spline interpolation to fit the
  data in X and Y, integrates the spline, and returns
  values of the integral evaluated at the points in XI.

  PPI=SPINTGRL(PP) returns the piecewise polynomial vector PPI
  describing the integral of the cubic spline described by
  the piecewise polynomial in PP. PP is returned by the function
  SPLINE and is a data vector containing all information to
  evaluate and manipulate a spline.

  YI=SPINTGRL(PP,XI) integrates the cubic spline given by
  the piecewise polynomial PP, and returns the values of the
  integral evaluated at the points in XI.

  See also SPLINE, PPVAL, MKPP, UNMKPP, SPDERIV

Index